PROBLEMS OF OUR PHYSICAL ENVIRONMENT

ENERGY TRANSPORTATION POLLUTION

JOSEPH PRIEST
Miami University, Oxford, Ohio

PROBLEMS OF OUR PHYSICAL ENVIRONMENT

ENERGY TRANSPORTATION POLLUTION

ADDISON-WESLEY PUBLISHING COMPANY
Reading, Massachusetts
Menlo Park, California • London • Don Mills, Ontario

This book is in the
ADDISON-WESLEY SERIES IN PHYSICS

Cover photograph: The sloop *Clearwater* passes an enormous junkyard as she sails down the Hudson River. She was designed for use as a waterborne stage during the presentation of riverside music festivals intended to draw public attention to the polluted condition of the water. Construction of the craft was promoted through the efforts of folk singer Pete Seeger and funded by his contributions and those of other ecology fighters. (UPI Telephoto.)

Copyright © 1973 by Addison-Wesley Publishing Company, Inc. Philippines copyright 1973 by Addison-Wesley Publishing Company, Inc.

ISBN 0-201-05972-X
BCDEFGHIJ-HA-7876543

PREFACE

A course in a physical science is required of all students in nearly every college and university. The actual course depends in large part on the philosophy of the school and the department involved. It may be a course in physics, chemistry or geology, some combination of courses in these areas, or, perhaps, a single entity called "Physical Science" which somehow embodies each discipline. A department often has a captive audience because of this requirement and the courses sometimes tend to be "diluted" versions of those offered for majors. There is, however, a plea for more relevance and participation. This book evolved from a segment of a new course structure at Miami University which responded to this plea.

Miami University students in the past decade were required to take 9 quarter hours (that is, 3 one-hour class meetings each week for 30 weeks) of a physical science course (chemistry, geology, or physics). The physics contribution was a course which bridged the scientific and humanity cultures. It was innovative and popular. Some 6000 students took the course and classes were consistently full. In 1971, the university requirement was changed to 9 quarter hours of a natural science which includes both the physical and biological sciences. Furthermore, a student is not locked into a full-year course but can take his 9 hours in any number of ways. The Physics Department responded to this challenge by organizing a series of 3 quarter-hour courses (3 fifty-minute class periods each week for 10 weeks) which could be used for the natural science requirement. These were

Physics and Society	Discoveries in Physics
Space Physics	The Atom and Its Nucleus
Physics and Our Environment	Symmetries in Physics
Physics of Music	Atmospheric Physics

"Physics and Society" is a prerequisite for any course in the sequence. It is a somewhat condensed version of the full-year "Physics for Humanists" course mentioned earlier. The set of courses evolved from a poll of 360 students who were taking the "Physics for Humanists" course. "Physics and Our Environment" was picked as the first, second, or third choice by 314 students. Students were undoubtedly attracted by the environmental theme rather than by the physics. Nevertheless, physics has a vital role in the understanding and appreciation of the problems (and beauty) of the environment and it is crucial that this be conveyed to students. One of the most direct applications of physics is to the problems of pollution and depletion of natural resources, which are the unfortunate consequences of the energy conversion processes in automobile engines, electric power plants, and the like. This text stresses such environmental considerations. It is intended for use by beginning college students and presumes no college course as a prerequisite. Some knowledge of algebra is assumed. Throughout the text, the scheme is first to present basic physical principles appropriate measuring units, and then apply them to a system of environmental interest. The model concept is used to a great extent to identify and quantify the problematical areas.

Approximately four 50-minute class periods were devoted to each of chapters 1 through 4, three each to chapters 5 and 6, and two each to chapters 7 through 9 during the 10-week term. I believe that it would be difficult to cover the entire book in 10 weeks if the students had no physics background. It would be well-suited for a 15-week semester schedule for students with no physics background. Liberal use was made of 35 mm slides. Student response to this feature was good. There are several good movies available on loan from various agencies and these were used from time to time. The questions at the end of each chapter were very useful for class discussion. The problems, many of which were too difficult for most students, were useful for expanding the text material. Suggestions for some more in-depth topics are given at the end of several chapters. These should be useful in classes where the size permits reports and class participation. I found it very meaningful to use pertinent demonstrations for illustrating a physical principle or an environmental effect.

The approach taken in this book is not unique nor particularly novel. It is not a comprehensive treatment of either physical science or the environment. Many will say that it overemphasizes the physical science aspect and, for their own tastes, they may be right. Others will say that important physical principles are omitted. I see the book as a way of teaching physical principles using an extremely important subject, the environment, as motivation. There is every opportunity for both teacher and student to expand on the topics presented and to introduce topics of their own liking. I would hope that they would be motivated to do so. My own experience is that students strongly desire an expansion over what they can deal with in the text by themselves. I feel that I have been able to alert students to problems of the environment and have made them better interpreters, evaluators, and voters through an understanding of some of the physical principles involved.

Oxford, Ohio
January 1973

J. P.

ACKNOWLEDGMENTS

The evolution of this book from an idea and a few sketchy outlines transpired only with the help and encouragement of many. The numerous students who researched much of the background material were particularly helpful. The careful, critical reviews of Professors John Shonle, Richard Fisher, Allan Evans, and Sam Austin were especially useful. Professors John Kraushaar, Delbert Devins, David Cowan, Vincent Llamas, Bruce Marsh, Ryoichi Seki, Robert O. Smith, John Trishka, and Verner Jensen provided very important input. I am grateful to Dr. Don Cunningham who instigated the first "Environmental Physics" course at Miami University and interested me in the subject. I greatly appreciate the help of my colleagues at Miami University, especially Professor David Griffing, who conceived the pattern for course restructure which figured heavily in the writing of this book. The typists, Juanita Killough, Brenda Persinger, Cathy Wright, Kathy Taylor, and Pam Rousseau did an excellent job. I owe a special debt of gratitude to my wife and children for their patience and encouragement during some very trying periods.

The author

TO THE STUDENT

This book has a dual purpose. The first is to alert students to problems of their physical environment and to make them better "voters." The second is to teach some physical principles and their applications. Nearly all teachers would agree that both of these are important in this day and age, but there would probably be much disagreement on how to achieve this goal. My philosophy in this has been influenced by a letter from a former student. Because of its eloquence I am reproducing it essentially as I received it.

November 8, 1971

Dear Dr. Priest,

 Life is certainly full of coincidences. I have just finished reading the book *How Children Fail* by John Holt and thinking about my years at Miami in light of this book.

 Holt's book deals mainly with the teaching of arithmetic but all of his thoughts are equally applicable to the teaching of physics. Physics seems just as mysterious and unreal to college students as arithmetic does to grade schoolers. I feel the book should be required reading for all teachers and students. If the other books by Holt are as good as this one, they should also be required reading.

 I feel there is just one basic fault to most physics education. In Holt's words, as teachers "we cut children (students) off from their own common sense and the world of reality by requiring them to play with and shove around words (equations) and symbols that have little or no meaning to them. Thus we turn the vast majority of our students into the kind of people for whom all symbols are meaningless; who cannot use symbols as a way of learning about and dealing with reality... The minority, the able and successful students, are very likely to be turned into something different but just as dangerous: the kind of people who can manipulate words (equations) and symbols fluently while keeping themselves largely divorced from the reality for which they stand". (Unfortunately I fall into the latter category.) All teachers should keep foremost in their minds that "numerical arithmetic (physics) should look to children (students) like a simpler and faster way of doing things that they know how to do already, not a set of mysterious recipes for getting right answers to meaningless questions." Physics teachers need to be extra careful not to assume that since physics by definition deals with the physical world, it also deals with reality as seen by the student. Reality for the student consists only of those phenomena which he has consciously experienced.

Whether I have achieved this goal remains to be seen. I have, however, consciously tried to follow this plan.

Miami University   J. P.
January, 1973

CONTENTS

PROBLEMS OF OUR PHYSICAL ENVIRONMENT

ENERGY TRANSPORTATION POLLUTION

1

ENERGY AND POWER: PRINCIPLES, DEMANDS, AND OUTLOOK

INTRODUCTION

A contemporary dictionary* calls physics "the science of *matter* and *energy* and of *interactions* between the two" and environment "the combination of external or extrinsic *physical* conditions that affect and influence the *growth* and *development* of organisms." Man is the main organism of interest in our environment, and the interaction of energy and man is a crucial element. Energy is the driving mechanism behind man and everything he does.† Energy from food is required just to keep us going. But subtleties such as the style of food preparation also depend in large part on the availability of energy. Eskimos, for example, eat raw meat because energy is unavailable for cooking. In general, however, man, uses energy for much more than mere sustenance. His transportation, industry, and homes use energy in a variety of forms (Fig. 1) which directly affect his style and standard of living. A measure of standard of living is difficult but the gross national product (GNP) is often used as an indicator. It is probably not surprising that the GNP per person in the United States is the largest in the world. Furthermore, the United States consumes 35% of the world's energy derived from fossil fuels even though it has only 6% of the total world population. Energy consumption in all forms amounts to nearly 100 times that necessary for survival (about 2500 food calories each day). This correlation of energy and GNP holds for most nations as illustrated in Fig. 2. One aspect of this enormous energy consumption that has become of great concern to the technologically advanced countries is pollution. And as the underdeveloped nations become industrialized and "move up" on the energy-GNP curve, pollution problems become worldwide. Pollutants of incredible varieties (chemicals, heat, noise, light, dirt, and so on) and magnitudes are adversely affecting man both physically and psychologically. It is ironic to find that the very technology which has brought about the control of many diseases and relief from starvation and has contributed in so many ways to the "good life" is the culprit. The internal combustion engine and fossil fuel burning electric generating plants are prime examples of the mixed blessings of technology, whose benefits, on balance, surely outweigh the detriments at this stage. And technology will no doubt help us solve our problems.

    The earth is a system which has been likened to a spaceship. It derives energy from the sun as well as from sources within the earth itself. In this sense it is no different from any craft; the fuel supply is limited. The system earth, though, has been operated as if the fuel supply were limitless and as if the rejects from its "exhaust" could have no effect on the delicate balance of natural processes driven

---

*The American Heritage Dictionary, American Heritage Publishing Company, Inc., New York.

†Two very interesting publications on the role of energy in society are: *Energy and Society*, by Fred Cottrell, McGraw-Hill Book Co., Inc. New York (1955), and *Energy and Power*, A Scientific American Book, W. H. Freeman and Company, San Francisco (1971).

Movie: "Multiply and Subdue the Earth," 2 reels, 67 min., color. Both reels are excellent, but the first constitutes a good introduction to environmental problems. Audio Visual Service, Indiana University, Bloomington, Indiana 47401

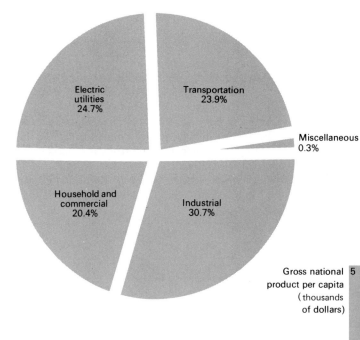

Fig. 1　Sector breakdown of energy consumption for the United States.

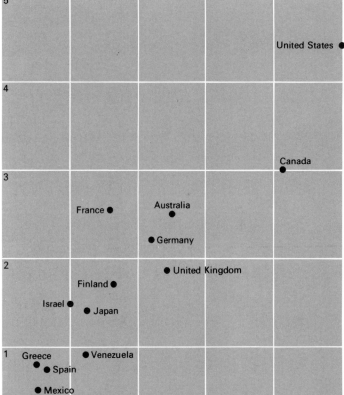

Fig. 2　Gross-national-product per capita versus per capita power consumption for selected nations. (Data for the graph was obtained from The Statistical Office of the United Nations, New York, N.Y.)

by the sun. This, however, is not true. The supply of fossil fuels is not only finite, but the presently known oil and natural gas deposits are estimated to last less than 100 years. This shockingly short period of grace is a consequence of both the limited nature of the fuel supplies and the increasing *rates* of depletion.

Today we are deluged by articles in the daily press, magazines, and professional journals that discuss man in relationship to his environment (Fig. 3). This book is intended to show you how physical science ties in with environmental concerns and how a knowledge of physical principles and concepts will therefore help you to understand, assess, and appreciate the environment so that ultimately you can participate intelligently in the hard decisions that will have to be made.

Environment
Environmental Science and Technology
Federal Documents—United States Senate Hearings Reports
Fortune
Life
Major Newspapers—New York Times, for example.
M.I.T. Technology Review
National Geographic
Science
Science News
Science and Public Affairs—Bulletin of the Atomic Scientists
Scientific American
Time
U.S. News and World Report

Fig. 3 Articles relating to the environment appear regularly in an incredible variety of journals and magazines. This is a list of sources which were particularly useful in the writing of this text.

RATES

We are all familiar with the idea of rate. For example, a 3-minute phone call costs a fixed amount of money, say $1, and this cost is referred to as a phone rate. The important concept involved is that of an amount (cost), time, and units (dollars for costs, minutes for time). Given a certain sum of money and the phone rate, we can readily figure out how many 3-minute phone calls we can make. But if the phone rate were to change with each call, then the answer would not come so easily. There are many environmental rates that are of great interest to us, and one of the major concerns is that these rates are increasing regularly. Since we know that resources are limited, like money for phone calls, it is fair to ask: "How long will the resources last?" It is this type of question which motivates this discussion on rates.

The U.S. Interstate Highway System is one of the engineering marvels of our time. It is a system designed to accommodate and encourage safe, high-speed travel. While enjoying and admiring the system, we cannot help but be impressed with the amount of land "consumed," perhaps 100 acres per day during construc-

A full directional interchange. Note the size of the system compared to a vehicle on one of the roads and the large amount of land "consumed." (Photograph courtesy of the Federal Highway Commission.)

tion (an acre is slightly smaller than the area of a football field).* If asked "How much land is taken up in a year?" nearly everyone would say 36,500 acres. Asked further "What principle did you employ?" the answer would probably be "amount equals rate times time," symbolized as

$$A = r \cdot t$$
$$= \frac{100 \text{ acres}}{\text{day}} \cdot 365 \text{ days}$$
$$= 36{,}500 \text{ acres}$$

You might not have put in the units and canceled them as if they were numbers. However, this step of including units is very important. The units make the equation physical, help us to avoid mistakes, and aid in figuring out some results which may not be so simple.

_____

*"How the Interstate Changed the Face of the Nation" by Juan Cameron, _Fortune_ **84**, 1 (1971).

A rate, in general, is a *change* in some quantity divided by the time required to produce the change.

$$\text{rate} = \frac{\text{change in a quantity}}{\text{time required to produce the change}}$$

The quantity in the example considered is acres of land. You'll encounter many different types of rates in physics and environmental problems. For example, speed, which is the change in distance per unit of time, expressed in mi/hr or m/sec, and the coal production rate expressed in tons/year. (The notation mi/hr means miles per hour, or miles divided by hours.)

The numerical calculation just considered tacitly assumes that the rate is the same at all times during the one-year time period. As mentioned at the outset, there is no reason why it has to be, and in fact, it probably is not in a real situation. Now asked "How much land is taken up in a year if the rate changes daily?," we find that the answer is not so obvious. However, we can solve this and many other similar problems fairly easily if we first draw a "picture" of the situation. The "picture" is a graph of rate versus time. If the rate is the same throughout the year, then every point plotted will be the same height above the horizontal time axis (Fig. 4). A line connecting these points is parallel to the horizontal axis. Note that the plot forms a rectangle with sides of "lengths" 100 acres/day and 365 days. The area of this rectangle, like that of any rectangle, is the product of the lengths of the sides. Thus

area = 100 acres/day · 365 days = 36,500 acres

But this result is just what was calculated from the expression A = rt which is not an accidental development. In general, the amount is merely the "area" under the rate versus time curve whether or not the rate is constant.

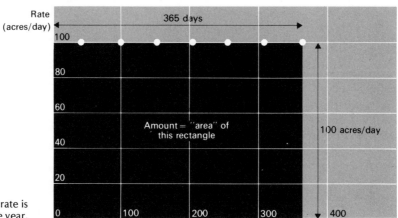

Fig. 4 Graph of rate versus time if the rate is 100 acres per day at all times during the year.

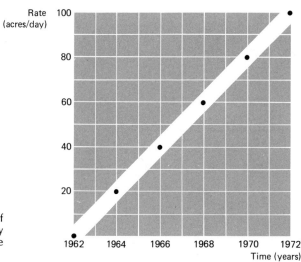

Fig. 5 A hypothetical way in which the rate of consumption of land by the interstate highway system may have changed throughout the years.

The rate of consumption of land by the interstate highway system has obviously changed since the inception of the practice. Figure 5 shows, hypothetically, how this rate may have changed throughout the years. Let us use this to illustrate the concept of "amount related to area" when the rate changes with time. Note that, unlike Fig. 4, the initial time for the plot does not start at zero; instead it begins at 1962. However it is the *change* in time which is important when determining amounts in some time period; hence the 3650 day *change* between 1962 and 1972 is the significant factor. Recognizing this and assuming that the area under the curve is simply that of a triangle with a base of 3650 units and height 100 units, we find that the amount, in acres, is

amount in acres = area of a triangle

$$= \tfrac{1}{2} \text{ base times height}$$

$$= \tfrac{1}{2} (3650 \text{ days}) (100 \tfrac{\text{acres}}{\text{day}})$$

$$= 182,500 \text{ acres}$$

Again note the importance of units. A wrong result would have followed had we used the time period in years.

Data plots and the ensuing straight line curve like that shown in Fig. 5 are fairly common. When this occurs for rate and time, we say that "rate is proportional to time." Mathematically, the rate would be related to time by the equation $r = ct$. Here c is called the proportionality constant and is a number with appropriate physical units. Simply stated, this relation means that the rate changes the same amount in equal time intervals. For example, the rate in Fig. 5 changes by 20 acres/day in every two-year time period. It also means that if the time interval doubles, then the rate also doubles. You should check this out for yourself using Fig. 5.

Figure 6 shows a generalized curve for rate proportional to time. Again, the amount is the "area" of a triangle. Thus

$$Area = \tfrac{1}{2} \, base \times height$$
$$= \tfrac{1}{2} \, t \cdot ct$$
$$= \tfrac{1}{2} \, ct^2$$

We could continue this process ad infinitum for various geometric figures, but would learn little more. A different geometry, which is meaningful, is given in Exercise 19. You are encouraged to work through it. The real challenge comes if the rate versus time is some peculiar curve, as is often the case (Fig. 7). Actual calculation of the amount is difficult, if not impossible, but the area can be found, for example, by plotting rate versus time on graph paper and counting the squares to get the area. Exercise 13 leads you through this procedure. Exercise 25 using data for the worldwide production of coal and lignite provides an opportunity to apply these ideas to a real situation of interest. You are encouraged to work through these exercises.

ESTIMATES OF LIFETIME OF FOSSIL FUELS

About 90% of the energy consumed in the United States is derived from the fossil fuels—coal, oil, and natural gas. The rates at which these fuels are produced are shown in Figs. 8, 9, 10. The production rates of oil and natural gas are increasing in contrast to coal production which has averaged about 550 million tons/year since 1910. Reading these data, we may be tempted at first to say "so what." But if we use the data to estimate the lifetime of our fuel resources, our attitude will change. Two pieces of information are required for this estimate: (1) an estimate of the remaining resources and (2) an estimate of the way the rates will change with time. Estimates of the remaining resources are subject to change simply because new deposits are discovered periodically (the fairly recent discovery of oil in Alaska is an example). The estimates in Table 1 are current and typical.*

Fig. 6 Plot of rate versus time if $r = ct$.

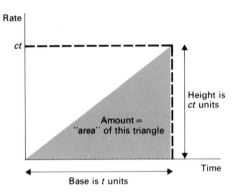

Fig. 7 Plot of rate versus time if the rate varies in an irregular fashion. The area can be obtained by counting the small squares.

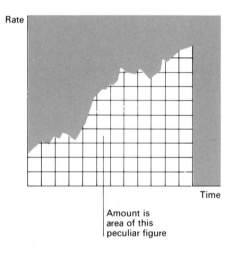

Table 1. Estimates of United States reserves of fossil fuels*

Fuel	Estimate of resources from beginning of time	Estimate of resources remaining	1968 production rate
Coal	1635 billion tons	1580 billion tons	0.550 billion tons/yr
Oil	165 billion barrels	74 billion barrels	3.4 billion barrels/yr
Natural gas	1075 trillion cubic feet	702 trillion cubic feet	19.3 trillion cubic feet/yr

*The oil and natural gas estimates do not include the Alaska discoveries. Coal deposits are given in short tons. For comparison with other units, 1 short ton = 2000 lb, 1 long ton = 2240 lb, 1 metric ton = 2200 lb.

*M. King Hubbert, "The Energy Resources of the Earth," *Scientific American* **224**, 60 (September 1971). M. King Hubbert, "Energy Resources," *Resources and Man*, W. H. Freeman and Co., San Francisco, 1969.

Fig. 8 United States annual bituminous coal production.

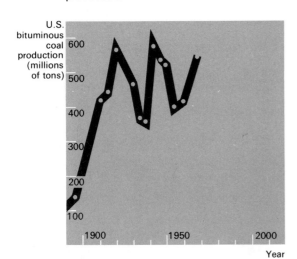

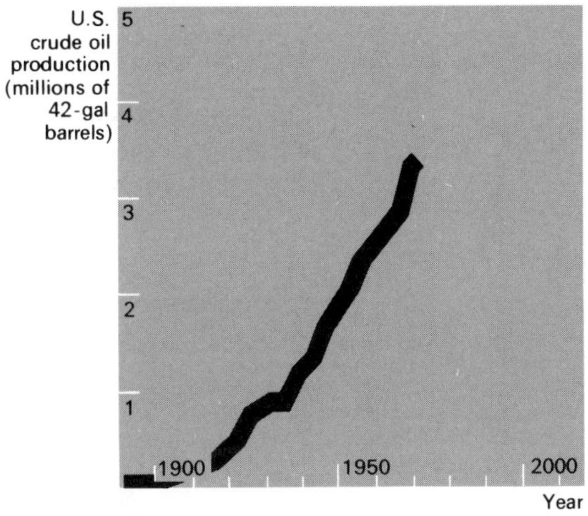

Fig. 9 United States annual crude oil production.

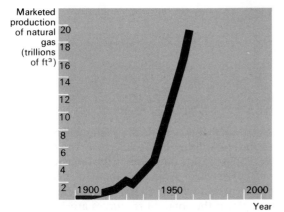

Fig. 10 United States annual marketed production of natural gas.

Assuming that the rates stay at the 1968 levels (see Fig. 9, for example), we find that the lifetime of the remaining resources will be

$$T_{coal} = \frac{amount}{rate} = \frac{1580 \text{ billion tons}}{0.55 \text{ billion tons/yr}} = 2900 \text{ yr}$$

$$T_{oil} = \frac{74 \text{ billion barrels}}{3.4 \text{ billion barrels/yr}} = 22 \text{ yr}$$

$$T_{gas} = \frac{702 \text{ trillion ft}^3}{19.3 \text{ trillion ft}^3/yr} = 36 \text{ yr}$$

Although the lifetime for coal is still fairly long, it is shockingly small for oil and gas. It is true that these results apply only to the United States, but the situation is not much different worldwide.

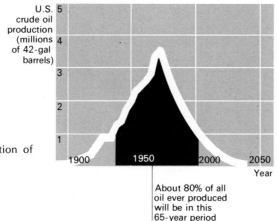

Fig. 11 Possible way that consumption of crude oil will progress in time.

It is, of course, unrealistic to assume that the production rates will remain at the 1968 levels. But so is saying that the rate will continue to increase. The rate must reach a peak and then decline (Fig. 11). If the rate decreases in the same way that it increased, then the production rate curve will be symmetric about the peak value. Where will the peak occur? If the rate curve is symmetric, it will peak when half the total resources are used. For example, if the total oil resources are 165 billion barrels, then we might expect the production rate to peak after about 82 billion barrels have been produced. From Table 1, 91 billion barrels were produced through 1968, so production should peak around 1968. The rate in future years might be as shown in Fig. 11. When the rate varies with time as shown in this figure, it is meaningless to talk about the lifetime for the reserves because these will never be exhausted completely due to the prohibitive cost of working marginal deposits. But the time required to produce a certain percentage of the total reserve can be determined. For example, about 80% of our oil resources will have been produced and consumed between 1940 and 2000—an incredibly short time compared with the millions of years it took for oil deposits to build up. In fact, this much of our oil will have been consumed in a time span which is small even in comparison with the age of the United States. Clearly, the fossil fuel reserves are limited and it is imperative to seek new technologies for energy production.

ENERGY

The measuring units of tons, barrels, and cubic feet associated with coal, oil, and natural gas are fairly common and present no conceptual difficulty. Likewise, we all know that gasoline for our automobiles is purchased by the gallon and that in the operation of the automobile it is depleted at some rate (often too high for our liking). However, it would probably be more meaningful to purchase gasoline in terms of its energy content since that is the usage for which it is intended. And although we have some intuitive feeling for what energy is, the concept involves ideas and measuring units unfamiliar to many of us. Yet if we want to understand more fully the environmental impact of the utilization of our energy resources, then it is important that we understand some of the physical principles and measuring units involved.

Obviously some compromise has to be made on the depth of the treatment because complete texts and courses can be built around this theme. If we establish our starting point as the idea of a force (push or pull) doing work on some object by moving it some distance, then it is logical to consider first the basic notions of speed, velocity, and acceleration which form a background for the concepts of forces and their action. These concepts are fundamental for several discussions in later chapters. Our treatment intends to illustrate only those *principles* and *units* which are useful in the study of our physical environment. Much supplemental reading material is available and you may want to delve into it. Several appropriate references are given at the end of this chapter.

Speed, Velocity, and Acceleration

Speed, velocity, and acceleration can all be categorized as rates. Speed is the most common of the three. Here the amount is a distance traveled in some time period so that

$$\text{speed} = \frac{\text{distance traveled}}{\text{time period}}$$

$$s = \frac{d}{t}$$

If a runner completes a 100-yard dash in 10 seconds, his speed would be

$$s = \frac{100 \text{ yd}}{10 \text{ sec}} = 10 \text{ yd/sec}$$

Every automobile is equipped with a speedometer which continuously displays the speed of the vehicle. The speedometer in an American automobile denotes the speed in miles/hour. Hence if the speedometer of a car reads 50 miles/hour and if this speed is maintained, then in a time period of 3 hours the car will travel a distance

$$d = s \cdot t$$

$$= 50\frac{\text{mi}}{\text{hr}} \cdot 3\,\text{hr}$$

$$= 150\,\text{mi}$$

If this speed is not constant, then the distance traveled must be found from the "area" under the speed versus time curve.

It is important to note that the concept of speed avoids all reference to direction. It is easy to imagine situations where the direction in which an object travels is just as important as the distance traveled. For example, a pilot who wants to know how long it will take to fly between New York and Chicago needs to know both the speed and direction of the plane. Velocity provides the means by which the direction of travel is included in the speed. A car traveling due North at 50 miles/hour would have the same speed as one traveling due East at 50 miles/hour. Their velocities, though, would be different.

Determining a distance traveled when the speed changes with time is not a particularly easy task. If the direction of the object also varies, you can imagine how complicated the problem becomes. Without delving into the problem, we can extract the essentials involved by considering the situation where an object moves only in a straight line. Thus we have only to account for two directions. If these two directions are characterized by the East-West geographic directions, then the object must be traveling due East, due West, or be stationary. This East-West nomenclature is common, but there are various other designations. The designation of the two directions as positive $(+)$ and negative $(-)$ is useful because it coincides with the positive and negative directions associated with graphs and also because the addition and subtraction of velocities are handled by the ordinary rules of algebra involving positive and negative numbers. An object traveling to the right with a *speed* of 50 miles/hour would have a *velocity* of $+50$ miles/hour.

−direction   Origin   +direction

Acceleration, like speed, finds everyday usage. Alvin Toffler in his best-selling book "Future Shock" talks about the "acceleration of change in our time." Every driver knows that he must depress the accelerator pedal to increase the speed of his automobile. All automobiles do not, however, respond in the same way to a depression of the accelerator. Some will change speed much faster than others. The acceleration response of a car is often characterized by the time it takes to achieve a certain speed, say 60 miles/hour, starting from rest. In the comparative evalua-

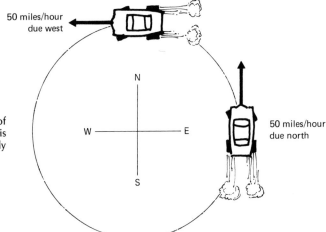

50 miles/hour
due west

N

W —————— E

S

50 miles/hour
due north

Fig. 12 A car traveling at a constant speed of 50 miles per hour on a circular racetrack is accelerating because its direction is constantly changing.

tion of two cars, the car achieving the designated speed in the *least* time would have the *greater* acceleration. All these everyday observations and ideas are incorporated in a simple rate expression

$$\text{acceleration} = \frac{\text{change in speed}}{\text{time required to produce the change}}$$

To be a little more rigorous, direction should be included and acceleration defined in terms of a change in velocity rather than change in speed. In other words, an object can be accelerated by changing either (or both) the speed and direction. For example, a car traveling at a constant speed of 50 miles/hour on a circular race track is accelerating because its direction is constantly changing (see Fig. 12). Just as for velocity, we will limit our discussion to straight-line motion. If the velocity of a car changes from +50 miles/hour to +100 miles/hour in 1 hour, then its acceleration is

$$a = \frac{(100 - 50)\ \text{mi/hr}}{1\ \ \text{hr}}$$

$$= \quad + \quad\quad 50 \quad\quad \frac{\text{mi}}{\text{hr}^2}$$

$$\quad\quad \text{direction}\quad \text{magnitude}\quad \text{units}$$

A positive acceleration means that the velocity has changed in the positive direction. For example, if a car traveling at +50 miles/hour began to accelerate at +50 miles/hour^2, then its velocity after 1 hour would have changed by +50 miles/hour and would be

$$+50\frac{\text{mi}}{\text{hr}} + 50\frac{\text{mi}}{\text{hr}} = 100\frac{\text{mi}}{\text{hr}}$$

If the velocity of a car changes from $+100$ miles/hour to $+50$ miles/hour in 1 hour, then its acceleration is

$$a = \frac{50 - 100}{1} = -50 \frac{mi}{hr^2}$$

This means that the velocity is changing in the negative direction. If a car traveling at $+50$ miles/hour began to accelerate at -50 miles/hour^2, then its velocity after 1 hour would have changed by -50 miles/hour and would be

$$= 50 \frac{mi}{hr} - 50 \frac{mi}{hr} = 0$$

In this case, we would say that the car is decelerating because its *speed* is decreasing.

Newton's Three Laws of Motion

Force is a word used in a variety of contexts, but it always implies a stimulus for motion of some kind. As examples: we are *forced* to get up for a 7:30 A.M. class, or he was *forced* off the road by another car. A push or pull is, perhaps, the simplest illustration of a force. Newton's three laws of motion provide both a verbal and a quantitative base for discussion of forces acting on some object. Newton's statement of the first law reads: "Every body continues in its state of rest, *or of uniform motion* in a straight line, unless it is *compelled* to change that state by *forces* impressed on it." Uniform in this context means constant and the motion refers to velocity. So if the velocity is constant (both in magnitude *and* direction), the acceleration is zero. And the velocity can be changed only by applying a force. Whenever an object is observed to be accelerating, Newton's first law suggests looking for the force responsible for the acceleration. Often this force which produces the acceleration is obvious; other times it is not. A batted baseball is accelerated by the force that the bat exerts on the ball. A ball dropped from the window of a building accelerates toward the ground. But what is the force that accelerates it? It turns out that the earth, though not in contact with the ball, pulls on the ball and accelerates it. This force is commonly called gravity.

Suppose that an experiment is conducted in which a moving automobile makes a collision with an automobile which is initially at rest. The struck automobile is accelerated by virtue of a force exerted on it by the moving automobile. Clearly we should expect that if the force increases, the acceleration also increases. And if the moving automobile always exerts the same force, then we should expect that the struck automobile always achieves the same acceleration. For the sake of argument, suppose that a force of 100 units produces an acceleration of 1 unit when there are no occupants in the struck automobile. Now suppose that we loaded the stationary automobile with passengers. Would we expect a force of 100 units to produce an acceleration of 1 unit? Surely we would not. And we would probably all agree that the acceleration of the loaded automobile would be less. You need not do this experiment if you remain unconvinced. You might rather try kicking different objects on the floor with the same force. Every object offers some resistance to being put into motion. This property of resistance is called inertial mass.

As for velocity and acceleration, force has an associated direction. There are simple experiments which illustrate why. If you want to push a car horizontally on a flat road, you do not push down on the top of the car. Likewise, if someone pushes on the rear of a car with the same force that you are pushing on the front of the car, then the two forces cancel and nothing happens to the car. If we again consider only motion in a straight line, then we can use the same procedure as for velocity and acceleration, and distinguish the two possible directions of a force as $+$ and $-$.

Newton's second law of motion, or the law of inertia as it is sometimes called, quantifies all these ideas and states that the acceleration experienced by an object is proportional to the *net* force acting on it and is inversely proportional to its inertial mass:

$$\text{acceleration} = \frac{\text{net force}}{\text{inertial mass}}$$

$$a = \frac{F}{m}$$

or

$$F = ma$$

The quantification is not complete until we assign units and methods of measurement for force and mass. The most universal system of units uses meters for length and seconds for time. Acceleration would then be expressed in meters per second per second, abbreviated m/sec^2. In this system, the unit of mass is the kilogram (kg).* The standard 1-kg mass is a cylinder made of platinum housed at the International Bureau of Weights and Measures near Paris. The mass of any arbitrary object is measured relative to this standard mass. In principle, this is done by applying the same force to the standard and arbitrary masses and then measuring the acceleration of each. The arbitrary mass is then calculated with the help of Newton's second law.

$F = ma$ (arbitrary mass)

$F = m_s a_s$ (standard mass)

Since the force is the same for both masses, then

$ma = m_s a_s$

$m = \dfrac{m_s a_s}{a}$

---

*The system using meters, kilograms, and seconds as the units of length, mass, and time is commonly called the MKS system. Whenever possible, this system will be used.

Using Newton's second law, we can now easily see what the unit of force is.

$$F = ma$$
$$= \frac{kg \cdot m}{sec^2}$$

(1)

One $kg \cdot m/sec^2$ is called a newton (N). If a 1-kg mass experiences an acceleration of 1 m/sec², then the force on the mass is 1 N. There are many secondary methods of measuring forces such as with a spring balance, but they are all consistent with this basic method.

Given the acceleration of an object, then one can sort of work backwards and find the velocity and the distance traveled by an object. Newton's second law is extremely important because it allows you to determine the acceleration of an object from an analysis of the forces acting on it. We will find this feature extremely useful, for example, when we study motor vehicles in Chapter 6.

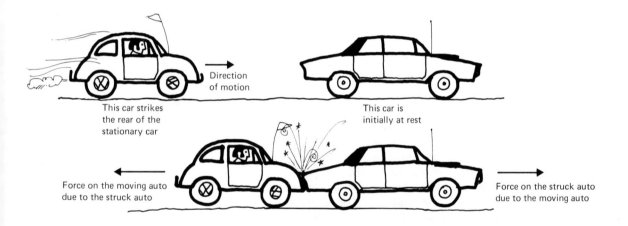

Fig. 13 Illustration of the forces on two colliding cars.

Newton's third law of motion states that if one object (call it A) exerts a force on another object (call it B), then B exerts an equal but oppositely directed force on A. Let us consider the colliding automobile example discussed in connection with Newton's second law. In our new context if the moving automobile exerts a force on the struck automobile which pushes it forward, then the struck automobile exerts an equal but oppositely directed force on the moving automobile which tends to push it backwards (see Fig. 13). Newton's third law does not seem like a particularly useful concept in terms of calculating the motion of objects or systems. It is, however, an extremely important concept and is particularly useful when trying to understand how things like atoms, molecules, and earth-satellite systems are held together.

As an illustration of these ideas, let us verbally and analytically examine the motion of a particle of ash released from a smokestack. Several forces act on the particle. First, the earth exerts a downward pull. This is just the force of gravity mentioned earlier. It is the same force that accelerates a ball to the floor when released from your hand. Second, the particle is buoyed upward just as a balloon in an air atmosphere. Third, there is a drag, or frictional, force similar to that exerted on an airplane as it flies through the atmosphere. The drag force is always opposite to the direction of motion. Finally, if there is a wind, it exerts a force. Figure 14 is a "picture" of this complicated force situation. The arrow denotes the direction of the force and the length of the arrow denotes its magnitude. Since the wind force is difficult to handle and not essential to the illustration, let us ignore it. Then if we call the direction up from and down toward the earth positive and negative, respectively, the net force is

net force = drag force + buoyant force − gravity

In symbols

$$F_{net} = F_D + F_B - F_G$$

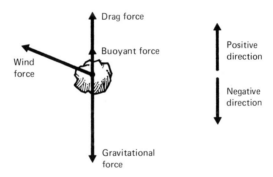

Fig. 14   An example of the forces acting on a particle released from a smokestack. The direction shown for the drag force presumes that the particle is falling toward the earth.

Newton's second law equates this net force to the mass of the particle times its acceleration.

$$F_D + F_B - F_G = ma \tag{2}$$

and

$$a = \frac{F_D + F_B - F_G}{m}$$

If the particle accelerates upwards, then there must be a net force in the upward direction which we have called the positive direction. Hence the sum of the drag and buoyant forces, $F_D + F_B$, must be greater than gravity, F_G. This is exactly what happens when a balloon rises. Conversely, the particle falls when the net force in the downward direction exceeds the net force in the upward direction. What does it mean if the net upward force *equals* the net downward force? The equation says

g = 9.832 m/sec² at the poles

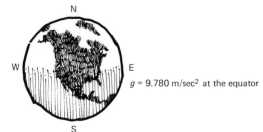

g = 9.780 m/sec² at the equator

Fig. 15 Illustration of the variation of g on the earth's surface.

the acceleration is zero. Does this mean that its velocity is zero? No, it means only that its velocity is constant. This is what happens when a parachutist achieves constant velocity. If there were no drag or buoyant forces, i.e. no air, then Eq. 2 would reduce to

$$a = -\frac{F_G}{m}$$

The minus sign means that the acceleration is always in the negative or downward direction. It turns out that this acceleration depends somewhat on the location of the mass relative to the earth, but is independent of the size and shape of the mass. This was dramatically demonstrated by the Apollo astronauts who showed that different objects dropped from the same height above the moon's surface struck the surface at the same time. This acceleration due to the gravitational force is given the symbol g (see Fig. 15). For all practical calculations, g is taken to be 9.8 m/sec². For example, the velocity of a particle starting from rest and subject to the gravitational force would be

velocity = acceleration × time
$$v = g \cdot t$$
$$= 9.8\, t$$

In 10 seconds the particle will achieve a velocity of

$$v = 9.8 \frac{m}{sec^2} \cdot 10 \text{ sec}$$

$$= 98 \text{ m/sec}$$

The force of gravity on an object is called its weight. From Eq. 1 we see that the force of gravity, that is, the weight, is

$$F_G = W = mg$$

Thus weight in the MKS of units is expressed in newtons. In the familiar British system of units, the unit of weight is the pound. For purposes of comparison, 1 pound is equivalent to 4.45 newtons.

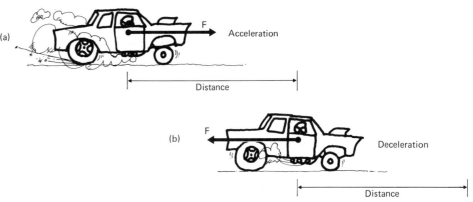

Fig. 16 (a) Force acts in the same direction that the car moves tending to increase its speed. This is a case of positive work.
(b) Force acts in a direction opposite to that in which the car moves tending to decrease its speed. This is a case of negative work.

Work

In everyday language, the word "work" is associated with labor or toil. In this sense, it can be thought of as "a physical or mental effort directed *toward* the production of something." From the physics standpoint, the effort is supplied by a force, and if a force acts on a mass while it moves through some distance (*d*), work is said to be done. The amount of work is defined as

$$W = Fd$$

Work has the units of force times distance, that is, $kg\text{-}m^2/sec^2$. One $kg\text{-}m^2/sec^2$ is called a *joule* (J). Although the physical definition of work retains much of the popular notion, there is a subtle but important difference. No matter how much force (or effort) is exerted, no work is done in the physics sense if the mass does not move. It is important also to recognize that a force can act either in the same direction as the mass is moving or opposite to the direction in which the mass is moving (Fig. 16). We distinguish between these two physically different situations by calling the work positive if the force is in the direction of motion, and negative if the force is opposite to the direction of motion. The *net* work on an object is the algebraic sum of the works done by each force acting on the object. For example, suppose that two forces of $+100$ N and -50 N caused an object to be moved a distance of $+2$ meters. Then the work due to the two forces, respectively, would be $(+100)(+2) = +200$ J and $(-50)(+2) = -100$ J. The net work would be $+200 - 100 = +100$ J. The reason for making such a distinction will become clearer later.

Energy and Work

Energy in almost any context can be thought of as the capacity for doing work. Thinking physically we might ask: "Under what conditions does something have a capacity for moving an object through some distance?" A car *A* in motion has this capacity because if it runs smack into the rear of car *B* stopped at a red light, *A* will surely move *B* some distance. The energy associated with motion is called *kinetic energy*. The kinetic energy of a mass *m* moving with speed *v* is defined as

$$E_k = \tfrac{1}{2}mv^2$$

The units are kg-m^2/sec^2, which are precisely those of work. If we retain the concept of energy as the capacity for doing work, it would have probably been more surprising had work and energy not had the same units. The work-energy principle ties these concepts together quantitatively by stating that the *change* in kinetic energy of a mass m is equal to the *net* work done on the mass.

$$W_{net} = E_k \text{ (final)} - E_k \text{ (initial)}$$
$$= \tfrac{1}{2}mv_f^2 - \tfrac{1}{2}mv_i^2$$

where v_f and v_i are the final and initial speeds. The reason for distinguishing positive and negative work is now clearer. If W is positive, the final kinetic energy is greater than the initial kinetic energy. The converse is true if W is negative. This also illustrates the notion of an energy conversion process which is so important in, for example, the electric generation process.

Potential Energy

A car at rest has no kinetic energy, yet it is *potentially* capable of moving another object *if* it can be put into motion. The potentiality arises from the intrinsic energy of the gasoline which, when burned in the automobile engine, can be *converted* to energy of motion. Water at the top of a hydroelectric dam is capable of exerting a force on the rotor of a generator *if* it is released (Exercise 15). Before release, the energy is potential. Once released and set into motion, the water possesses kinetic energy. There is potential energy in coal, oil, and gas because *if* ignited, heat energy is released. There is also potential energy in the mass of the nucleus of the atom which can be released *if* an appropriate reaction is initiated.

The potential energy of dammed water is converted to electric energy at the Grand Coulee hydroelectric station. Although no pollutants are produced by this method, it is not without environmental problems because of disruptions that sometimes result from damming a river. (Photograph courtesy of the United States Department of Agriculture, Soil Conservation Service.)

Energy Conversion and Efficiency

Energy put to *use* is kinetic. An automobile is *useful* when it is in motion. We are *useful* when we are doing things. Heat energy involves the kinetic energy of atoms and molecules. Electricity, probably the most *useful* form of energy, results from the motion of electrical charges (Chapter 2). "Useful" in this context means convenient, abundant, economical, easily convertible, and clean (in the environmental sense). The basic forms of available energy are electromagnetic (sun, for example), chemical, nuclear, thermal (heat), mechanical, and electrical. Useful energy results from conversions of one form of energy into another. Let us look, for example, at the production of electricity in a coal burning power plant: first the chemical energy of the coal is converted into thermal energy which is converted, in turn, to mechanical energy by a steam turbine which drives an electric generator (Fig. 17).

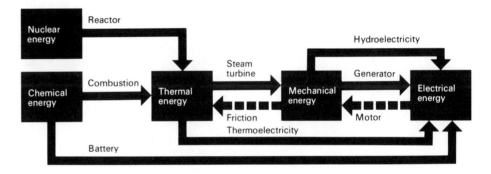

The principle of conservation of energy which states that the amount of energy is the same before and after conversion applies to all these processes. This does *not* mean that all the energy released is converted into one desired form. Rather it means that if all *forms* of energy are accounted for, the total amount of energy before and after the conversion is the same. For example, a steam turbine converts about 50% of the thermal-energy input to mechanical energy. The remaining thermal energy is released to the environment.

The notion of *efficiency* as a quantitative indicator is significant in these processes. For our purposes, it suffices to say that

$$\text{efficiency} = \frac{\text{useful energy out}}{\text{total energy converted}}$$

In symbols [the Greek epsilon (ϵ) is often used to denote efficiency]

$$\epsilon = \frac{E_{\text{out}}}{E_{\text{converted}}}$$

The function of an incandescent light bulb is to convert electric energy into visible radiant energy. Yet the efficiency of this process is only about 5%. This

Fig. 17 Schematic diagram of the common energy conversion processes.

means that if 5×10^6 Joules of electric energy are supplied to a light bulb, the amount converted to light energy is

light energy = total energy converted × efficiency

$$= (5 \times 10^6 \text{ Joules}) \times 0.05$$

$$= 2.5 \times 10^5 \text{ Joules}$$

The remainder of the energy that was converted, 4.75×10^6 Joules, emerges as thermal (or heat) energy.

Power

The overall mileage (distance) an automobile travels during its lifetime probably hasn't changed significantly in the last 40 years. If this is true, then the total work done by an automobile is essentially the same. However, the *rate* at which work is done has changed because the speeds have increased over the years. The rate of doing work is thus a significant quantity. This rate is called power, defined quantitatively as

$$\text{power} = \frac{\text{amount of work done}}{\text{time required to do work}}$$

$$P = \frac{W}{t} \frac{\text{joules}}{\text{sec}} = \frac{W}{t} \text{watts}$$

where W is the work done in the time interval t. The horsepower (hp) rating of an engine is a measure of the rate at which the engine does work. One horsepower is equivalent to 746 watts (W). So if an engine develops 200 hp, it is doing work at the rate of

200 hp $\times$ 746 W/hp = 1.492×10^5 J/sec

An electric motor of a household appliance develops only a fraction of a horsepower, say $\frac{1}{6}$ hp. This is equivalent to about 120 W.

Other Common Energy and Power Units

Words associated with units of measurement are sometimes entertaining, historically enlightening, and provide challenges for crossword enthusiasts. A noggin (1/32 of a gallon) is a good example. However, life would certainly be simpler if there were only one universal measuring system. For instance, the joule and watt are really all the units needed for energy and power. There are, however, a variety of other measuring systems in use, and hence we have to learn how to convert from one system to another. Table 2 is provided to help in this chore. Most of these units are defined in subsequent chapters.

Table 2. Some Common Energy and Power Units

Energy

kilowatt-hour (kW-hr) = 3,600,000 joules (J) = 3.6 × 10⁶ J
calorie (cal) = 4.186 J
kilocalorie (kcal) = 4186 J
British thermal unit (Btu) = 252 cal = 252 (4.186) J = 1050 J
electron volt (eV) = 1.602 × 10⁻¹⁹ J
million electron volt (MeV) = 1,000,000 eV = 10⁶ eV
food calorie = kcal = 1000$_{cal}$

Power

3,413 Btu/hr = 1 watt (W)
kilowatt (kW) = 1000 W = 10³ W
megawatt (MW) = 1,000,000 W = 10⁶ W
gigawatt (GW) = 1,000,000,000 W = 10⁹ W

Energy Equivalents

1 gallon gasoline = 126,000 Btu = 1.26 × 10⁵ Btu
1 pound bituminous coal = 13,100 Btu = 1.31 × 10⁴ Btu
1 cubic foot natural gas = 1,050 Btu = 1.05 × 10³ Btu
1 42-gallon barrel of oil = 5,510,000 Btu = 5.51 × 10⁶ Btu

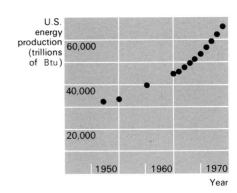

Fig. 18 Total annual energy consumption in the United States.

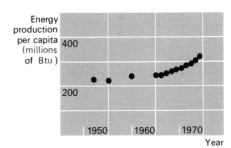

Fig. 19 Total annual energy consumption per capita in the United States.

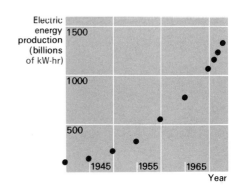

Fig. 20 United States annual electric energy production.

DEMANDS FOR ENERGY

As we have discussed at the beginning of this chapter, the ravenous energy appetites of the United States have given rise to grave concern about pollution and depletion of energy stores. If the energy consumption were diminishing, then the alarm could be unwarranted; but the data in Fig. 18 show that the opposite is true. Anxieties about population growth suggest that it is responsible for these ever-increasing energy demands. We can examine the effect of population growth by dividing energy production by population and replotting the resulting per capita production as shown in Fig. 19. When we compare the production per capita and energy production growth curves, we see that both rise, although the former at a smaller rate, as expected. Therefore, the fact that energy consumption per capita is increasing means that as individuals we are requiring energy at an ever-increasing rate.

Are there demands for particular forms of energy? Yes, specifically for electric energy* and energy from fossil fuels. This is dramatically illustrated in Fig. 20. There are many reasons for this increase in energy requirements, but most of them can be reduced simply to our desire for convenience. We have a passion for gadgets that mix, toast, roast, open garage doors, trim hedges, and above all, cool homes in the summer. It is not necessarily the energy used to operate

*The principles and concepts of electric energy are discussed in Chapter 2.

some of these devices that is of concern, but the energy that goes into making them. Aluminum, which is widely used, especially in packaging, requires much energy in the refining process. Synthetic fibers like nylon and rayon also are important energy consumers. Besides consuming great quantities of energy, these materials are extremely hard to dispose of appropriately. Witness, for example, the aluminum containers around our parks and highways.

PREDICTING THE GROWTH OF ELECTRIC ENERGY IN THE UNITED STATES

The data in Fig. 20 illustrate with great impact the dramatic increase in electric energy production in the United States. Using ideas already discussed, we could

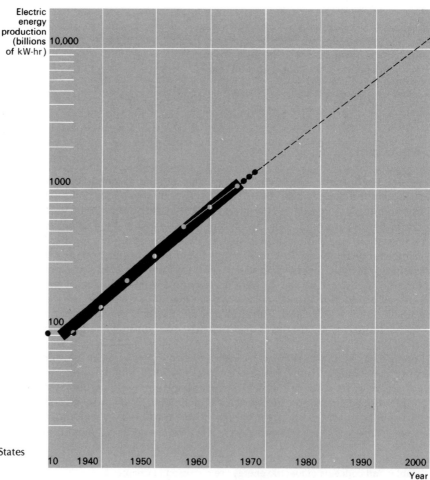

Fig. 21 Semilogarithmic plot of United States electric energy production versus time.

determine the total production in the given time period, obtaining a useful quantitative result. There are, however, more interesting and fundamental questions to be answered. For example, what is a reasonable estimate of the energy production ten years from now and why does it increase in the observed manner? Making the estimate seems easy because it involves only an educated guess of the progression of the graph. But, since the graph is not a straight line (it is nonlinear) a quick guess may be rather deceptive. Fortunately, data can often be plotted in another way which will yield a straight line and make estimations more reliable. One of the most useful techniques is to plot the logarithm of the number rather than the number itself. Actually, because this method is so widespread, logarithmic and semilogarithmic graph paper is available, where one or both of the axes are proportional to the logarithms, saving one the chore of looking up logarithms in a table.* Figure 21 shows the results of plotting electric energy production versus time on semilogarithmic graph paper. We can certainly agree that the data are reasonably well described by a straight line over the entire period and very well described from 1955 to 1968. The rate of growth has decreased, but not very much, for the 1955–1968 period. Extending the straight line and estimating the energy production for 1980 and 2000 yields 3000 and 12,000 billion kW-hr. This compares favorably with more sophisticated estimates of 2700 and 8000 billion kW-hr often cited in the literature.

Extending a curve in this manner is called *extrapolation*. Extrapolation is somewhat like walking out on a tree limb that is in danger of breaking; there is a limit to how far one can go without getting into trouble. Straight-line graphs are the most desirable for predictions but there is no guarantee that the line will continue to be straight far into the future. The more the line is extended, the less reliable the predictions. Some of the dangers of "walking too far on the tree limb" are illustrated in Exercises 11 and 22.

EXPONENTIAL GROWTH

"The art of projecting technological requirements may be likened to a feat of archery. One sets down on a century-wide chart the past 70-year record (*logarithmic graph paper* is essential because of the *exponential character* of the 20th century's technology), and the archer then lines up his arrow shaft with the progression of historical points. The arrow is sent upward on a straight line course and some celestial observer notes where it crosses the terminus of the century. If this sounds somewhat zany, it really is not because most of the points on the chart fall on a *straight* line. They reveal man's monotonously ravenous appetite for power in all forms. The only real variation in the chart is a pronounced dip at

---

*Logarithms and techniques of plotting data are discussed in Appendix 1 to help those not familiar with these ideas. A complete understanding of logarithmic-type graphs is not necessary. However, you should at least be able to "read" them. For example, in Fig. 21 you should be able to determine the energy production in, say, 1953.

© 1970 United Feature Syndicate, Inc. Reprinted with permission.

the time of the Great Depression in the 1930's. Incidentally, kilowatts and the Gross National Product marh in lock step over the course of the past 70 years."*

The key words in Dr. Lapp's eloquent statement are logarithms, straight line, and exponential. The meaning and usefulness of logarithmic and straight lines have already been discussed. "Exponential" is a word encountered very early in the mathematics of calculus. It is widely used to characterize the growth and decay processes of many systems, including biological ones. If "something" grows or decays exponentially, it simply means that a plot of the logarithm of that "something" versus time is a straight line. The line points up for growth and down for decay. Since the logarithm of electric energy production versus time is a straight line pointing up, production is said to be growing exponentially. More quantitatively, if "something" grows exponentially, it also means that "something" will double its value at regular time intervals. For example, Fig. 21 reveals that electric energy production in the United States has doubled about every 10 years since 1935. Many other quantities of environmental interest are increasing exponentially and this is the root of many of our concerns because staggering amounts can be recorded in incredibly short time periods when something grows at an exponential rate. A simple calculation involving the growth of electric energy production in the United States illustrates this dramatically.

The present electric generating capacity of the United States is about 300 million kilowatts. A typical power company produces about 1 million kilowatts so the total power could be produced by 300 typical companies. Assuming that a typical company occupies a square piece of land 0.2 miles (about 1000 feet) on a side, then the 300 companies would occupy an area of 12 square miles. Approximating the United States as a rectangle 1000 miles by 3000 miles, its total area would be 3 million square miles. Now ask, "If the present rate of electric energy production continues, how long will it be until the entire country is covered by power plants?" The answer: after about 16 "doublings" or about 160 years if the doubling time is 10 years. There is no question that this will happen *if* the growth rate continues and current technology for producing electric energy is used. Figure 22 shows the expected growth in steam generating centers for the period 1970 to 1990. The predominant environmental message is that we cannot allow these exponential growths to continue unabated.

To put this idea of exponential growth on a more common basis, suppose that you invested $100 in a savings plan in which your money doubled every 5 years. After 5 years you would have accumulated $200. After 10 years you would not be particularly wealthy, but would have $400. If you continued investing for 50 years or 10 doubling periods, you would accumulate $102,400 and this is significant. Now a savings firm does not tell you how long it takes to double your investment. Rather you are told the investment rate. For the example above, the investment rate is 14 percent per year, abbreviated 14%/yr. If you are familiar with ordinary savings firms, you know that this is a fairly high rate. Still 14%/yr does not seem

---

*From Ralph E. Lapp, "Where Will We Get the Energy." *The New Republic*, July 11, 1970. Reprinted by permission of The New Republic, © 1970, Harrison-Blaine of New Jersey, Inc.

Fig. 22 (a) The major steam generating centers in the United States in 1970. (b) Expected major steam generating centers in the United States in 1990. (From "Power Generation and the Environment," by Rolf Eliassen, *Bulletin of the Atomic Scientists*, September, 1971.)

like a large quantity. Interestingly, the growth rate and doubling time are approximately related by the simple expression

doubling time = 70 ÷ growth rate in %/yr

For example, if the growth rate is 14%/yr, then the doubling time is

$$\text{doubling time} = \frac{70}{14} = 5 \text{ years}$$

which agrees with the example cited earlier. It is clear that both the doubling time and exponential growth rate are useful quantities. Once something has been identified as growing exponentially from a semilogarithmic plot of the data, it is a simple matter of getting the doubling time from the graph. However the growth rate involves a calculation which, if done rigorously, is beyond the intended level of the reader. Nonetheless it can be estimated using common ideas. The fact that it is a percent suggests a fractional change multiplied by 100%. For example, if the weight of a person increases from 125 to 150 pounds the percent change is

$$\frac{25}{125} \times 100\% = 20\%$$

↑
fractional
  change

In the example considered, the investment increased from $100 to $200 in the first five-year period. The percentage change from the beginning is thus 100%. However, the percentage change implied in exponential growth is based on the average between the initial and final values. For the investment example, the average value is between 100 and 200. We can estimate this in the same way we would figure an average grade for two test scores of 100 and 200:

$$\text{average} = \frac{\text{initial value plus final value}}{2}$$

$$= \frac{100 + 200}{2}$$

$$= 150$$

Then the average percent change is

↙change
$$\frac{100}{150} \times 100\% = 67\%$$
↖
  average accumulation in the 5-year period

The rate aspect is achieved by dividing the average percent change by the time period over which the change occurred. Thus

$$\text{growth rate} = \frac{67\%}{5 \text{ years}} \cong 14\%/\text{year}$$
↑
approximately

Let us now do a similar calculation for electric energy production by picking two representative data points from Fig. 21.

Year	Energy (billion kW-hrs)	Average energy	Change in energy	Change in time
1945	222.49			
		275.81	106.65	5
1950	329.14			

$$\text{growth rate} = \frac{106.65}{275.81 \ (5)} \times 100\%$$

$$= 7.7\%/\text{yr}$$

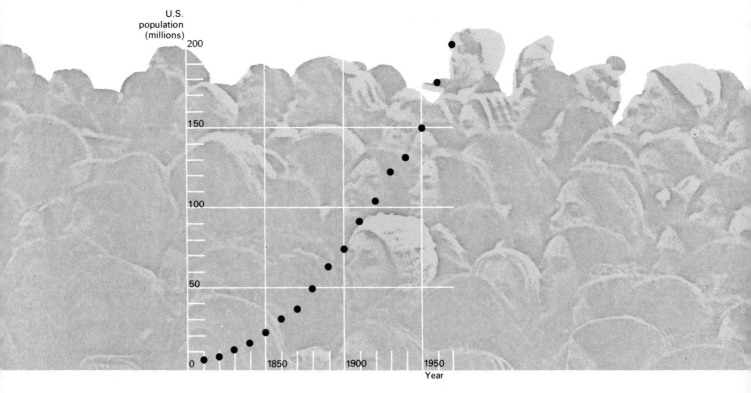

This growth rate does not seem large. It is, in fact, very much like an ordinary savings firm rate. Yet it means that electric energy production doubles about every 10 years which puts an enormous strain on our energy resources and production capabilities.

Fig. 23 United States population versus time.

POPULATION GROWTH IN THE UNITED STATES

Today's concern with environmental problems has made population growth a burning issue. We shall deal with it here not only because of its direct effect on energy requirements but because we want to find out a little more about why populations grow as they do.

   Figure 23, in which the United States population is plotted as a function of time, shows that the population grows in fairly regular but nonlinear fashion much as electric energy production does. Figure 24 shows the same data plotted on semilogarithmic paper. It is clear that the population does not grow exponentially over the entire time period shown, but from, say 1810 to 1870 and 1920 to 1970, the growth is nearly exponential. An analysis (Exercise 20) similar to that for electric energy yields a growth rate of 1.3%/year for the 1920–1970 period. (Compare this with a growth rate of 7.7%/year for electric energy.) What does a population growth rate of 1.3%/year mean? Well, it took until 1916

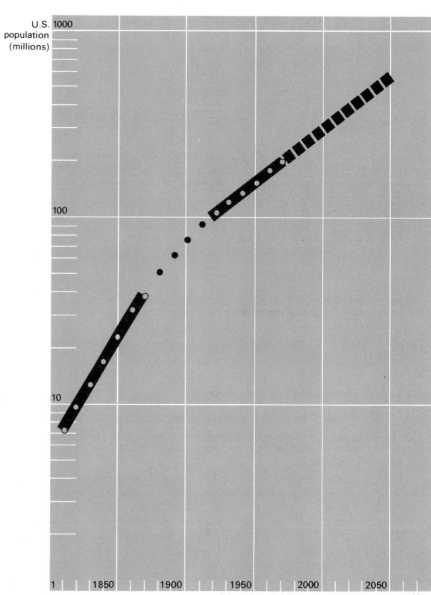

Fig. 24 Semilogarithmic plot of United States population versus time.

for the population in our country to reach 100 million, but it took only 50 additional years to reach 200 million. If the growth continues at this rate, it will take only 30 more years for the population to reach 300 million. The fact we must seriously be concerned with is that exponential growth produces enormous and consequential changes in very short time periods once it starts from a large-number base (see Exercise 21).

Predicting future population is difficult because of the effects of immigration, wars, economy, etc. Any professional attempt would take these factors into consideration. But if we ignore them and predict on the basis of the way things have been going since 1920, we would get something like 305 million for 2000. Such a prediction compares very favorably with more professional ones.

All the properties of exponential growth can be summarized in the expression

$$\frac{A}{t} \div \overline{A} = \text{constant} = \lambda$$

A is the change in the amount in the time period t and $\overline{A}$ is the average value of the amount in the time period t. Regardless of the time period considered, this quantity always yields the same number which is the constant denoted by the Greek lamda (λ). This expression describes many phenomena in physics (see Exercise 26 for example), and physicists instinctively want to know why. They would like to be able to calculate λ starting from first principles. One might say that "all the physics is contained in the parameter λ." All too often the calculation of this innocent looking parameter is extremely difficult. Regarding population growth, it might be said that "all the sociology is contained in the parameter λ" and indeed this is difficult to understand because of such factors as religion, education, economics, etc. Yet it is clear that if the population is to be stabilized, then λ must be reduced to zero. That is why the Zero Population Growth (ZPG) organization first adopted the slogan "Stop at Two" and later "One Will Do." (Exercise 27 illustrates some of the fallacies of the "Stop at Two" reasoning.)*

REFERENCES

Introductory Physics

Physics: A Descriptive Analysis, Albert J. Read, Addison-Wesley Publishing Company, Reading, Mass. (1970).

Elementary Physics: Atoms, Waves, Particles, George A. Williams, McGraw-Hill Book Company, New York (1969).

An Introduction to Physical Science, J. T. Shipman, J. L. Adams, Jack Baker, and J. D. Wilson, D. C. Heath and Co., Lexington, Mass. (1971).

Conceptual Physics, Jae R. Ballif and William E. Dibble, John Wiley and Sons, Inc., New York (1969).

---

*For a short, interesting article on population growth, see "Predicting and Preventing Population Problems" by Robert J. Trotter in *Science News*, **100**, No. 7, August 14, 1971.

Physics: An Ebb and Flow of Ideas, Stuart J. Inglis, John Wiley and Sons, Inc., New York (1970).

Statistical Data

U.S. Bureau of the Census, *Statistical Abstracts of the United States 1970.* (91st Edition) Washington D.C., 1970.

U.S. Bureau of the Census, *Historical Statistics of the United States, Colonial Times to 1957*, Washington, D.C., 1960.

World Almanac.

Federal Power Commission.

The User's Guide to Protection of the Environment, Paul Swatek, Ballantine Books, Inc. (New York), 1970.

QUESTIONS

1. Alvin Toffler in the book "Future Shock" writes "The acceleration of change in our time is, itself, an elemental force." What analogies can you draw between this statement and Newton's second law of motion?

2. A kilowatt-minute is not a commonly used unit. If it were, what physical quantity would it measure?

3. Discuss some of the ways that energy affects man's lifestyles and civilizations.

4. Compare the concept of efficiency as applied to energy conversion with its meaning as used in everyday life.

5. What are some rates of interest that have a direct bearing on our environment?

6. Follow a major city newspaper for a week and note the number of articles that pertain to our environment.

7. A front page article in a prominent newspaper stated that population in the United States is growing at the rate of 10% per year. Do you believe this? Explain. What is your own feeling about population problems in the United States?

8. Discuss some of the problems that arise in trying to estimate fossil fuel reserves.

9. Suppose that you had a fixed amount of money, say $1 million for the sake of argument, and you started spending this at an increasing rate. Why must this rate eventually peak and about how much money would you expect to have left when the rate peaks?

10. Suppose that a natural resource is being consumed exponentially and that the "doubling time" is 10 years. How does the amount consumed in any 10-year period compare with all that was consumed prior to the beginning of this period?

11. Exponential growth occurs when the rate of change of some thing at some time is proportional to the amount of the things present at that time.

$$\frac{N}{t} \alpha N$$

This is why exponential growth is so common. Think of some things which might grow exponentially and then, if possible, get data to check your conjectures.

12. Every benefactor of a student in a college or university is aware of a tuition rate expressed in dollars/year. What feature of this rate makes it difficult for a benefactor to predict the total cost of a four-year education?

EXERCISES

Easy to Reasonable

Exercise 1. A car traveling with a velocity of -50 mi/hr changes its velocity to -100 mi/hr in 10 sec. What is its acceleration in mi/hr²?

Exercise 2. Suppose that speed expressed in mi/hr is proportional to time expressed in min. What are the units of the proportionality constant? What physical quantity would characterize this constant?

Exercise 3. The foot, second, and pound are the common units of length, time, and force in the British system of units. What would the common unit of acceleration be in this system? What would the unit of mass be in this system?

Exercise 4. Suppose that a car obtains 15 miles per gallon of gas when it averages 45 miles per hour. One might also want to know how many gallons per hour the car uses at this speed. Determine this by combining the units, mi/gal and mi/hr, in a way that gives units of gal/hr. How would you have worked the problem otherwise?

Exercise 5. The human population derives energy from food at the rate of about 2500 food calories per capita per day. Show that this corresponds to a per capita consumption rate of about 100 thermal watts.

Exercise 6. Simply as an exercise in conversion of units, show that a kilowatt-hour is equivalent to 3413 Btu.

Exercise 7. Perform the following experiment. Take a sheet of 8½″ by 11″ paper, typically 0.003 inches thick, and cut it into two equal pieces. Then take the two pieces and cut them into four equal pieces. Continue this process of doubling the number of pieces a total of eight times. Are you surprised that the final stack is about one inch thick? Calculate how thick the stack would be if you continued for a total of 25 steps. Answer: 3 miles! This is a good example of how a system grows if it doubles at regular intervals.

Start

First cut

Second cut

Third cut

Exercise 8. On the average, a person uses about 150 gal of water per day for all purposes. Such a figure seems high but it can easily be accounted for. The water supply on a household toilet is a box of approximate dimensions 18 inches long, 7 inches wide, and 12 inches deep. A gallon of water occupies a volume of 231 cubic inches. How many times would a toilet have to be flushed to account for 50 gal? Estimate the amount of water used in showering if the rate of flow of water is 4 gal per min. You will have to estimate how long it takes you to shower.

Exercise 9. Using any method that you desire, show that the electric energy production has doubled about every 10 years since 1930.

Exercise 10. A certain resource of environmental interest is being depleted exponentially at the rate of 7% per year.

    a) How long does it take to use one-half of the amount that is available at some given time?

    b) If there were 120 million tons of this material available in 1970, how much would you expect to be available in the year 2000?

Fig. 25 A typical automobile traffic pattern near a city during rush-hours. Pick out the automobile in which the number of passengers, including the driver, can be identified and determine the average number of passengers per car. This photograph also shows how the median strip of a freeway can be utilized by buses or trains to facilitate mass transit of people. (Photograph courtesy of U.S. Department of Transportation.)

Exercise 11.   The average number of passengers in an automobile (including the driver) in 1970 was about 1.3 (Fig. 25). In 1950, the corresponding figure was closer to 2.0. Using these numbers, extrapolate to 1990 and determine the number of passengers per car trip. What lesson do you learn about the dangers of extrapolation?

Exercise 12.
  a) A typical household electric clothes dryer consumes electric energy at a rate of 6000 W. Estimate the time required to dry a load of clothes and figure the total cost if the rate is 8¢/kW-hr.
  b) A night light in a child's bedroom typically uses a 5 W bulb. Estimate the cost per year if the bulb is left on continuously.

Exercise 13.   The graphs in Fig. 26 are plotted on the same scale. For (a) the distance can be calculated easily because the rate is constant.

distance = rate × time
       = 50 mi/h̶r̶ · 10 h̶r̶ = 500 mi

For (b) the distance cannot be calculated easily because the rate is not constant. But for (a) or (b) the distance is proportional to the area under the curve which is directly related to the number of squares in the enclosed area. For (a) there are 50 squares (count them). Estimate the number of squares in (b) and by comparing your result with the results for (a) estimate the distance traveled in (b).

Reasonable to Difficult

Exercise 14.   The right-of-way for an electric transmission line may extend a few hundred feet on either side of the line.

  a) Look up or estimate how many acres of land are now used for right-of-ways for electric transmission lines (1 acre = 43,560 sq. ft.).
  b) If power requirements continue to double every ten years, estimate the time required for the United States to be completely taken up by right-of-ways for transmission lines.

Exercise 15.   A mass of water at the top of a dam has potential energy because of its advantageous position. It seems reasonable that the amount of potential energy would be proportional to the mass (call it m) and the height it is above the dam (call it height h).

potential energy is proportional to mass × height

P.E. α mh

Potential energy must have the units of work. If the equation above is written as an equality, then

P.E. $= kmh$

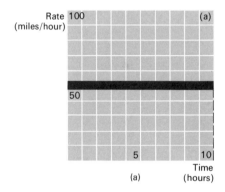

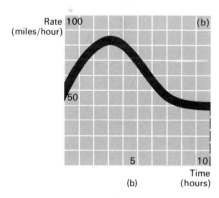

Fig. 26   (a) A constant rate versus time curve. (b) A situation in which the rate varies with time.

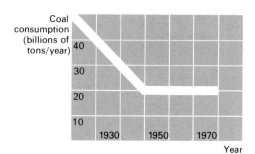

Fig. 27 A hypothetical coal consumption curve.

where k is the proportionality constant. Then kmh must have the units of work. If m is in kg and h is in meters, what are the units of k? What physical constant do you know of that also has these units? Do you suspect any connection?

Exercise 16. Suppose that coal is consumed at the rate shown in Fig. 27.

a) How much is consumed between 1910 and 1970?

b) Measuring from 1910, at what year will half of the coal be consumed?

Exercise 17. Arthur Schlesinger, an adviser to President Kennedy, in a public address spoke of the "velocity of development of technology". As an example, he considered the distance that man could travel in 3 hours at various times in history. Limited to walking he could go about 10 miles and with domestication of the horse, about 25 miles. Consider the effect of the railroad, piston, jet and SST airplanes and spacecraft, and estimate these distances as a function of the time they appeared in history. If this trend continues, estimate the distance man might travel in 3 hours in the year 2000. Can you think of some possible upper limit for this distance?

Exercise 18. The first large vacuum-tube computer was completed in 1951. Seven years later the first transistorized computer appeared with a computing speed ten times faster. Today's computers which were first introduced in 1965 compute ten times faster than the first transistorized computers. As computing speed has increased, the physical size of the computer has decreased by nearly the same proportion. It appears that computer technology will continue to develop at this pace. Make some estimates and conjectures for the consequences of this in the year 2000. When you have done this you might want to read "Man in Space and Chips in Space" by Robert Jastrow in *The New York Times Magazine*, January 31, 1971. (See also "Computer Pollution," *Time* **98**, Aug. 16, 1971.)

Exercise 19. The equation $r = at$ for rate proportional to time "says" that the rate is zero when time is zero, that is, when the clock is started. Often the rate is not zero at this time. The equation

$$r = at$$

can be altered to include the initial rate by writing

$$r = \text{initial rate} + at$$
$$= r_0 + at$$

The notation r_0 is used for the initial rate. The subscript 0 is a reminder that it corresponds to zero time. A plot of r versus t now appears as shown in Fig. 28. The amount is still the area under the curve which is simply the area of a triangle plus the area of a rectangle. Show that

$$A = r_0 t + \tfrac{1}{2}at^2.$$

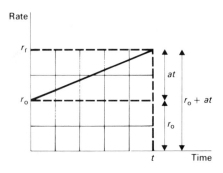

Fig. 28 Uniform rate versus time curve if the initial rate is not zero.

Exercise 20.   The United States population from 1920 to 1968 is given in Table 4.

a) Using a procedure like that used in the text for the growth of electrical energy production, show that the population grows at the rate of about 1.3% per year.

b) Make a graph of the log of the population versus time and deduce the time required for the population to double.

c) Using your rate deduced in part (a), calculate the theoretical doubling time, or $T = 70 \div$ growth rate, and compare your result with that obtained from the graph.

d) Estimate the United States population for the year 2050 and compare with some estimate you can find in the literature.

Table 4

Year	Population (millions)
1920	105.71
1930	122.78
1940	131.67
1950	150.70
1960	179.32
1968	201.17

Exercise 21.   An estimated 500,000 Pakistanis were tragically killed by a cyclone in November 1970 (*Science* **171**, #3971, 12 February 1971). Surprisingly, perhaps, this population loss was repaired in about 40 days. Look up the 1970 population and growth rates for Pakistan and verify this figure by a calculation assuming exponential growth.

Exercise 22.   A spring when stretched tends to restore itself to its unstretched (or equilibrium) position. X is the extension from the unstretched position. A plot of restoring force versus extension might be as shown in Fig. 29.

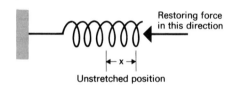

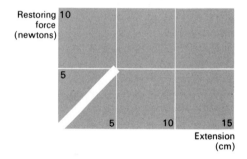

Fig. 29 A typical restoring force–displacement relationship for a spring.

a) Based on the graph shown, estimate the restoring force if the spring is extended 10 cm.

b) Suppose, in actuality, that the restoring force is 8 N. What has happened?

c) Suppose that when the extension is 12 cm the restoring force is zero. What has happened? What lesson is learned? Ligaments in the knee, for example, also behave very much like this.

Difficult to Impossible

Exercise 23.   The production of natural gas in the United States is shown in Fig. 10. Because the reserves are limited, the production rate will peak sometime after 1970. Sketch a curve which you feel will reasonably approximate the production rate in the years following 1970 and estimate the fraction of the total amount produced in the 65-year period around the peak.

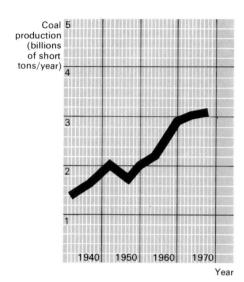

Fig. 30 World production of coal, lignite, and anthracite.

Exercise 24. The equation which characterizes exponential growth is

$$A = A_0 e^{\lambda t}$$

If $A = 2A_0$, then the quantity has doubled. The time for this is called the doubling time, t_D. Substituting yields

$$e^{\lambda t_D} = 2$$

Show by taking logarithms of both sides that $t_D = 0.69/\lambda$. Values of the logarithms can be found in Appendix 1.

Exercise 25. The world production of coal and lignite is shown in Fig. 30.

  a) Determine the total amount produced for the time period shown.

  b) Measuring back from 1968, determine the time required to produce one half of the total amount produced up to 1968.

Exercise 26. Several systems which decrease exponentially either with time or distance will be encountered. This would mean that a plot of the log of whatever is decreasing versus time is still a straight line, but a line which slopes in a direction opposite to that for exponential growth.

Table 5

	Time (min)	Decays/sec
	0	4617
	0.6	3963
	1.2	3332
	1.8	2904
	2.4	2542
The accepted	3.0	2114
half-life is	3.6	1820
2.55 min	4.2	1541
	4.8	1279
	5.4	1151
	6.0	960
	6.6	789
	7.2	728
	7.8	603

It also means that $(S/t) \cdot (1/S)$ is negative. The decay of a sample of radioactive nuclei typifies this behavior. Some data for the decay of radioactive cesium is given in Table 5.

  a) Plot the log of decays per second versus time.

  b) For a system which decreases exponentially, it is appropriate to talk about the time required for the system to decrease by a factor of 2 (this is analogous to the doubling time for exponential growth). This time is called the half-life when the system is radioactive nuclei. Determine the half-life of the radioactive cesium using your graph from part (a).

  c) Determine the decay rate in %/min. Calculate the half-life and compare your result with that deduced from the graph.

Exercise 27. Any attempt at predicting population growth requires (1) the distribution of people according to age and (2) birth and death rates. This information is known, available, and, no doubt, fairly complex. For the sake of illustration, suppose that at some given time a particular distribution is as shown in Fig. 31.

a) Determine the total number of people at this given time. Now suppose that no one dies until age 65. Secondly, suppose that only those women presently in the 20–35 age group have children and that each has two in a 15-year period. Assume that one-half the people in the 20–35 age group are women.

b) Make a sketch like Fig. 31 showing the population distribution in 15 years.

c) Determine the total population and deduce whether the population has grown.

d) The purpose of this exercise is to show that a population can grow for a while even though each mother has only two children. It has been stated that if this were in effect now, that is, if married couples have only two children and if all women married and gave birth, the population would still grow for about 70 years. Try to figure out why this is so.

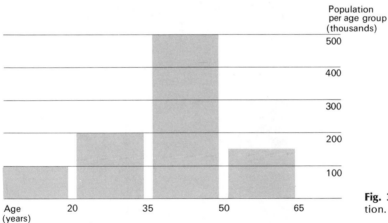

Fig. 31 A hypothetical population distribution.

2

ELECTRIC ENERGY: PHYSICAL MODELS, BY-PRODUCTS, AND ALTERNATIVES*

Photograph by EPA newsphoto/Editorial Photocolor Archives.

*"The Energy Game" is a film which discusses the electric power crisis in the U.S. It is available on free loan from AAAS, Office of Communications Programs, 1515 Massachusetts Ave., Washington, D.C. 20005.

INTRODUCTION

A pollutant is any substance added to the environment in sufficient concentration to affect something that man values. There are many things that man values to a greater or lesser degree; some he may even sacrifice. That which he is most reluctant to compromise is his health and well-being. Table 1 lists the air pollutants considered as health hazards by the Environmental Protection Agency (E.P.A.).

The major polluters fall into three predominant categories: transportation; stationary fuel combustion sources (this category includes electric power plants); and industry. Strictly on a weight basis, transportation accounts for about 50% of all air pollution. However, various pollutants produce different environmental effects and a small amount of one may be much more detrimental than a large amount of another. Thus when setting maximum allowable amounts of pollutants, the effect must also be considered. Interestingly, fuel combustion sources and industry are mainly responsible for sulfur oxides and particulate matter while transportation is mainly responsible for carbon monoxide, nitrogen oxides, and hydrocarbons. From this delineation it is logical for us to progress to an examination of the technology of each and the pollutants which follow. The procedure governing our examination will follow the general theme of "physical models, by-products, and alternatives."

Table 1. Sources of air pollution (1969)—millions of tons annually. (From *Statistical Abstracts of the United States: 1971.* (92 Edition) U.S. Bureau of the Census, Washington, D.C. 1971.

	Carbon monoxide	Sulfur oxides	Nitrogen oxides	Hydro-carbons	Particulate matter	Total
Transportation	111.5	1.1	11.2	19.8	0.8	144.4
Fuel combustion (Stationary Sources)	1.8	24.4	10.0	0.9	7.2	44.3
Industrial Processes	12.0	7.5	0.2	5.5	14.4	39.6
Refuse disposal	7.9	0.2	0.4	2.0	1.4	11.9
Miscellaneous	18.2	0.2	2.0	9.2	11.4	41.0
						281.2

CONCEPT OF A MODEL

The word, model, has many connotations. To an aeronautical engineer it might mean a scaled-down version of an aircraft or to a mother it might denote the son whose behavior is ideal. But this type of meaning is not implied in topics like "The Bohr Model of the Atom," "Models of the Ocean," "A Model of Los Angeles Smog," or "Economic Models for Transportation." In such cases, we mean a quantitative, numerical description of a physical system.* The formulation is based on

*See *Computers, BASIC, and Physics*, Herbert D. Peckham, Addison-Wesley, 1971, for a good description of a mathematical model.

(a)

(a) Macroscopic view of a transistor. (Photographs courtesy of RCA *Electronic Age*). (b) Microscopic views of the surface structure of the transistor material.

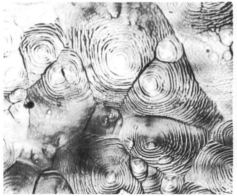 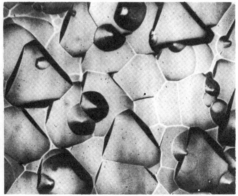 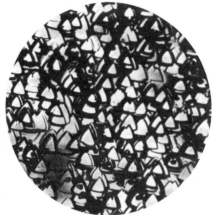

(b)

measurements taken of the system or assumptions made about the system. The basic idea is to develop a formalism which describes the behavior of the system to an extent which allows predictions of what the system will do when certain features are changed. For example, a model of Los Angeles smog would be useful for predicting the change in smog conditions as a result of control devices for automobile exhaust emissions.

A model is developed along the lines of macroscopic (implying large-scale) and microscopic (implying small-scale) approaches. The former is based on measurements of the external parts of the system; the latter on assumptions about

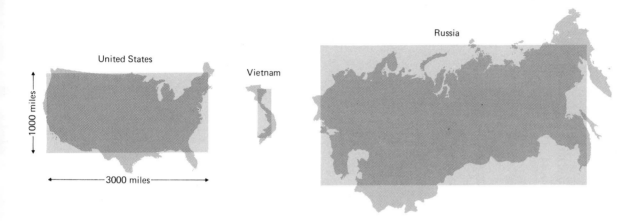

Fig. 1 First step in the development of a model of the world.

the internal parts of the system. It is obvious that the microscopic approach must necessarily involve assumptions when you consider that the inside of the system may never really be seen. For example, if we want to model the behavior of a gas in a container (the air in an automobile tire, for example), we can take external measurements of the pressure, volume, and temperature and try to deduce mathematical relations between the variables *or* we can make assumptions about the interactions of the atoms in the gas and try to relate the properties of the individual atoms to pressure, volume, and temperature. The second approach is more fundamental, but also more difficult.

The astronauts and cosmonauts are the only people who have had an actual macroscopic view of the earth. The lucid photographs they obtained are probably one of the real bonuses of the space program; the photographs dramatically illustrate the limited nature of our system. A model of the earth from this macroscopic viewpoint might be thought to consist of parcels of land of various size, called nations, which are approximated by rectangles (Fig. 1). Looking inside these countries with the intention of gaining a more microscopic view, we ask, "What distinguishes one nation from another?" The list of distinguishing factors would be long, but would include such things as politics, religion, sports, composition of the population, and, as already seen, standard of living and energy consumption. Both of our descriptions would serve useful purposes. The macroscopic rectangular-nation model allows us to estimate the time required for the United States to become covered by electric power plants if present consumption growths continue. The microscopic model illustrates the important role of energy in our society.

FUNCTION OF THE ELECTRIC GENERATOR

The device which ultimately supplies electric energy to the transmission lines leading to our homes and factories is called an electric generator. It is a device which converts rotational mechanical energy to electric energy. All automobiles with batteries have such a device to replace electric energy drawn from the battery. Knowing what the generator does and not what it looks like, we might model it macroscopically as in Fig. 2. In the United States, the rotational energy is supplied primarily by a steam turbine (85.0%) or by the water from a river (14.9%).* A contemporary generator will convert 99%† of the mechanical energy to electric energy which is derived from work done by electric charges. Let us now look a little more microscopically at the physical principles involved in this process. These principles will be useful in many other contexts throughout the book.

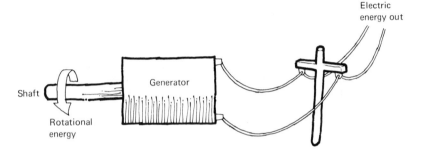

Fig. 2 Symbolic representation of conversion of mechanical energy to electric energy.

PHYSICAL PRINCIPLES

Gravitational Force and Gravity

Any two objects having the property of *mass* will exert an attractive force on each other. This attraction is due to the gravitational force which, along with the electromagnetic, and "weak" and "strong" nuclear forces, is one of the four fundamental types in nature. The gravitational force between the earth and a satellite keeps a satellite accelerating in orbit about the earth.

m_1 is attracted toward m_2

m_2 is attracted toward m_1

Two masses designated m_1 and m_2 are attracted toward each other by the gravitational force between the two.

Federal Power Commission Oversight, a report covering the principal policy questions now facing the Federal Power Commission, Serial 91-58, U.S. Government Printing Office, Washington, D.C. (January 30, 1970).

†"The Conversion of Energy," Claude M. Summers, *Scientific American*, **224**, 3, 148 (September 1971).

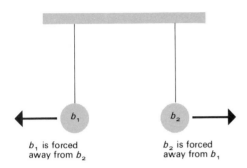

b_1 is forced
away from b_2

b_2 is forced
away from b_1

Fig. 3 Two balloons designated b_1 and b_2 which possess a net charge of the same type will repel each other. This is due to the electric force between the two.

Electric Force and Electricity

If two balloons are first rubbed on a person's hair and then suspended alongside each other with thread, they will repel each other (see Fig. 3). This is due to the electric force and is attributed to the property of *charge*. Each balloon acquired a net charge as a result of the rubbing process.

The balloons are comprised of atoms which individually have mass; and the atoms, in turn, are built up from still smaller entities, electrons and nuclei,* which have both mass and charge. There exist in nature two types of charges which are distinguished by the different effects they produce on each other. The charges could be distinguished by calling one *A* and the other *B*, for example, but because the charge of one type can be neutralized by a charge of the other type, it is appropriate to call the charges positive and negative. The unit of charge in the MKS system of units is the coulomb (C). The charge of the electron is -1.602×10^{-19}C and there is no confirmed smaller unit of charge. The net positive charge on the nucleus just balances the negative charge of all the electrons. The overwhelming mass of the atom is in the nucleus and is confined to a volume approximating a sphere having a radius of about 10^{-13} centimeters. The electrons surround the nucleus, the closest one being about 10^{-8} centimeters from the nucleus. In perspective, if the radius of the nucleus were about the thickness of a dime, the nearest electron would be about the length of a football field away. In some sense, the atom is like our solar system in which the nucleus is the sun and the electrons are the planets. Extending this model for the atom to our experiment with the balloons, we can think of the rubbing process as having initiated a transfer of electrons from the hair to the balloons. The balloons acquire electrons and become negatively charged; the hair loses electrons and is left with a deficiency of negative charge and an excess of positive charge. The balloons repel each other because they both have a net negative charge. They would also repel each other if they both had a net positive charge. But they would attract each other if one had a net positive charge and the other had a net negative charge.

Any charge, positive or negative, in motion constitutes an electric current. However, electric current in the metallic wires of our toasters, light bulbs, etc., is due only to the motion of electrons. Moving charge in a conductor is much like water flowing in a pipe. We can talk about a rate of flow of charge just as we can talk about a rate of flow of water. The rate of flow of charge is a quantitative measure of electric current. Formally

$$\text{current} = \frac{\text{amount of charge flowing by some position in a conductor}}{\text{time recorded for the flow}}$$

$$I = \frac{Q}{t} \frac{C}{\text{sec}} = \frac{Q}{t} \text{ amps}$$

A coulomb per second is called an ampere (amp). If 100 coulombs of charge flow by some position in a conductor in 10 seconds, we would say the current is 10

*This electron-nucleus model of the atom is adequate for understanding electricity and the basic structure of molecules. It will be necessary to expand it later (Chapter 4) to include the internal structure of the nucleus.

amps. A force must be exerted on the electrons to cause them to flow through a conductor just as a force must be applied on water to keep it moving through a pipe. The force does work on the electrons just as work is done on an automobile when it is forced some distance. It is the role of the electric power companies to supply this electrical pressure. This electrical pressure is called the potential or voltage and can be defined in terms of the amount of work done on a charge as it moves between two points.

$$A \rightarrow Q \quad B$$

Work W is done in moving charge Q from A to B.

Electric potential $= \dfrac{\text{work}}{\text{charge}}$

$$V = \frac{W}{Q} \frac{\text{joules}}{\text{coulomb}} = \frac{W}{Q} \text{ volts}$$

A joule per coulomb is called a volt (V). If 100 joules of work are done on 1 coulomb of charge when it moves between two points A and B, the potential difference or voltage between A and B is 100 volts. The potential difference between the two "holes" in the electric outlets in a house is typically 115 volts. A battery supplies either 6 or 12 volts for an automobile.

Electrons in a conductor acquire kinetic energy as a result of the work done on them. This is just the work-energy principle. The electrons dissipate energy in a variety of ways such as heat, light, and mechanical energy. Just as in the mechanical case discussed in Chapter 1, the rate at which the electrons, or any charges, dissipate energy is called electric power or, for short, just power.

$$P = \frac{W}{t}$$

Since electric work is also expressed in joules and time in seconds, electric power would be expressed in joules per second, or watts.

From the definition of power, we can write

$$\begin{array}{ccc} W & = & P \quad \cdot \quad t \\ \text{energy} & & \text{power} \quad \text{time} \end{array}$$

So any proper unit of power multiplied by a unit of time yields a unit of energy. Kilowatts and hours are proper units of power and time. Thus the kilowatt-hour, the unit in terms of which we pay the utilities company, is a unit of *energy*. It is a peculiar unit because the units of time do not cancel those in power.

$$W = P \text{ (kW)} \cdot t \text{ (hr)}$$

$$= P \frac{\text{kJ}}{\text{sec}} \cdot t \text{ (hr)}$$

Units do not cancel.

The energy consumed by a 1000-watt light bulb lit for 10 hours would be

$W = 1000 \text{ W} \cdot 10 \text{ hr}$
$= 10,000 \text{ W-hr}$
$= 10 \text{ kW-hr}$

Typically, electric energy may cost 5¢ per kW-hr. Hence the cost of operating the 1000-watt bulb for 10 hours would be

$\text{cost} = 10 \text{ kW-hr} \cdot 5¢/\text{kW-hr}$
$= 50¢$

The cost is not prohibitive. On the other hand, the cost of operating such things as electric clothes dryers, stoves, and air conditioners may be significant.

Electric power can also be expressed in terms of voltage and current. Since

$$P = \frac{W}{t} \quad \text{and} \quad W = QV,$$

then

$$P = \frac{VQ}{t}.$$

But Q/t is simply the current I. So $P = VI$. Volts multiplied by amperes is a measure of electric power. An automobile light bulb rated at 10 watts operated from a 12-volt battery would use a current of

$$I = \frac{P}{V} = \frac{10}{12} = 0.83 \text{ amp}$$

The terminals of a battery used in an automobile are designated as positive and negative. This designation is referred to as the polarity. When the battery is connected to an element such as a light bulb, electrons flow out of the negative terminal and through the bulb (see Fig. 4). This type of current is called direct, and is denoted d.c. The voltage at the outlet in a house changes (or oscillates) both in size and polarity at regular intervals of time. Typically, the size and polarity at the outlet repeats 60 times each second. Such voltage is called alternating, or a.c., because when it is connected to an element, the direction of the current alternates with time. One of the main reasons for using this type of voltage is that energy losses in transmission from the generating facility can be minimized. It is also much easier to change the size of an a.c. voltage which is the function performed by the transformer that you often see on a utility pole or in a distributing substation in a city.

Most electric devices will operate only on a specific type of voltage, that is, a.c. or d.c. The motor on a refrigerator will operate only on a.c. However, those devices which generate heat such as an iron will function on either a.c. or d.c. and the relation for power, $P = VI$, applies to both types. If a 1000-watt iron is connected to a 115-volt outlet, it would use a current of

$$I = \frac{P}{V} = \frac{1000}{115} = 8.7 \text{ amp}$$

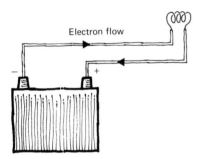

Electron flow

Fig. 4 When a device such as a light bulb is connected to a battery, electrons flow out of the negative (−) terminal, through the device, and into the positive (+) terminal. This type of current is called direct current and is denoted d.c.

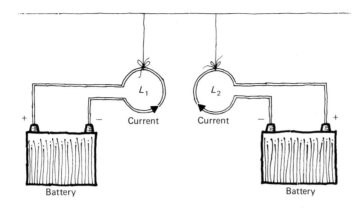

Fig. 5 If two loops of wire are suspended and a current is caused to flow in each, they will experience mutual forces as a result of a magnetic force created by the moving charges in the wires.

Magnetic Force and Magnetism

We can replace the balloons in our original model by two loops of wire carrying a current, and we would find that the loops experience mutual forces (see Fig. 5). This is due to the magnetic force and is attributed to the property of a *charge in motion*. The magnetic effects seen in a simple toy magnet are due to motions of charges in the atoms comprising the magnet. Experimentally, one finds that the magnetic effects in a long or bar-type magnet are concentrated near the ends (called poles) of the magnet and that different magnetic forces are produced by the different ends.

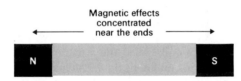

Simple bar magnet.

If we distinguish the poles as N and S, then we find that two N or two S poles will repel each other, while an N and an S pole will attract each other (Fig. 6). Because of this similarity to the behavior of charges, it would be appropriate and more rational to call these poles positive and negative. However, as we all probably know, there is a magnetic concentration near the north and south geographic poles of the earth and the notion of N and S poles has evolved from this correlation.

A gravitational field is said to exist in any region of space in which a *mass* experiences a gravitational force. For example, a gravitational field exists in a room; if we hold a pencil above the floor and release it, it is accelerated downward by the gravitational force. Similarly, electric and magnetic fields are said to exist in a region of space if charges and magnets experience electric and magnetic forces. For example, a magnetic field exists around a bar magnet because another magnet will experience a force if brought into its vicinity.

If a charge moves in a magnetic field, it experiences a force due to a magnetic interaction between a *moving* charge and a magnetic field. This is the principle

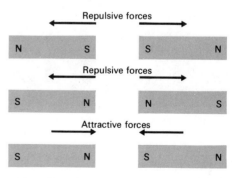

Fig. 6 Illustration of the mutual forces on two magnets for various orientations of the magnets.

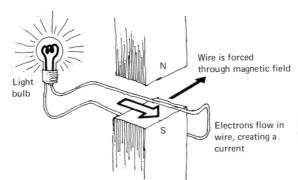

Light bulb

Wire is forced through magnetic field

Electrons flow in wire, creating a current

Fig. 7 A current is induced in a wire when it is forced through a magnetic field.

Electric power transmission lines do not require inordinate land space for construction. However, significant air space is required. And as seen in this photograph, land is sometimes degraded in the construction process. (Photograph by Ned. A. Hood, U.S. Department of Agriculture.)

employed to force electrons through a conductor like a metallic wire to create an electric current. An ordinary wire used to connect an appliance to an electric outlet contains vast numbers of negative (electrons) and positive (nuclei) charges. The positive charges are "tied down" by the physical structure of the conductor and are not free to move. In fact, only the outermost electrons of the individual atoms can be freed easily. If the wire is forced through a magnetic field (see Fig. 2.7), all charges experience a force by virtue of being moving charges in a magnetic field; but only those outermost electrons move. It is the motion of these electrons that create an electric current in the wire. It takes mechanical energy to force the wire through the magnetic field. In a practical electric generator used by a commercial electric power company, this mechanical energy is supplied by a steam turbine or moving water. The mechanical energy is converted to electric energy at a certain efficiency.

A practical electric generator would not use a single wire as depicted in Fig. 7. Rather, a coil, called the armature, containing a large number of turns of wire is rotated in a magnetic field. The magnetic field is created by a current in a stationary coil which surrounds the rotating armature.

ENERGY MODEL OF A FOSSIL FUEL ELECTRIC POWER PLANT

The steam turbine is the major source of mechanical energy for electric generators. The turbine operates by forcing steam from vaporized water onto blades attached to the shaft of the turbine (Fig. 8). The water is vaporized by the heat generated from burning fossil or nuclear fuels. With this basic model in mind, we can expand the system to that shown in Fig. 9. Several of the major components of this process can be identified in the commercial facility shown in Fig. 10. Although these diagrams function as a flow chart for energy and are useful, it is quantitative information that is required. So let us "dump" 1000 pounds of coal into the hopper and see how much electric energy comes out and what by-products evolve along the line. At this stage, sulfur oxides, particulate matter, and heat are of most interest. The release of carbon dioxide is discussed later.

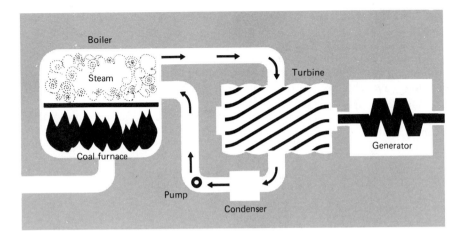

Fig. 8 Schematic representation of generation of electricity.

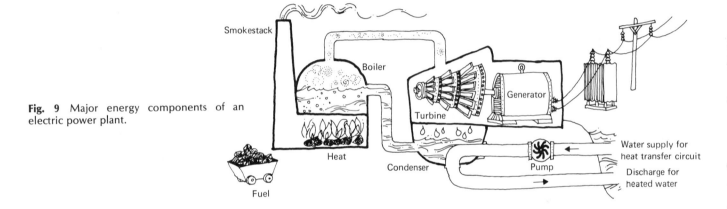

Fig. 9 Major energy components of an electric power plant.

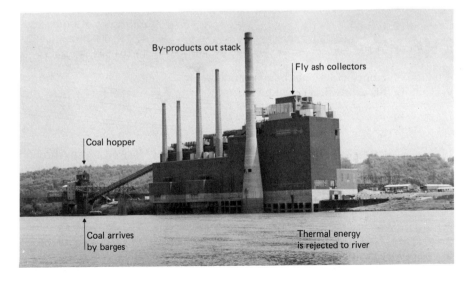

Fig. 10 The Cincinnati Gas and Electric Company Beckjord Station located on the Ohio River.

Finding coal for the hopper or, more appropriately, data for the model is no problem.* There is, however, a wide variation in the characteristics, in particular: the energy available per pound (heat of combustion); the sulfur content; and the ash content. A "typical" coal might be one from midwestern United States. For no particular reason, let us choose a coal from Ohio. The properties of this coal are given in Table 2. Let us now follow the 1000 pounds of coal through the power plant.

Table 2. Characteristics of coal from Jefferson County, Ohio

Heat of combustion (h) = 12,720 Btu/lb
Fractional sulfur content by weight (f_s) = 0.051
Fractional ash content by weight (f_a) = 0.107

Boiler

Energy available at the boiler = energy available per pound of coal × number of pounds

$$= 12{,}720 \frac{\text{Btu}}{\text{lb}} \cdot 1000 \text{ lb}$$
$$= 12{,}720{,}000 \text{ Btu}$$

If we wanted to symbolize this for possible use with a computer program, we could call E_{in} the energy available at the boiler and write

$$E_{in} = C \cdot h$$

where C and h are the amount of coal and heat of combustion.

Energy out of the boiler and into the turbine $(E_{turbine})$ = energy in the boiler (E_{in})
× energy conversion efficiency of boiler (ϵ_b)
$$= 12{,}720{,}000 \text{ Btu} \times 0.88$$
$$= 11{,}194{,}000 \text{ Btu}$$

A typical efficiency of 0.88 has been assumed.†
In symbols, this expression could be written:

$$E_{turbine} = E_{in} \cdot \epsilon_b$$
$$= Ch\epsilon_b$$

The energy that doesn't go into the turbine escapes, or is released, mostly through the smokestack.

Handbook of Chemistry and Physics, Chemical Rubber Co., 2310 Superior Ave., N.E., Cleveland, Ohio, 35th Edition, pp. 1758–1759.

†The efficiencies used in these calculations were taken from "The Conversion of Energy" by Claude M. Summers, *Scientific American* **224**, No. 3, 148 (September 1971).

Energy out the smokestack (E_{stack}) = energy into boiler (E_{in}) − energy into
turbine ($E_{turbine}$)
= 12,720,000 − 1,194,000
= 1,526,000 Btu

In symbols

$$E_{stack} = E_{in} - E_{turbine}$$
$$= Ch - Ch\epsilon_b$$
$$= Ch\,(1 - \epsilon_b)$$

The sulfur content in the coal is converted to sulfur oxides by chemical reactions.
There is no simple way to figure how much sulfur will be converted to oxides. Engineering formulas are used in making the estimations; one such formula is*

amount of sulfur oxides (lb) = 38 × percent sulfur in coal
× number of tons of coal

For the case being considered, the amount of coal is $\frac{1}{2}$ ton. Hence

amount of sulfur oxides = 38 (5.1) ($\frac{1}{2}$)
= 97 lb

Now these oxides do not necessarily escape out the smokestack if there is some
mechanism installed to extract them. There are very few power plants equipped to
remove even part of the sulfur oxides produced. We will assume, then, that all
97 pounds go out the smokestack and into the atmosphere.

The fly ash is simply an unburnable product in the coal.

amount of fly ash produced = amount of coal × fractional content of coal that
is ash
= 1000 lb × 0.107
= 107 pounds

Most power plants do have facilities for removing about 99% of the fly ash before it
goes out the smokestack. Hence 1% goes out the smokestack.

amount leaving smokestack = 107 (.01) = 1.1 pounds

Turbine

Energy out the turbine and into generator ($E_{generator}$) = energy into turbine ($Ch\epsilon_b$)
× energy conversion efficiency of turbine (ϵ_t).
= 11,194,000 (0.47)
= 5,261,000 Btu

---

*Selected Materials on Environmental Effects of Producing Electric Power, U.S. Government Printing Office, 1969, p. 256.

In symbols

$$E_{generator} = Ch\epsilon_b\epsilon_t$$

A typical conversion efficiency of 0.47 has been assumed for ϵ_t. The low value for this efficiency is probably the most difficult thing to accept in this discussion. However, there is a practical limit to the efficiency of a steam turbine which is imposed by the physical laws of thermodynamics. This is discussed in detail in Chapter 3.

Condenser

$$\text{thermal energy rejected to condenser } (E_{thermal}) = \text{energy into turbine } (Ch\epsilon_b)$$
$$- \text{energy into generator}$$
$$(Ch\epsilon_b\epsilon_t)$$
$$= 11,194,000 - 5,261,000$$
$$= 5,933,000 \text{ Btu}$$

In symbols

$$E_{thermal} = E_{turbine} - E_{generator}$$
$$= Ch\epsilon_b - Ch\epsilon_b\epsilon_t$$
$$= Ch\epsilon_b (1 - \epsilon_t)$$

Generator

$$\text{energy out of generator } (E_{electric}) = \text{energy into generator } (Ch\epsilon_b\epsilon_t) \times \text{energy}$$
$$\text{conversion efficiency of generator } (\epsilon_g)$$
$$= 5,261,000 (0.99)$$
$$= 5,209,000 \text{ Btu}$$
$$= 5,209,000 \text{ Btu } (1050 \frac{\text{joules}}{\text{Btu}}) \cdot \frac{1}{3,600,000} \frac{\text{kW-hr}}{\text{joules}}$$
$$= 1520 \text{ kW-hr}$$

In symbols

$$E_{electric} = Ch\epsilon_b\epsilon_t\epsilon_g$$

With these numbers, the model of Fig. 9 can be quantified as in Fig. 11.

This procedure accomplishes several things.

1. Numerical results are obtained for a given set of conditions. This gives a feel for the emissions from a typical power plant.

2. The origin of the pollutants in the energy conversion process has been established.

3. Most importantly, a model and mathematical formulas applicable to any system of this type have been deduced. Only the numbers are different for another system of the same type.

The magnitude of the pollution cannot be fully appreciated because the element of time has not been considered. A 1000-megawatt unit operating at

Microscopic Origin of Pollutants 55

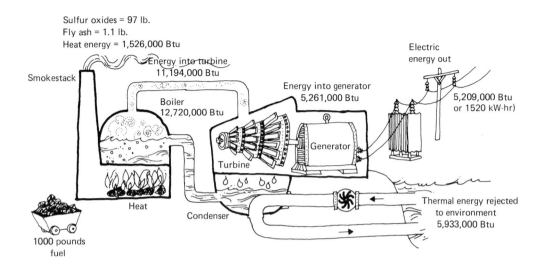

Sulfur oxides = 97 lb.
Fly ash = 1.1 lb.
Heat energy = 1,526,000 Btu

Smokestack

Energy into turbine
11,194,000 Btu

Boiler
12,720,000 Btu

Energy into generator
5,261,000 Btu

Electric
energy out

5,209,000 Btu
or 1520 kW-hr)

Generator

Turbine

Heat

Condenser

Thermal energy rejected
to environment
5,933,000 Btu

1000 pounds
fuel

capacity would use the 1000 pounds of coal in about 5 seconds. This means that about 10,000 tons of coal are used per day and this would release 1000 tons of sulfur oxides to the atmosphere! Already the sulfur oxides are approaching the tolerance levels set by some states.

Fig. 11 Typical outputs for conversion of 1000 pounds of coal into electric energy.

MICROSCOPIC ORIGIN OF POLLUTANTS

Consideration of the origin and magnitude of particulates, sulfur oxides, and thermal energy followed naturally from the energy model of an electric power plant. To look at the origin of the pollutants more microscopically requires an extension of the basic ideas of the electron-nucleus model of the atom.

An atom is one of a large class of physical systems appropriately called *bound systems* or *bound* states. These systems are bound in the sense that they are held together by some binding force. For example, an orbiting satellite and the earth constitute a bound system which is held together by the gravitational force between the two (Fig. 12). The electrons and nucleus of an atom are bound together by the attractive electric force that exists between any oppositely-charged particles.

The number of electrons in a neutral atom is called the atomic number and it is this atomic number which determines the chemical character of an atom. Table 3 shows the first 18 atoms in the order of increasing atomic number of electrons. As the atomic number increases, the mass of the atom also increases. This is due both to the additional electrons and to an increase in the size of the nucleus. There are a total of 92 naturally-occurring chemical species, but the air pollutants of interest now are primarily associated with atoms from the species shown in Table 3. It is possible to have atoms with the same number of electrons, but with nuclei of different mass. These are called isotopes and we will see later just what

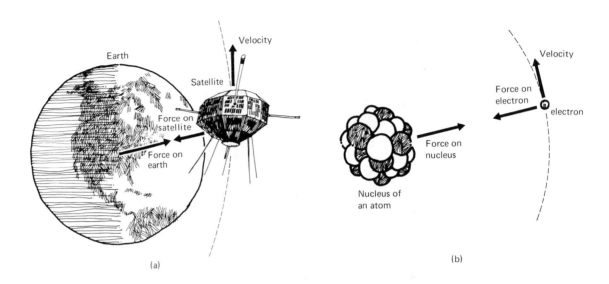

Fig. 12 (a) An orbiting satellite and the earth. (b) A nucleus and an electron. Both (a) and (b) constitute bound systems tied together by gravitational and electric forces, respectively.

is the origin of their differences in mass. Since isotopes have the same number of electrons, they share the same chemical characteristics.

A molecule is a bound system of atoms which is also held together by electric forces. This seems contradictory because an atom alone is electrically neutral and therefore it would appear that no electric force should exist. But what happens is that the electronic charge, especially the charge normally farthest from the nucleus, redistributes in such a way that net attractive electric forces are created. For example, gaseous hydrogen normally exists in a molecular state in which each molecule consists of two hydrogen atoms. The atoms are close enough in the molecule that the electron of one atom may temporarily be transferred to, or shared by the other. When such redistribution of charge occurs, the result is a net attractive force which holds the atoms together. Because of the motion of the electrons, the distribution of charge continually changes, but, on the average, there is some net attractive force. The type of binding we have described is called covalent. There are various types of fairly complex binding (or bonding) which in chemistry go under names like ionic, metallic, or hydrogen bonds. The details of these bonds are important, but not necessary for this discussion.

The simplest and stablest oxygen molecule consists of two oxygen atoms. A symbolism (or model) for the formation of this molecule is

$$O + O \rightarrow O_2$$

The breakup of O_2 into two oxygen atoms is symbolized by

$$O_2 \rightarrow O + O$$

Table 3. The first 18 atoms in order of increasing atomic number and mass. The mass is that of the most abundant isotope and is based on a scale of exactly 12 for the most abundant isotope of carbon.

Atom	Chemical symbol	Number of electrons	Relative mass
Hydrogen	H	1	1.007825
Helium	He	2	4.00260
Lithium	Li	3	7.01600
Beryllium	Be	4	9.01218
Boron	B	5	11.00931
Carbon	C	6	12.00000
Nitrogen	N	7	14.00307
Oxygen	O	8	15.99491
Fluorine	F	9	18.99840
Neon	Ne	10	19.99244
Sodium	Na	11	22.9898
Magnesium	Mg	12	23.98504
Aluminum	Al	13	26.98153
Silicon	Si	14	27.97693
Phosphorous	P	15	30.99376
Sulfur	S	16	31.97207
Chlorine	Cl	17	34.96885
Argon	Ar	18	39.96238

Oxygen is required for the combustion of fossil fuels. It constitutes about 21% of "air" and normally occurs in this typical molecular state. The reactions from which energy is derived in the combustion of coal* are

$C + O_2 \rightarrow CO_2$ (carbon dioxide)
$2C + O_2 \rightarrow 2CO$ (carbon monoxide)
$S + O_2 \rightarrow SO_2$ (sulfur dioxide)
$N_2 + O_2 \rightarrow 2NO$ (nitric oxide)
$N_2 + 2O_2 \rightarrow 2NO_2$ (nitrogen dioxide)

This represents the microscopic origins of sulfur dioxide and carbon monoxide which are considered pollutants in the sense of direct damage to plants and animals; of carbon dioxide which, although nontoxic, is of concern because of its possible long-term effects on the weather (see Chapter 3); and of nitrogen oxides which are important in the formation of photochemical smog (see Chapter 6).

ENERGETICS OF CHEMICAL REACTIONS

The symbolic representation for the chemical reaction plays much the same role as the pictorial models (Figs. 9 and 11) did for the electric power plant. And, just as in the examination of the power plant, the next step is to make our findings

---

*A 20-minute color film "Principles of Combustion" is available on loan from: National Coal Association, Coal Building, 1130 17th Street, N.W., Washington, D.C., 20006.

quantitative. For the reaction, we draw on the principles of physics and chemistry. One already invoked, but not specifically mentioned, was the conservation of atoms: the end products of a reaction contain the same number and type of atoms as the initial reacting particles. Another principle which is vital to this discussion is the conservation of energy: the *total* energy is the same before and after the reaction. In words

total energy before = total energy after

In symbols

$E(\text{before}) = E(\text{after})$

Care must be exercised, though, to ensure that all *forms* of energy are included, in particular the energy equivalence of mass.

The hydrogen atom provides a good, simple illustration of this concept. The hydrogen atom has one electron and, to balance with it, one unit of positive charge in its nucleus. The formation of the atom from these two constituents is depicted according to our scheme as

p^+	+	e^-	→	H
nucleus with one posi- tive unit of charge		electron with one nega- tive unit of charge		neutral hydrogen atom

Now if the masses of the nucleus and the electron are added and compared with the mass of the neutral atom we would find that the mass of the atom is 0.0000000146 amu* *less* than the combined masses of the constituents. A first guess might be that this is just due to uncertainties in the measurements of the masses. But it is not. It is due to one of the most exciting aspects of physics—the equivalence of mass and energy. Early in the twentieth century, Albert Einstein proposed that mass and energy were equivalent and related by

$E = mc^2$

where c is the speed of light in a vacuum. This means that anything having the property of mass is a *potential* source of energy. It says nothing about the *form* of that energy. It also means that energy can be transformed to mass through the relation

$$m = \frac{E}{c^2}$$

When the hydrogen atom is "assembled" from a nucleus and electron, its mass decreases because mass has been transformed to energy which binds the atom together. Similar results would be obtained for all atoms and, in fact, for any bound system assembled from small entities and held together by some appropriate force.

---

*1 amu = 1.6604×10^{-27} kg

Recognizing the equivalence of mass and energy, it is appropriate to include energy in the chemical reaction symbolism and write, for example,

$$C + O_2 \rightarrow CO_2 + energy$$

If thermal energy is liberated as is the case with this reaction as well as with those for the formation of CO and SO_2, the reaction is said to be exothermic. Thus we have pinpointed the origin of energy at the microscopic level. If thermal energy is required to make the reaction proceed, the energy term is negative and the reaction is said to be endothermic.

There are many other aspects of physics and chemistry that can be built into this model for a reaction. Our treatment has been sufficient to see the microscopic origin of the problematical by-products and thermal energy. Let us proceed to examine the environmental and technological consequences of them.

PARTICULATE MATTER*

General Properties

The particulate matter emerging from the smokestack of a coal-burning electric power plant enters the atmosphere in the form of fine, solid particles. There are, of course, many other sources of particle-like pollutants. Often pollutants are in the liquid state and it is appropriate to consider these along with the solids. The word particulate is intended to cover both states.† Particulates are classified according to their size. If all particulates were spherically shaped, the classification could be easily made in terms of the diameter (or radius) of the sphere. Or, if the particulates were all cubic, the length of a side would be sufficient. However, particulates will generally have peculiar and imprecise geometric shapes. Examine, for example, a large piece of dust. In spite of the fact that particulates are peculiarly shaped, they are classified according to diameter (or radius), which is determined by *assuming* that a particulate is a sphere. To see how this works, we need the concept of mass density. Objects of the same size do not necessarily have the same mass (or weight). For example, a quart of air has substantially less mass than a quart of milk. Mass density is a measure of how much mass is contained in a prescribed volume. Formally

$$mass\ density = \frac{mass}{volume}$$

$$\rho = \frac{M}{V}$$

(The Greek letter rho (ρ) is the symbol normally used for mass density.) Thus mass density would have units of mass divided by volume. Typical units are grams per

A fossil-fuel-burning electric power plant such as the one shown here first converts heat energy into mechanical energy. In the process, significant amounts of heat energy are rejected to the environment, usually to a stream, river, or lake. This has led to concerns for thermal pollution of these bodies of water. (Photograph courtesy of Billy Davis, *Louisville Courier Journal* and the Environmental Protection Agency.)

---

*The publication *Air Quality Criteria for Particulate Matter*, E.P.A. Publication AP-49 was used extensively as a source of material for this section.

†The word aerosol is often used interchangably with particulate. Some workers, though, prefer to consider aerosols as particles with diameters less than some particular value, for example 10^{-4} meters. (See, for example, the definition of aerosol in Reference 2.)

cubic centimeter (g/cm³) and kilograms per cubic meter (kg/m³). Water has a mass density of 1 g/cm³. This means that, regardless of geometric shape, 1 cm³ of water has a mass of 1 g. So if you measure the mass of a certain amount of water to be 10 g using, say, a balance like that used to "weigh" a letter for mailing, you can say that its volume is 10 cm³. Similarly, if you know that the density of particulate matter is 2 g/cm³, then knowing its mass from some measurement you can calculate its volume from the expression

$$V = \frac{M}{\rho}$$

For example, if a given particulate has a mass of one-millionth of a gram (10^{-6}g) and a density of 2 g/cm³, it would have a volume of

$$V = \frac{10^{-6}\ g}{2\ g/cm^3} = 5 \times 10^{-7}\ cm^3$$

If we assume that it is a sphere, it has a volume $V = 4/3\ \pi\ (D/2)^3$, where D is its diameter. Equating the volume of this assumed shape to its actual volume, we have

$$\underbrace{\frac{4}{3}\ \pi \left(\frac{D}{2}\right)^3}_{\substack{\text{volume of} \\ \text{a sphere}}} = \underbrace{5 \times 10^{-7}}_{\substack{\text{actual volume} \\ \text{of particulate}}}$$

The diameter is then

$$D = \sqrt[3]{\frac{24(5) \times 10^{-7}}{4\pi}}$$

$$= 0.98 \times 10^{-2}\ cm$$
$$= 98\ \text{micrometer}$$
$$= 98\ \mu m$$

(μm symbolizes micrometer meaning 10^{-6} meters or 10^{-4} centimeters). Particulate diameters will range from about 0.0002 μm to 500 μm.

Particulate concentrations are usually reported as a mass density which is the total mass contained in a cubic meter of air. Typical units are millionths of grams (micrograms) per cubic meter, symbolized μg/m³. These mass concentrations range from about 10 μg/m³ in remote nonurban areas to 2000 μg/m³ in heavily polluted areas. The arithmetic average for cities in the United States is about 105 μg/m³. Even though the concentrations vary substantially throughout the country, the distribution according to size is essentially the same everywhere. (Fig. 13).

Particulates with diameters greater than 0.1–0.2 μm account for about 95% of the total mass. But about 95% of the particulates have diameters less than 0.1–0.2 μm (Exercise 21). This fact should not be overlooked because, environmentally, the numbers in which the smaller-sized particles exist may be more important

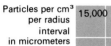

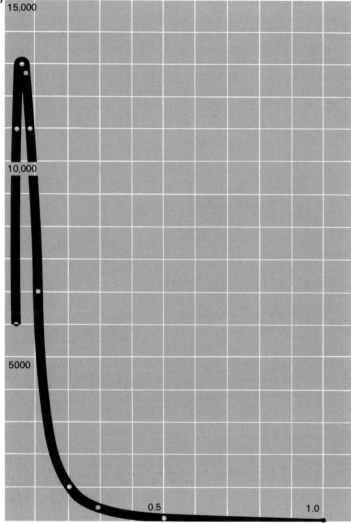

Fig. 13 Typical measured size distribution of atmospheric suspended particles. This curve will be higher or lower or shifted left or right depending on the location. Its shape, though, will always be very much like that shown.

than the fact that the bulk of the total mass is composed of the larger-sized particles. Particulates evolve from a number of natural and man-made sources. Those smaller than 1 μm in diameter come principally from condensation and combustion processes. Those with diameters between 1 μm and 10 μm come from such things as industrial combustion products and sea sprays. Those with diameters greater than 10 μm result from mechanical processes such as wind erosion and grinding. The sizes and origin of atmospheric particulates are shown in Fig. 14.

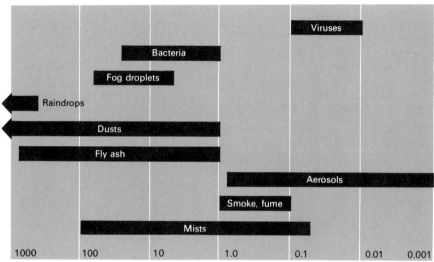

Fig. 14 The sizes and origin of atmospheric particulates. Note that this source chooses to call aerosols particles with diameters less than one micrometer.

There are several environmental concerns over particulates in the air.

1. The dirt they produce after settling to the ground is annoying and requires energy and money for removal.

2. They produce effects detrimental to materials, plants, and animals, including man.

3. There is concern over their effect on the heat balance of the earth through reflection and absorption of solar radiation.

These effects are complicated by the fact that many of the particulates remain aloft for unusually long periods and in some cases become permanently airborne. Let us examine the reasons for this and then elucidate the environmental effects.

Settling of Particulates in the Atmosphere

If two solid spheres of different sizes made from the same material are released from rest in a fluid like oil, for example, both will eventually drop at a constant speed. However, the larger one will achieve the greater speed. Particulates in the atmosphere are not spheres and their motion is influenced greatly by winds, but the larger ones do settle faster for exactly the same reason. The smaller particulates often settle so slowly that they become permanently airborne. A simple calculation using the rate concept shows why. Table 4 shows the relation between settling velocity and particulate diameter. Let us determine the time it takes a micrometer-diameter particulate to settle 1 kilometer (0.62 mi). A 1-micrometer diameter late has a settling velocity of 4×10^{-3} cm/sec (Table 4). Thus

$$\text{settling time} = \frac{\text{distance traveled}}{\text{settling velocity}}$$

$$= \frac{1 \text{ km}}{4 \times 10^{-3} \text{ cm/sec}} \cdot 10^5 \frac{\text{cm}}{\text{km}}$$

$$= \frac{10^8}{4} \text{ sec}$$

$$= \frac{10^8 \text{ sec}}{4 \, (3600) \, \dfrac{\text{sec}}{\text{hr}} \quad 24 \, \dfrac{\text{hr}}{\text{day}}}$$

$$= \frac{10^8}{4 \, (.36) \times .24 \times 10^6} = 2.9 \times 10^2$$

$$= 290 \text{ days}$$

Table 4. Approximate settling velocities in still air at 0°C and 760 mm pressure for particles having a density of 1 g/cm³ (From *Air Quality Criteria for Particulate Matter*, National Air Pollution Control Administration Publication No. AP-49.)

Diameter (micro- meters)	Settling velocity (cm/sec)
0.1	8×10^{-5}
1	4×10^{-3}
10	3×10^{-1}
100	25
1000	390

Particulates with diameters less than 1 micrometer settle so slowly that they tend to migrate thousands of miles before settling to the ground. It is not uncommon to find particles characteristic of the state of Arizona to have migrated as far east as the state of New York. Many particles actually become permanently airborne and it is these which cause concern about long-term weather effects.

The relationship between particulate diameter (or radius) and settling velocity can be understood by simply extending the smoke particle example used to illustrate Newton's second law in Chapter 1. The only additional information needed is an explicit relation for the drag force. This can be made plausible with a little thought about the behavior of spheres falling in a thick fluid.

If we neglect the buoyant force on a particulate, but include the drag force, then the force picture is as shown in Fig. 15.

net force = drag force − gravitational force

$$F_{\text{net}} = F_D - F_G$$

$$= F_D - mg$$

$$= ma$$

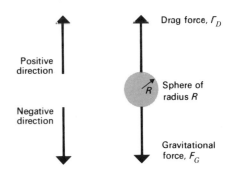

Fig. 15 A particulate in the atmosphere is pulled toward the earth by gravity. Its motion toward the earth is retarded by a drag force due to contact between the particulate and constituents of the atmosphere.

We have used the result $F_G = mg$ from the earlier discussion and have equated the net force, $F_D - mg$, to mass times acceleration, ma, in accordance with Newton's second law of motion. The fact that the particle achieves a constant, limiting velocity means that its acceleration is zero, and, therefore, that the net force is zero. Hence,

$$F_D = mg$$

Now the particle accelerates initially until it reaches the limiting velocity. Consequently the drag force must somehow depend on velocity. Let us assume that the drag force is directly proportional to the velocity.

$$F_D \propto v$$
$$= Cv$$

C is a factor which at least makes both sides of the equation have the units of force; it would also contain the other effects that may have been neglected so far. Proceeding, we find that

$$Cv = mg \quad \text{or} \quad v = mg/C$$

This latter relation explains the observed effect, namely, that the velocity increases as the mass increases. Since the mass is proportional to volume and the volume of a sphere is proportional to the cube of the radius, we might go on to assume that the velocity is also proportional to the cube of the radius.

$$v \propto R^3$$

This is not quite true because the drag force also depends on the size of the sphere. In the simplest sense, the drag force is proportional to the radius, so that $F_D = CRv$ rather than $F_D = Cv$. This would make the velocity proportional to the square of the radius:

$$v \propto R^2$$

For example, if the radius increased by a factor of 10, the velocity would increase by a factor of 100. Table 4 shows some actual particulate velocities. Note that the suggested velocity-radius dependence is not quite true, but that the dependence of velocity on radius is quite strong.

Environmental Effects

Health Effects

As mentioned earlier, particulate concentrations average about 100 μg/m^3 in American cities. If the concentration were close to this value at all times, then it would be difficult, if not impossible, to assess particulate effects on human health. But if for some reason the concentrations rose to some abnormally high value and were accompanied by an unusual number of health effects, then some association is possible. Still there is considerable margin for error because high

Cincinnati City Hall photographed in 1963 shows the accumulation of 34 years of dirt. (Photograph courtesy of the Environmental Protection Agency.)

particulate levels are normally accompanied by high sulfur oxide levels and it is difficult to separate the effects of each. In fact, the two pollutants together can produce effects which neither is capable of doing alone. This phenomenon is referred to as synergism (Question 3). There are several incidents of unusually high pollution levels brought on by peculiar weather conditions which produced stagnation of the air.* In some areas, particularly London, the conditions were complicated by heavy fogs. Indeed many were accompanied by significant increases in health effects. The most cited incidents of this type occurred in the Meuse Valley, Belgium (1930), Donora, Pennsylvania (1948), London (1952), and New York City (1953, 1966). The London episode produced some 4000 excess deaths. Some 5910 persons (42.7% of the total population) were affected to some degree in the Donora incident. Based on studies such as these, certain conclusions have been reached. They are not absolute but they are helpful as guidelines for setting standards. Two typical fairly reliable conclusions are as follows:

1. AT CONCENTRATIONS of *750 $\mu g/m^3$* and higher for particulates on a 24-hour average, accompanied by sulfur dioxide concentrations of 715 $\mu g/m^3$ and higher, *excess deaths* and a considerable *increase in illness* may occur.

---

*The physical origin of these conditions is discussed in Chapter 3.

2. IF CONCENTRATIONS ABOVE 300 $\mu g/m^3$ for particulates persist on a 24-hour average and are accompanied by sulfur dioxide concentrations exceeding 600 $\mu g/m^3$ over the same period, *chronic bronchitis* patients will likely suffer *acute worsening of symptoms.*

More of the details leading to these and several other similar conclusions are given in *Air Quality Criteria for Particulate Matter*, E.P.A. Publication AP-49. The points to be made here about health effects are

1. There are documented correlations between high particulate concentrations and human health effects.

2. Damage is done primarily to the respiratory system. Those suffering from bronchitis and other respiratory ailments are particularly affected.

Donora, Pennsylvania as it appeared on a polluted day in 1949. Air pollution was the cause of a major epidemic in this city in 1948. (Photograph courtesy of the Environmental Protection Agency.)

3. Sulfur oxides must also be considered when assessing effects of particulates because of synergistic effects.

More will be said about these effects in the discussion of sulfur oxides.

Effects on Materials

The fallout of particulates produces an obvious unsightliness on buildings and streets. But apart from this, particulates, especially when accompanied by sulfur oxides and moisture can literally attack structural materials. For example, steel and zinc samples corroded three to six times faster in New York City than in State College, Pennsylvania, where the particulate concentrations were about 180 and 60 $\mu g/m^3$, respectively.

Effects on Vegetation

There is little evidence of the direct effect on plant life of the mixture of particulates typically found in the air. There is, however, evidence that specific types of particulates do produce damage. One of many examples in the case in which a marked reduction in the growth of poplar trees one mile from a cement plant was observed after cement production was doubled.

Two New York City scenes recorded one day apart. The heavy smog is a result of heavy smoke concentrated by a stagnant air condition. (Photograph courtesy of the *New York Daily News*.)

Visibility

The natural beauty of a red sunset or a clear, blue sky is a treat for all to enjoy. And although the astronauts who explored the moon enjoyed many exciting views that we on earth cannot, they saw no red sunsets or blue skies. The reason is that these effects are produced by the scattering of light by atoms and molecules in the atmosphere and there is no such atmosphere on the moon. White light is a mixture of the colors of the rainbow. The amount of light scattered depends very strongly on the color. Blue light is scattered much more than red. Sunsets are red because the "blues" are scattered out of the line of sight of the viewer by atoms and molecules. The sky is blue because the blues are scattered toward the earth by the atmosphere.

Particulates in the atmosphere will also scatter and absorb visible light and other radiations not perceived by the human eye; this is of concern because of the amount of particulates that are accumulating. The effects are often seen as dense hazes in areas of high particulate concentrations. The hazes limit the maximum distance from which objects can be distinguished from their background. This maximum distance is called the visual range. There is a distinct correlation between visual range and particulate concentration which is often expressed as

$$(\text{visual range}) \times (\text{concentration}) \cong \text{constant}$$

is approximately

The relation is approximate because other factors contribute to the visual range and, also, because there is some intrinsic error in making the measurements. The number associated with the constant is 750 miles per ($\mu g/m^3$). So if the particulate concentration were 100 $\mu g/m^3$, the visual range would be approximately

$$\text{visual range} = \frac{750}{100} = 7.5 \text{ mi}$$

Solar Radiation

The fact that visibility is reduced by scattering and absorption of radiation means also that energy associated with the radiation is diminished. The notion of turbidity is introduced as a quantitative indicator of this energy loss. The word turbidity comes from turbid which implies cloudy, smoky, or hazy conditions.

Suppose that a source of radiation impinges on a volume of air having a suspension of particulates. Let X denote the thickness of this volume in the direction of the radiation and let E_0 denote the energy striking each square centimeter of the volume each second. The energy emerging decreases as the thickness and particulate concentration increases. It turns out that the decrease is exponential, precisely like the exponential behaviors already discussed except that the attenuating parameter here is thickness with units like meters rather than time with units like seconds. The emerging and impinging radiation are related by

$$E = E_0 \, e^{-TX}$$

where e is equal to 2.72. T is called the turbidity. It accounts for the type of radiation and the size and concentration of the scattering and absorbing particles; it is a number which varies from zero to infinite. Zero means that no radiation is removed; infinite means that all radiation is removed. Solar radiation encounters a "thickness" of air (that is, the atmosphere) in its path toward the earth. There will be a complex distribution of particulates mixed with the air in the atmosphere which will vary for different positions on the earth. So the turbidity and energy loss will also vary with position on the earth. There is, of course, some energy loss even if there are no particulates; particulates only enhance this loss.

Particulate concentrations have become high enough in some areas that sunlight has been significantly reduced at the earth's surface. Analyses of the radiation received in these areas have led to the following conclusion: For concentrations varying from 100 $\mu g/m^3$ to 150 $\mu g/m^3$, where large smoke turbidity factors persist, in middle and high latitudes direct sunlight is reduced up to *one-third* in summer and *two-thirds* in winter. Furthermore, it is estimated that the total sunlight is reduced five percent for every doubling of the particulate concentration.

An astronaut orbiting the earth outside the earth's atmosphere would be in a position to actually measure the amount of radiant energy proceeding from the sun to the earth. He would also observe that some of the radiation is returned back into space as a result of reflection and scattering in the atmosphere and on the earth. The ratio of the energy reflected back into space to the incident energy is called the albedo. Various estimates of the albedo average around 0.4, meaning that about 40% of the solar radiation returns to space. Particulates would, of course, contribute to the albedo both through reflection and absorption. If particulate concentrations became significant and if reflection predominated over absorption, then the earth would be deprived of solar energy and the temperature of the earth would decrease. If absorption predominated, the earth would warm. There is no real evidence that either effect is occurring. However, the worldwide air temperature has been decreasing since 1940 and this may be the reason.

Federal Standards for Particulate Concentrations

The determination of an unacceptable pollutant level was a decision left to the state or city until the enactment of the 1970 Clean Air Act. The local standards varied from none to some at least comparable with the new federal standards, and enforcement was difficult because the legal machinery was usually not provided. However, it appears that there are sufficient legal provisions for the federal standards. For example, federal agents have the power to close down factories if the EPA deems it necessary. The federal standards for particulates are:

The maximum average 24-hour concentration which is not to be exceeded more than once per year is 260 $\mu g/m^3$.

The annual maximum mean concentration is 75 $\mu g/m^3$.

Note that the maximum allowable annual mean concentration is less than the average for most American cities. Clearly, adjustments by the major contributors, namely industry and power plants (Table 1), are necessary.

In calculating the particulate emissions from a power plant, it was assumed that only 1% of the particulates generated actually escaped to the environment.

Illustration of acute pollution in St. Louis. (Photograph courtesy of the St. Louis Post Dispatch.)

This is only true for the most up-to-date plants and the percentage of particulate emission is somewhat larger on the overall national average. For example, power plants emit about 3 million tons of particulates and burn about 300 million tons of coal annually. So if the average ash content of coal is 10%, there would be about 30 million tons of particulates produced. If 3 million tons escaped, then the average efficiency is more nearly 90% rather than the 99% for the best technology. As new power plants are built and older less efficient systems are replaced, these emissions will diminish to an expected 0.82 million tons in 1980.* Similar efforts will be required for other industrial sources. The technology for reducing emissions is available, although it is costly, and there should be no problem in meeting the federal standards for allowable concentrations. Let us now examine the principles of the collection mechanisms.

Particulate Collection Devices

Particulate collection devices are categorized as gravitational, cyclone, wet collectors, electrostatic precipitators, fabric filtration, afterburner (direct flame),

---

*W. W. Moore, *Reduction in Ambient Air Concentrations of Fly Ash—Present and Future Prospects*, Proceedings: The Third National Conference of Air Pollution, U.S. Public Health Service Publication No. 1649, U.S. Government Printing Office, Washington, D.C. (1966).

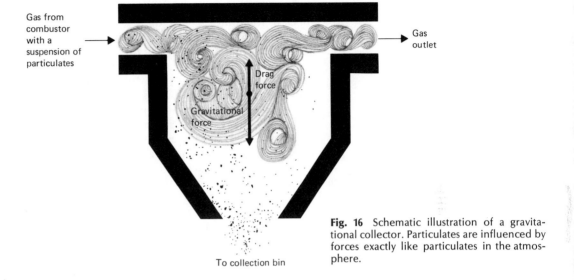

Gas from combustor with a suspension of particulates

Gas outlet

Drag force

Gravitational force

To collection bin

Fig. 16 Schematic illustration of a gravitational collector. Particulates are influenced by forces exactly like particulates in the atmosphere.

and afterburner (catalytic). The details, including advantages and disadvantages, are given in Reference 3. The most common and appropriate for insight into principles involved are the gravitational, cyclone, and electrostatic precipitator.*

Gravitational Collector

The principle of the gravitational collector is shown in Fig. 16. The gases containing a suspension of particulates from the fuel combustion enter a large chamber. Like particulates in the atmosphere, they are influenced by the gravitational force of the earth pulling down and a drag force which retards their motion. Some of the particles will settle enough so that they will fall through the bottom of the container and be collected. As expected from our discussion of particulates settling in the atmosphere, the larger ones will settle fastest and are most likely to be the ones collected. Gravitational collectors function efficiently only for particulate diameters greater than about 50 micrometers. They are useful, though, because of their simplicity and ease of maintenance.

---

*A 20-minute, color film "Collection of Particulate Matter" describing the operation of five collection devices is available on loan from Technical Audiovisual Branch, Office of Technical Information and Publications, Research Triangle Park, North Carolina, 27709.

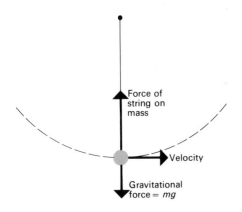

Fig. 17 A mass attached to a string and revolving in a vertical circle is acted on primarily by gravity and a force exerted on the mass by the string. These forces combine to keep the mass traveling in the circular path.

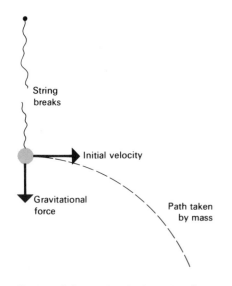

Fig. 18 If the tension in the string shown in Fig. 17 exceeds the breaking strength of the string, the string breaks and the mass flies outward from its circular orbit.

Cyclone Separator

Any rapidly rotating air mass is loosely referred to as a cyclone. Such a condition is created with the combustion gases in a cyclone separator. The result is that the particulates are forced to the outer edge of the cyclone where they can be collected. The principle is as follows. Any mass in circular motion is accelerating and, therefore, has some net force acting on it. For example, if a mass attached to a string is swung in a circle, the force of the string on the mass combined with the gravitational force on the mass provide a net force which keeps the mass in its circular path (Fig. 17). By Newton's third law, the mass exerts an opposite force on the string which produces a tension in the string. If the tension exceeds the breaking strength of the string, the mass flies outward from its circular orbit (Fig. 18). If a gas with a suspension of particles is rotated by some sort of cyclonic wind action, then the particles will subscribe to the circular motion only if there is some force available to accelerate it. This must be a force exerted on the particles by the gas. If the force is insufficient, the particle moves outward. In a cyclonic separator, the gas containing the particulates is forced into cyclonic motion down through a tapered cylinder. (Fig. 19) The particulates move toward the walls of the cylinder. Some strike the walls and fall into a collection bin. This type of device is somewhat better than the gravitational system for collecting particulates of a smaller diameter. Still, the lower limit for the diameter is about 5μm for efficient collection.

Electrostatic precipitator

The electrostatic precipitator is the most efficient device for removing small diameter (less than 5 μm) particulates. It is based on the principle that objects with unlike charges attract each other. The precipitator is, in principle, a hollow container, say a cylinder, with wires located within it (Fig. 20). The wires are maintained at a high negative electric potential of the order of 50,000 volts. The strong electric field created by this potential pulls electrons off the central wire by an effect termed field emission. Some of the gas molecules, such as oxygen (O_2), will capture an electron and acquire a net negative charge (O_2^-): These molecules will be accelerated toward the wall of the cylinder because its potential is positive relative to the suspended wires. During transit, the O_2^- ion may attach itself to a particle and the particle-O_2^- composite will migrate toward the wall. Once the composite strikes the wall, it becomes neutralized and the electric force vanishes. The particulates are then removed from the walls of the container.

The electrostatic precipitator is the most efficient of the systems discussed and is capable of removing 99% of the total *mass* concentration. However, the efficiency for particulates less than 0.1 μm in diameter is poor. Thus even though the electrostatic precipitator may remove 99% of the mass, only about 5% of the *number* of particulates are accounted for because there are many more of the smaller particulates in the gas effluent (see Fig. 13). The reason the electrostatic precipitator has difficulty in extracting the small particulates is that the probability of capture of an ion decreases with the *effective* size of the particulate. The effec-

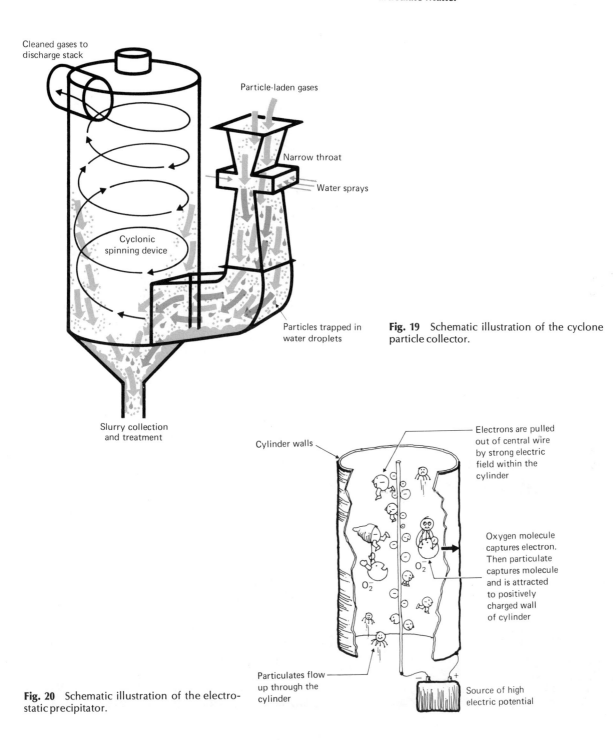

Cleaned gases to
discharge stack

Particle-laden gases

Narrow throat

Water sprays

Cyclonic
spinning device

Particles trapped in
water droplets

Fig. 19 Schematic illustration of the cyclone
particle collector.

Slurry collection
and treatment

Cylinder walls

Electrons are pulled
out of central wire
by strong electric
field within the
cylinder

Oxygen molecule
captures electron.
Then particulate
captures molecule
and is attracted
to positively
charged wall
of cylinder

O_2^-

O_2^-

Particulates flow
up through the
cylinder

Source of high
electric potential

Fig. 20 Schematic illustration of the electrostatic precipitator.

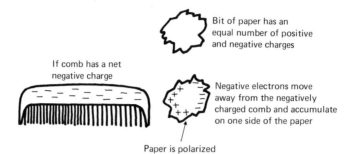

Bit of paper has an equal number of positive and negative charges

If comb has a net negative charge

Negative electrons move away from the negatively charged comb and accumulate on one side of the paper

Paper is polarized

Fig. 21 A bit of paper can be electrically polarized by bringing it near a comb which has been charged by combing one's hair.

tive size is determined by the degree of *polarization*. Polarization, in general, means a separation of groups of interests because of conflicting or contrasting positions. Where electric charges are concerned, it means a separation of the two types of charges. The positive and negative charges in a piece of electrically neutral material can be separated by placing the material in an electric field. This is what happens when a comb is charged by combing and then placed near a bit of paper (Fig. 21). Particulates in the gas effluent are polarized because of the large electric field established within the precipitator. As a result, the effective size for capture of an ion is increased. However, since the small particulates are extremely difficult to polarize, they are very difficult to remove. It is difficult to assess the effect of not being able to remove these small particulates, but, as mentioned earlier, it may be the actual *numbers* of particulates rather than the *mass* which is ultimately of importance.

The effectiveness of electrostatic precipitators for removing particulates emanating from a smokestack is illustrated in these pictures taken without and with the electrostatic precipitators in operation. (Photographs courtesy of Eastman Kodak Company.)

The Disposal Problem

The collection of fly ash has solved one problem but created another—disposal of the collected matter. Nearly 20 million tons of fly ash are collected annually and, for the most part, are pumped away from the precipitators at a cost of from 50¢ to $3.00 a ton into ponds where it settles to the bottom. Some commercial use is made of the by-product but it amounts to only a few percent of that collected. Some of the uses are as an aggregate for building bricks, improved traction material for car tires, fire-quenching material for coal mines, and fertility improver for soils.

Sulfur dioxide injury to a white birch leaf. (Photograph courtesy of the U. S. Department of Agriculture.)

SULFUR DIOXIDE

General Properties

All fossil fuels contain a certain amount of sulfur. However, coal has the greater proportion. The sulfur content, by weight, of commercial coal varies from about 0.3 to 5%. Nearly all the coal mined from the vast coal deposits east of the Mississippi River contains more than 1% sulfur. Sulfur dioxide (SO_2) and sulfur trioxide (SO_3) are the major oxides produced when the coal is burned. Still, the production of sulfur dioxide is some 40 to 80 times more probable than sulfur trioxide. Sulfur dioxide is a nonflammable, nonexplosive, colorless gas. Most people can taste it in concentrations of about 1 ppm in air.* It produces a pungent, irritating odor for concentrations greater than about 3 ppm in air.

It is easy to estimate how much SO_2 is produced from burning a ton of coal with a given sulfur content (see Exercise 1). However, because of the large variation in sulfur content, it is difficult to estimate how much is being produced in the entire country. An often quoted amount is 20 million tons for 1970 which is expected to reach 94.5 million tons in 2000†. Sulfur dioxide constitutes about one-fifth of all atmospheric pollutants and it is estimated that damage due to air pollution amounts to about $65 per capita each year.‡

Environmental Effects

Health Effects

A number of experiments have been conducted on the effects of sulfur dioxide on animals. Such research is meaningful in the environmental sense to the extent that the results can be extrapolated to humans. The studies show that it is the respiratory system which is most susceptible to damage. This information is important and has some ultimate use when trying to set standards. It is, however, the effect of sulfur dioxide in an atmosphere containing a variety of other pollutants which is of real significance because the sulfur dioxide gets involved chemically and produces other species more toxic than itself. For example, sulfur dioxide can be converted to sulfuric acid (H_2SO_4) which is highly corrosive and attacks almost anything. One might think of this as happening by first converting SO_2 to SO_3 by the reaction

$$2\,SO_2 + O_2 \rightarrow 2\,SO_3$$

using oxygen from the atmosphere. This would be followed by

$$SO_3 + H_2O \rightarrow H_2SO_4$$
$$\text{water}$$

---

*Parts-per-million, symbolized ppm, means the number of molecules of a given type out of a total of one million molecules. Historically, this is the most common unit for measuring pollution concentrations. However, the actual number of molecules in a given volume of air depends on the pressure and temperature. There is a trend, instigated by the E.P.A., toward reporting all pollution concentrations as mass concentrations as is the case for particulates. It is possible to relate a mass concentration in say $\mu g/m^3$ to ppm if the pressure and temperature is specified. Normally, the pressure is taken to be that of the atmosphere and the temperature is about that of a room in a house (78° Fahrenheit) The mechanics of how this is done is outlined in Exercise 30, Chapter 3. It is fairly difficult but you would be rewarded by working through it. Where it is important, concentrations in both ppm and $\mu g/m^3$ will be given.

†Sulfur Oxide Control: A Grim Future," *Science News* **98**, 187 (August 29, 1970).

‡*Waste Management and Control, a Report to the Federal Council for Science and Technology by the Committee on Pollution*, Publication 1400, Washington, D.C., National Academy Science–National Research Council (1966).

using water vapor from the atmosphere. This does in fact happen but only slowly because the formation of SO_3 by this proposed mechanism is slow. A more probable method of forming SO_3 is through a catalyst which modifies the reaction rate but is not consumed in the process. There are many possible catalysts. One is nitrogen dioxide (NO_2) which is also a combustion product.

$$NO_2 + SO_2 \rightarrow NO + SO_3$$

The NO_2 used to form SO_3 is regenerated by the reaction

$$2\,NO + O_2 \rightarrow 2\,NO_2$$

The role of the many catalysts in a polluted atmosphere in the formation of sulfuric acid is an extremely important environmental problem.*

The effects of SO_2, like particulates, on human health are obtained from epidemiological studies. Although these effects are difficult to assess quantitatively, they do have the distinct advantage of being derived from cases involving human subjects; in this sense, they are extremely valuable for deciding on allowable concentrations of various pollutants. Three fairly reliable conclusions from studies of this type are:

1. At concentrations of about 500 $\mu g/m^3$ (0.19 ppm) of sulfur dioxide (24-hour mean), with low particulate levels, *increased mortality rates* may occur.
2. At concentrations of about 715 $\mu g/m^3$ (0.25 ppm) of sulfur dioxide (24-hour mean), accompanied by particulate matter, *a sharp rise in illness rates* for patients over age 54 with severe bronchitis may occur.
3. At concentrations of about 120 $\mu g/m^3$ (0.046 ppm) of sulfur dioxide (annual mean), accompanied by smoke concentrations of about 100 $\mu g/m^3$, *increased frequency and severity of respiratory diseases* in school children may occur.

More of the details leading to these and several other similar conclusions are given in *Air Quality Criteria for Sulfur Oxides*, E.P.A. Publication AP-50. It is important to note that: (1) the elderly, young, and persons suffering from respiratory ailments such as bronchitis and emphysema are particularly bothered; and (2) there is an established synergistic effect between sulfur dioxide and particulates.

Effects on Vegetation†

Plants, in general, are easily damaged by sulfur dioxide. There is, however, a wide variation in the amount of SO_2 needed to damage different types of plants. For example, acute injury occurs in alfalfa (a common food crop for livestock) at con-

---

*"When Smog Turns Into Airborne Sulfuric Acid," *Science News*, March 4, 1972, p. 151. "Acid Rain," *Environment* **14**, No. 2, 33 (March 1972).

†An excellent 20-minute, 16 mm, color movie "Air Pollution and Plant Life, TF-102" is available on free loan from National Audiovisual Center, N.A.R.S., GSA, Washington, D.C., 20409.

centrations of 1.25 ppm for 1 hour. But 20 ppm for 1 hour is the concentration required to produce an equivalent damage in privet (a type of shrub commonly used for hedges). Damage has been reported for chronic exposures as low as 0.03 ppm. The formation of sulfuric acid, as described earlier, which attacks plants is also a problem as are synergistic effects of sulfuric dioxide with ozone and nitrogen dioxide.

Effects on Materials

As mentioned earlier, combinations of particulates and sulfur dioxide do significant damage to materials. This is especially true if conditions are favorable for the formation of sulfuric acid. Building materials and statues often are discolored and disfigured from attacks by sulfuric acid. In addition, metals corrode and things such as cotton, nylon, rayon, and leather are damaged by sulfur dioxide.

Federal Standards for Sulfur Dioxide Concentrations

The federal standards for sulfur dioxide concentrations are

> The maximum average 24-hour concentration which is not to be exceeded more than once per year is 365 micrograms per cubic meter (this amounts to 0.14 ppm).

> The maximum average annual concentration is 80 micrograms per cubic meter (this amounts to 0.03 ppm).

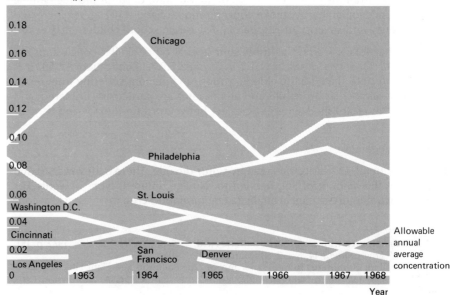

Fig. 22 Trends in sulfur dioxide concentration for eight cities. (From *Air Quality Criteria for Sulfur Oxides*, NAPCA Publication AP-50. Available from Environmental Protection Agency, 1626 K Street N. W., Washington, D.C., 20460.)

The question is "How closely do the urban sulfur dioxide concentrations correspond to the maximum permissible concentrations set by the E.P.A.?" The answer is: the concentration in some cities already exceeds the maximum limits and in many cities it is approaching the maximum limits. This is shown in Table 5 and Fig. 22.

Table 5. Sulfur dioxide concentrations at selected cities, 1968. (From *Air Quality Criteria for Sulfur Oxides*, National Air Pollution Control Administration Publication AP-50.)

	Maximum 24 hour (ppm)	Yearly average (ppm)
Chicago	0.51	0.12
Cincinnati	0.08	0.02
Philadelphia	0.36	0.08
Denver	0.05	0.01
St. Louis	0.16	0.03
Washington, D.C.	0.18	0.04

Control Methods for Sulfur Oxides

Thus far (1972), the only practical scheme for limiting sulfur oxide emissions is the use of fuel with less than 1% sulfur. This procedure was instigated in New York City in 1965. The results shown in Fig. 23 are dramatic. The reserves of low-sulfur fuel in the United States are large, but most of the fuel is located at a distance from most power plants which makes it expensive to obtain. Another serious problem is that the precipitators used for removal of fly ash are not as efficient when low-sulfur content fuels are used. For example, the efficiency is reduced from 99% to 98% if 1% sulfur coal is burned.

There are a variety of methods for reducing, not eliminating, sulfur oxide emissions. A process that is both economical and environmentally acceptable is at least three to eight years away according to a report by the National Research Council (1971). Two general approaches for removing the sulfur oxides are taken.

1. Sulfur oxides in the stack are removed by appropriate chemical reactions, and

2. removal of the sulfur from the coal before combustion.

In the first method, a salable product such as elemental sulfur or sulfuric acid is produced which tends to offset the cost of the sulfur oxide removal (Table 6). This is important because sulfur is used at a rate of 100 pounds per person per year in the United States. This would also alleviate some of the environmental problems associated with the mining of sulfur. The second method of removing the sulfur before combustion turns out to be extremely difficult because much of the sulfur is chemically bound to the coal.

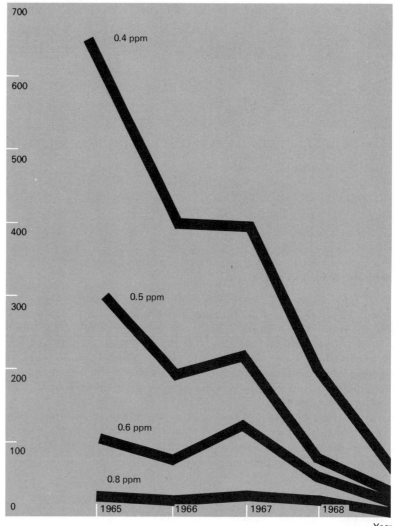

Fig. 23 Illustration of the effectiveness of sulfur dioxide abatement procedures in New York City. (From *Science* **170**, 707 (13 November 1970).)

Table 6. Characteristics of current technologies for removing SO_2 from stack gases. (From *Cleaning our Environment, The Chemical Basis for Action*, American Chemical Society, Washington, D.C. (1969) and *Selected Materials on Environmental Effects of Producing Electric Power*, U.S. Government Printing Office, Washington, D.C. (1969).)

Process	Principle	Efficiency	Capital cost 800mW plant	Salable by-product	(Total cost of removal per ton of coal)	
					Without by-product resale	With by-product resale
Limestone dolomite	Production of calcium and magnesium sulphates	40–50% dry collection. Better than 90% with dry plus scrubbing process		None		$0.36–0.63
Catalytic conversion	Production of sulfuric acid mist $SO_2+O\rightarrow SO_3$ $SO_3+H_2O\rightarrow H_2SO_4$	90% for plant using 3–3.5% sulfur coal	$16.5 million	Sulfuric acid Moderately concentrated $9/ton Fully concentrated $12/ton	$1.75	$0.72 0.38
Alkalized alumina	Absorption of sulfur oxides by alumina spheres	90% for plant using 3% sulfur coal	$8.5 million	Sulfur $25/ton or $40/ton	$1.54	$0.86 0.32

Another possibility is the conversion of coal to a burnable gas or a liquid with the sulfur removed in the process. The chemistry of these processes is known and it is only a matter of economics. Pilot plants are already in operation and much promise is held for this process.[*]

Historically, electric generating facilities have been located as close to the consumers as is possible so as to minimize energy losses in transit. Pollution in the cities is one of the environmental prices paid for this feature. There is a trend, however, toward moving the generating facilities to the source of fuel. Many of the fuel sources are located in southwestern United States and there are plans for facilities in Montana and Wyoming.[†] Although this arrangement alleviates pollution in the cities, it creates environmental problems by polluting and altering these vast scenic areas.[‡]

Two of the major goals of the federal government's drive for clean energy[§] are (1) a technology for removal of stack gases from the power and industrial plants burning fossil fuels and (2) coal gasification.

---

[*]"The Competitive Comeback of Coal", *Science News* **99**, p. 84 (Jan 29, 1971).

[†]"A Home on the Range for a Vast Industry" *Science News* **101**, 156 (March 4, 1972).

[‡]See, for example, "Hello Energy—Goodbye, Big Sky" *Life* **70**, 61 (April 16, 1971).

[§]See, for example, *Science* **172**.

COMMENT ON THERMAL POLLUTION

The energy model for generation of electric energy (Fig. 11) shows that about 45% of the energy extracted from combustion is rejected as thermal energy to the environment. Such a type of pollution is of major concern and must be studied. It is important, though, to try to first understand why so much thermal energy has to be rejected. This requires some basic knowledge of the thermodynamics of the thermal to mechanical energy-conversion process. It is the quest for this understanding which motivates the next chapter "Thermodynamics: Principles and Environmental Applications."

REFERENCES

Environmental Considerations

Selected Materials on Environmental Effects of Producing Electric Power, Joint Committee on Atomic Energy, Congress of the United States (91st Congress, First Session), August 1969. Superintendent of Documents, U.S. Government Printing Office, Washington, D.C., 20402.

Air Quality Criteria for Particulate Matter, National Air Pollution Control Administration Publication No. AP-49, January 1969. Available from Environmental Protection Agency, 1626 K Street N.W., Washington, D.C., 20460.

Control Techniques for Particulate Air Pollutants, National Air Pollution Control Administration Publication AP-51, January 1969. Available from Environmental Protection Agency, 1626 K Street N.W., Washington, D.C. 20460.

Air Quality Criteria for Sulfur Oxides, National Air Pollution Control Administration Publication AP-50. Available from Environmental Protection Agency, 1626 K Street N.W., Washington, D.C., 20460.

Nuclear Model of the Atom

Elementary Physics: Atoms, Waves, Particles, George A. Williams, McGraw-Hill Book Company, New York (1969), Chapter 25.

An Introduction to Physical Science, J. T. Shipman, J. L. Adams, Jack Baker, and J. D. Wilson, D. C. Health and Co., Lexington, Mass. (1971), Chapters 10–12.

The Atom and the Universe, James S. Perlman, Wadsworth Publishing Co., Inc., Belmont, California, Chapters 19–23.

Sulfur Oxide Pollution

Selected Materials on Environmental Effects of Producing Electric Power, U.S. Government Printing Office, pp. 105–109.

"Sulfur Oxide Control: A Grim Future," *Science News* **98**, 187 (Aug. 29, 1970).

"Environmental Protection in the City of New York," *Science* **170**, 706 (13 Nov. 1970).

"Air Pollution Standards Proposed", *Science News* **99**, 96 (1971).

Cleaning our Environment—The Chemical Basis for Action, American Chemical Society, Washington D.C., 1969, Chapter 1.

Episode 104, Environment **13**, 2 (1971).

"Progress in Abating Air Pollution," *Science* (March, 1970).

QUESTIONS

1. An Electric heater used to heat a house converts electric energy to thermal energy with nearly 100% efficiency. The conversion efficiency for the burning of fossil fuel is somewhat less than 100%. It is, however, substantially cheaper to heat a home by burning a fossil fuel than by using electric heat. Why?

2. Home heating using electric energy is often advertised as "clean." Comment on the rationale of this statement.

3. The concept of synergism goes on around us in a number of ways. A parent knows that the net mischievousness of two children is often far greater than the sum of the effects that each might produce alone. The combination of nylon stockings and a lady's legs produce a synergistic effect. Try to think of some other examples of synergism in our everyday lives.

4. A citizen outside a university community tries to understand the overall behavior by observing what the community does. He might try to understand it more fully if he were to look at the student-student, student-staff, and staff-staff interactions. In what ways are these approaches similar to the ways we employ in devising a model of a physical system?

5. A probability of something happening can be thought of as the number of times some event occurs divided by the number of attempts at trying to achieve the event. For example, if a coin is flipped 1 million times and a head comes up 500,000 times, the probability for obtaining a head on a single flip Is $\frac{1}{2}$. In what ways are the ideas of probability and efficiency similar?

6. What are some of the economic and environmental considerations for siting a fossil-fuel electric power plant?

7. Discuss some of the ramifications of rationing electricity as a mechanism for preventing blackouts.

8. What are some of the geographical factors considered when choosing a site for an industrial city? How do some of these factors contribute to pollution problems?

9. Most cities have a small electric energy margin for meeting peak demands, and blackouts have occurred in several areas. In the face of this, many power companies have continued, through advertising, to encourage the use of more electricity. What is your reaction to this policy?

10. Air pollution in cities by electric power plants could be minimized by moving the plants many miles away. What other problems, environmental and economic, would this produce?

11. Explain why a parachutist achieves a limiting speed with his "chute" on much sooner than if his "chute" is not on.

12. Somewhat of an environmental problem is being created by the large quantities of salt (NaCl) that are put on streets and highways in the winter to remove ice. Try to figure out the chemistry of this process and explain how the process works. Why might it be an environmental problem? (See "Plants and Salt in the Roadside Environment" by Arthur H. Westing in *Understanding Environmental Pollution* by Maurice A. Strobbe, The C. V. Mosby Co., St. Louis (1971).

EXERCISES

Easy to Reasonable

Exercise 1. How many pounds of sulfur are contained in 1 ton of coal having 5% sulfur content?

Exercise 2.
a) What is the thermal energy content of 5000 pounds of a typical coal (see Table 2)?
b) If this thermal energy is used to feed a steam turbine and the conversion efficiency is 85%, how much energy gets into the turbine?
c) What happens to the energy that doesn't get into the turbine?

Exercise 3. Show that 5,209,000 Btu are equivalent to 1520 kW-hr.

Exercise 4. If a railroad coal car can transport 100 tons and an electric generating facility requires 10,000 tons of coal per day, how many cars are required to supply the coal for one day's operation? What insight does this give you into the importance of coal transportation facilities and long-term labor agreements?

Exercise 5. Calculate the overall efficiency for the electric power plant model used in the text.

Exercise 6. An electric clothes dryer typically requires 6000 W of electric power. How much current does such a dryer use when connected to a 115-V outlet?

Exercise 7. In 1900 it took more than 20,000 Btu of heat to produce 1 kW-hr of electricity. Today it takes less than 9000 Btu. What are the corresponding efficiencies? (Be careful of units). What would you suspect is responsible for this two-fold increase in efficiency?

Exercise 8. An oxygen atom is nearly half as massive as a sulfur atom. Knowing the composition of the sulfur dioxide molecule, explain why nearly 100 pounds of sulfur dioxide are produced from the oxidation of 50 pounds of sulfur.

Exercise 9. A certain particulate has a mass of 10 μg and a density of 2.5 g/cm^3. What is the *radius* of the particulate assuming that it is a sphere? What would a length of one of its sides be if it were a cube?

Exercise 10. Table 1 shows that 12 million tons of particulates were emitted in 1966. To get a feel for the magnitude of such a quantity, consider the following. Suppose that a highway 50 ft wide extended for 3000 mi across our rectangular United States (Fig. 1). If these 12 million tons were spread uniformly over this highway, how deep would it be? Assume that one cubic foot of particulates weighs 100 lb.

Exercise 11. In 1967, about 10^{12} kW-hr of electric energy were produced by thermal energy conversion. The population in 1967 was about 200,000,000. Assuming that thermal energy was converted with a 35% efficiency and that 1 pound of coal produces 13,000 Btu of energy, calculate how much coal per person was required to produce the electric energy.

Reasonable to Difficult

Exercise 12. The average electric power plant requires about 10,300 Btu of thermal energy to produce 1 kW-hr of electricity. Estimate the quantity and cost for the coal required. Estimate, or find out, the cost of electricity in your area and figure out how much the utilities company charges to convert the coal to 1 kW-hr of electric energy.

Exercise 13. The basic carbohydrate molecule is CH_2O. Carbohydrates are produced in photosynthesis through the reaction of water, carbon dioxide, and sunlight.

  a) Figure out the chemical reaction of this process.

  b) Is the reaction exothermic or endothermic? Explain.

  c) What happens when the reaction is reversed?

Exercise 14. The total electric energy production in 1968 was 1.3 trillion kW-hr. Assume that this was produced by a coal-burning power plant with an overall efficiency of 40%. Estimate how much coal was burned. Assuming that the coal contained 2% sulfur, estimate the amount of sulfur dioxide produced. Compare with that quoted in Table 1.

Exercise 15. Using the data in Table 4, calculate the time required for particles of diameters of 0.1, 1, 10, 100, and 1000 micrometers to settle a distance of 1 km under the conditions given in the table.

Exercise 16. It was shown in the text that the burning of 1000 pounds of coal produces 1520 kW-hr of electric energy. Show that it takes a 1,000 MW power plant only about 5 sec to produce this 1520 kW-hr of energy and to burn this 1000 lb of coal.

Exercise 17. It was shown in the text that the burning of 1000 lb of coal produces 1520 kW-hr of electric energy. If the coal contains 5% sulfur, this produces about 1000 lb of sulfur oxides. The total amount of electric energy produced in 1966 was about 10^{12} kW-hr. Figure out how many tons of sulfur oxides would be produced assuming 5% sulfur content in the coal and compare your result with that shown in Table 1 for power plants. What factors contribute to the difference in your result and that shown in Table 1?

Exercise 18. The lifetime of a battery depends on the amount of current that is drawn from it. The more the current, the shorter the lifetime. The quality of a battery is often characterized by the number of ampere-hours (that is, amperes multiplied by hours) that it can provide.

a) What physical quantity would an ampere-hour measure?

b) If a battery is rated at 100 ampere-hours, how long would it last if it delivers 0.25 amperes?

Exercise 19. Electric power is related to voltage and current by $P = VI$. In a device like a toaster, voltage and current are related by $V = IR$, where R is called the resistance.

a) Show that the power can also be written as $P = I^2 R$ or $P = V^2/R$.

b) Would you expect the heating element of a toaster to have high resistance or low resistance? Explain.

c) If an electric toaster uses 1000 W when connected to a 115-V household electric outlet, what is its resistance?

Exercise 20. Surely we all would agree that the rate of flow of water from a faucet would increase if the water pressure increased. Likewise, the current in a wire would increase if the voltage across it increases. We might even suspect the voltage to be proportional to current; $V = CI$, where C is the proportionality constant.

a) What are the units of C?

b) Why would you expect this proportionality constant to depend on the type and size of the wire?

c) This constant is referred to as resistance, symbolized as R, and the expression $V = IR$ is called Ohm's law. Why do you suppose it is characterized as resistance?

Difficult to Impossible

Exercise 21. Figure 13 shows how particulates are typically distributed in the atmosphere according to their size. The total number in this distribution would simply be the total area under the curve. The number with radii from 0 to 0.2 μm and 0.2 to ∞ would be the area under the curve from 0 to 0.2 μm and 0.2 to ∞ respectively.

a) Estimate these areas and show that more than 90% of the particulates have radii less than 0.2 μm.

b) Try to give an argument as to why those with radii less than 0.2 μm account for only about 10% of the mass.

Exercise 22.   The amount of sulfur dioxide emitted per ton of coal burned was calculated from the expression Amount $= 38S$, where S is the percentage sulfur content in the coal. Sulfur dioxide is produced by the reaction

$$S + O_2 \rightarrow SO_2$$

Assuming that all the sulfur in the coal is free and goes into SO_2, figure out your own expression for the amount of SO_2 produced per ton and compare with the one used. Why might you expect your expression to differ somewhat from the one used?

3

THERMODYNAMICS: PRINCIPLES AND ENVIRONMENTAL APPLICATIONS

Steel plants pouring out clouds of smoke near Pittsburgh on the Ohio River. This contributed to Pittsburgh's name as the Smoky City before controls removed the pollutants from the air. (Photograph courtesy of the Environmental Protection Agency.)

MOTIVATION

For a very large fraction of the driving public, an automobile engine is a mysterious device that derives energy from gasoline and converts some of it to energy of motion. Many drivers know how many miles the car travels for each gallon of gasoline consumed, but few realize that only about 25% of the energy is converted to energy of motion. Most drivers also know that oil has to be added periodically to an automobile engine and that the oil serves as a lubricant to reduce friction between moving parts. Friction might be suspected as the cause for the low efficiency of the engine; we might suppose that if we were sufficiently ingenious or if we had a better lubricant, we could make an automobile engine with an efficiency close to 100%. The same conclusion might be reached when we read that the efficiency of a steam turbine used to convert thermal energy to mechanical energy for an electric generator is, at best, 50%. Friction is indeed a consideration, but a more fundamental cause is found in the thermodynamics of the energy conversion process.

The steam turbine, like all heat engines, extracts thermal energy from a heat source (a container of steam, for example), converts some of it to mechanical energy, and releases thermal energy to a heat sink of some sort (a lake, for example). This operation is depicted schematically in Fig. 1. Typically, as much thermal energy is released as is converted to mechanical energy. Because there are so many large steam turbines in operation, especially in the electric power industry, concerns are being registered for the biological health of our lakes, rivers, and streams that carry the burden of these heat loads.

Our goals in this chapter are several. First, we want to learn enough thermodynamic principles to understand why the efficiency of a commercial steam turbine is intrinsically low. Second, we want to use these principles to understand the methods being used to dispose of thermal energy in more environmentally acceptable ways. Third, we want to understand from the thermodynamic viewpoint the stable atmospheric conditions that prevent the dispersal of pollutants released from the combustion of fossil fuels. Finally, we want to investigate the possibility of increasing the average temperature of the earth as a result of the "greenhouse" effect produced by the accumulation of carbon dioxide in the atmosphere.

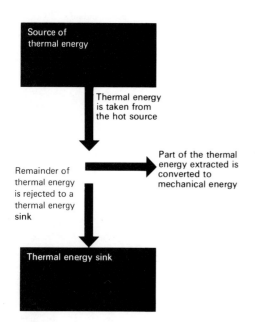

Fig. 1 An energy model for a heat engine.

BASIC THERMODYNAMICS

Pressure, Volume, Temperature

Both the steam turbine and the steam engine are devices that convert the energy of an expanding gas to useful mechanical energy. This is shown schematically in Fig. 2. The turbine has largely replaced the engine as a practical device. However, discussing the steam engine is much easier than discussing the turbine, and it is an ideal way to introduce some basic thermodynamics. Happily, our conclusions will also apply to the steam turbine.

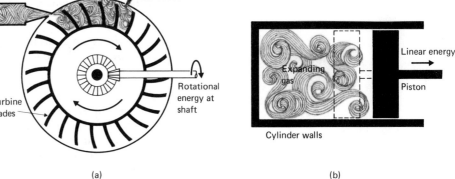

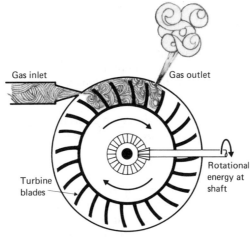

Fig. 2 (a) Schematic illustration of a steam turbine. Gas or vapor exerts a force on the blades of a turbine producing rotational kinetic energy. (b) Schematic illustration of a steam engine. Gas or vapor exerts a force on a piston contained in a cylinder producing linear kinetic energy.

Gas inlet

Gas outlet

Turbine blades

Rotational energy at shaft

(a)

Expanding gas

Linear energy

Piston

Cylinder walls

(b)

We start with the simple observation that an expanding gas pushing against a piston does mechanical work on the piston. We ask, "How can the work be calculated?" In answering this question, we use a macroscopic model of the gas. The appropriate variables are pressure P, volume V, and temperature T (see Fig. 3). Although a formal definition of these terms may not be familiar, their usage is common. Pressure implies a force. When we are pressured to study for an exam, it suggests that we are forced to do so. Automobile drivers are conscious of the air pressure in the tires. And when a balloon bursts, it is because the air pressure within the ballon exceeds the breaking strength of the balloon. Anything that

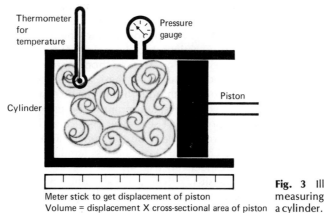

Thermometer for temperature

Pressure gauge

Cylinder

Piston

Meter stick to get displacement of piston
Volume = displacement X cross-sectional area of piston

Fig. 3 Illustration of an operational way of measuring the thermodynamic variables of a gas in a cylinder.

supports the weight of some object has a pressure exerted on it. The surface of the earth supports the weight of the air atmosphere above it; therefore the atmosphere exerts a pressure on the earth's surface or anything on the earth's surface. It might seem from these examples that force and pressure could be used interchangeably. There is, however, an important basic difference. Pressure accounts for the area over which the force acts. Formally,

$$\text{pressure} = \frac{\text{force on a surface}}{\text{area over which the force acts}}$$

$$P = \frac{F}{A}$$

Thus pressure has units of force divided by units of area. Typical units are newtons per square meter (N/m^2) and pounds per square inch (lb/in^2). For example, atmospheric pressure at sea level is about 15 lb/in^2. This means that each square inch on the surface of the earth at sea level experiences a force of 15 pounds. Note that the pressure can be very large if the contact area is very small. A very large pressure can be exerted with a thumbtack with only a small force because the contact area of the tack point is very small.

We have already used volume in connection with the concept of density. Here the volume is that which is occupied by the gas. As the piston moves, the volume changes. We designate a change in volume by the symbol ΔV. (The capital Greek delta, Δ, is commonly used to designate a change in a physical quantity.)

There is probably no scientific word more familiar than temperature. Nearly all houses have a thermostat which regulates the temperature within the house. Air temperature is a regular feature of TV newscasts and newspapers. Temperature is, however, a much subtler concept than pressure and volume. Unlike pressure and volume, which are ordinary mechanical quantities, temperature is a measure of the sensation of hot and cold. A thermometer is a measuring device used to assign a number related to the degree of the sensation of hot and cold. Thermometry is a science in itself and there are many ways to measure temperature. The basic idea is to find some physical quantity that changes when temperature changes. For example, since the height of a column of liquid such as mercury in a glass tube is dependent on temperature, a number designating the temperature can be assigned to the height of the mercury column when it is placed in some environment. This method requires that two reference temperatures be established. Customarily, these are the ice and steam points. The ice point is the temperature of an equilibrium mixture of ice and water at atmospheric pressure at sea level. The steam point is the temperature at which water boils at atmospheric pressure at sea level. The ice and steam points are assigned numbers 32 and 212, respectively, on the familiar Fahrenheit scale. They are assigned 0 and 100 and 273 and 373 on the less familiar, but more scientific, Celsius and Kelvin scales, respectively (Fig. 4).

If Celsius, Kelvin, and Fahrenheit mercury-type thermometers were placed in the same environment, the liquid would rise to the same level. But, since the

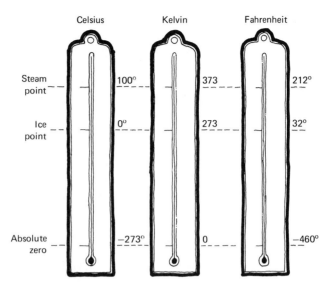

Fig. 4 Reference temperatures for the three common temperature scales.

scales differ as we have seen, each thermometer would register a different number for the temperature. Yet the numbers must be related to each other because they are derived from the same physical situation. The relations are

$$K = C + 273$$
$$F = \frac{9}{5}C + 32$$

You can check the validity of these equations by calculating the temperature at the ice point on the Kelvin and Fahrenheit scales knowing that this temperature is 0 on the Celsius scale.

Work Done by the Force of an Expanding Gas

A gas pushing against and moving a piston does work which, in principle, is no different from a person pushing on and moving a box on the floor (Fig. 5). The work, by definition, is the product of the force F and distance x the object is moved.

$$W = Fx$$

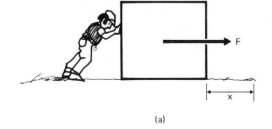 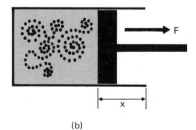

Fig. 5 (a) Work done by a person exerting a force F on a box. (b) Work done by an expanding gas pushing on a piston.

(a)   (b)

Since force is related to pressure by $F = PA$, the expression for work can also be written

$W = PAx.$

Now if the piston moves a distance x, then the volume will have *changed* by an amount

$\Delta V = Ax$

where A is the cross-sectional area of the cylinder. Hence the expression for work can be written

$W = P\,\Delta V$

and be expressed verbally by

work = pressure multiplied by *change* in volume

This expression has all the characteristics of the previously encountered relation

amount = rate multiplied by time

If pressure is plotted versus volume, then the area under the curve represents the work done. For example, if the pressure is constant, then the work is the area of a rectangle (Fig. 6).

There is no assurance that the pressure of a gas will be constant, and in any practical engine it probably will not be. This complicates the determination of the work done. Given the proper mathematical tools, the work can usually be calculated. Still there are cases where there is no recourse but to find the area under the pressure versus volume curve.

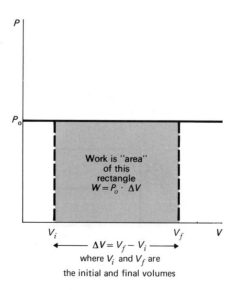

Fig. 6 Determination of work done by a gas on a piston if the pressure is constant.

Heat

Heat, like work, is a difficult topic to introduce because of preconceived ideas. When someone says "The house is cold, we need heat," the request is clear. Similarly, we know that exercise produces both warmth and an increase in body temperature and that energy is derived from food. So we make some intuitive connection between heat, energy, and temperature. The confusion between the everyday concept and the scientific concept arises when heat is thought of as some sort of energy *content* in a substance. Certainly there is inherent energy in something like a container of hot water, but in the scientific sense it would not be heat, it would be internal energy. The idea of internal energy seems to be rather all-encompassing, but we can narrow it down to just the ordinary kinetic and potential energy of the constituents of the substance. If energy is transferred from, say, the hot water to a potato in the water, we say that heat has been *transferred*. The point is that heat has meaning only in terms of energy in transit. For now, it suffices to consider heat as energy added to, or removed from, an object to raise or lower its temperature. Heat energy can be transferred only if a temperature difference exists, and the energy will flow from the object with the higher temperature. The calorie (cal) and British thermal unit (Btu) are units commonly used to quantify this concept. If the temperature of 1 g of water is raised from 14.5° to 15.5°C, the water is said to have absorbed 1 cal of heat. The kilocalorie (Kcal) or ordinary food calorie is equal to 1000 cal defined in this way. A Btu is the heat required to raise the temperature of 1 lb of water from 63°F to 64°F.

It is also meaningful to know how much heat is required to raise the temperature of one g of any substance one degree Celsius. This is called the specific heat, usually symbolized by C. The specific heat depends on temperature and pressure, but for most calculations it is assumed to be constant. The specific heats of some common substances are shown in Table 1. The concept of specific heat is also important in discussing gases. One talks about the heat required to raise the temperature of a gas containing a certain number of molecules. The number chosen is not important here. It is important, though, to recognize that if heat is added to a gas, both its volume and its pressure can change. For this reason, two gaseous specific heats are used. The specific heat at constant pressure, symbolized by C_p is the heat required to raise the temperature of the gas one degree Celsius (or one Kelvin) while keeping its pressure constant. The specific heat at constant volume, symbolized by C_v, is the heat required to raise the temperature of the gas one degree Celsius (or one Kelvin) while keeping its volume constant. The ratio of these two specific heats, C_p/C_v, is symbolized by γ and plays a very important role in thermodynamics. For example, it appears in the expression for the efficiency of a conventional gasoline engine in an automobile (see Exercise 24).

Table 1 Specific heat of some common substances

Substance	C (cal/g − °C)
Water	1.00
Aluminum	0.22
Iron	0.11
Copper	0.093

First Law of Thermodynamics

Rudolph Clausius, who probably did as much as anyone in the formal development of thermodynamics, stated in 1865 that "the total energy of the universe is

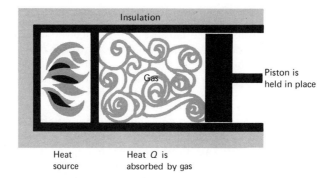

Fig. 7 The gas in a cylinder with a fixed piston absorbs thermal energy Q. No work is done by the gas, but the internal energy and temperature of the gas increases.

constant." This statement has come to be known as the first law of thermodynamics. Literally, it means that although one type of energy can be transformed into another, there is no change in the net amount of energy. However, the statement does not mean that the energy will be as accessible after the transformation and this is a crucial factor in the consideration of useful sources of energy.

In systems involving thermal energy, the first law of thermodynamics recognizes heat as energy in transit and formulates the concept in a quantitative way. To deduce this formal statement, consider the following experiments. A cylinder with a fixed piston is placed in contact with a heat source (Fig. 7). The heated gas in the cylinder presses against the piston; however, since the piston is held in place, no mechanical work is done. Where did the energy go then? It simply was absorbed by the gas, increasing its internal energy U. Formally

heat absorbed = *change* in internal energy
$$Q = \Delta U$$

This says nothing about how the energy is distributed in the gas or in what form it exists.

Now suppose that the heat source is removed and the cylinder is insulated from the surroundings so that heat can neither enter nor leave the cylinder. If the piston is released, it will move and mechanical work W is done. Where does the energy for this work come from? It can only come from the internal energy of the gas, since the cylinder is thermally insulated from its surroundings. Therefore, the internal energy must decrease. Formally

change in internal energy = work done on the piston
$$\Delta U = -W \qquad \text{(minus sign indicates a \textit{decrease} in internal energy)}$$

If heat is added *and* work is done, then we can write for the general situation

change in internal energy = heat added minus work done
$$\Delta U = Q - W \tag{1}$$

This is the formal statement of the first law of thermodynamics. Some time will be well spent if you consider what changes have to be made with Eq. 1 if the gas is compressed by forcing the piston in.

Efficiency of a Heat Engine

In a cyclic process, a system starts from some initial state of affairs, proceeds through a series of other states, and ultimately returns to the conditions of the initial state. A clock is a simple cyclic system. The clock is the system and a given time defines a state. Any given state (time) is repeated every 12 hours. A steam engine is also a cyclic system. A state of this system during any part of the cycle is specified by the pressure, volume, and temperature of the gas. The cycle involves the extraction of thermal energy (steam, for example) from a hot source, the conversion of thermal energy to mechanical energy, release of thermal energy to a cold heat sink when the vapor condenses to a liquid, and a return of liquid back to the boiler where it is again converted to steam (Fig. 8). There are complicated energy transfers throughout the cycle. However, if the system returns to its initial thermodynamic state with the same pressure, volume, and temperature, then there is no change in the internal energy in the cycle. If there is no change in the internal energy, the first law of thermodynamics says that the work done is just the *net* heat transferred. That is

$$\Delta U = \Delta Q - W$$
$$0 = \Delta Q - W$$
Therefore $$W = \Delta Q$$

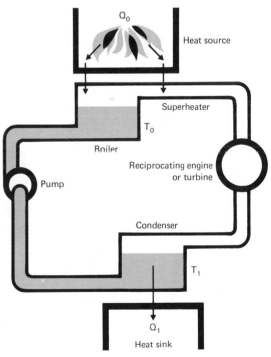

Fig. 8 (a) Schematic diagram of a cyclic process. (b) Schematic diagram of processes in a reciprocating steam engine or turbine.

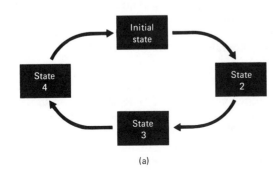

(a)

(b)

The net heat transferred is just the difference between the heat taken in and the heat released

$$\Delta Q = \text{heat absorbed} - \text{heat released}$$
$$= Q_{hot} - Q_{cold}$$

The efficiency for converting thermal energy to mechanical energy is

$$\epsilon = \frac{\text{work done}}{\text{thermal energy put into system}} \times 100\%$$

$$= \frac{W}{Q_{hot}} \times 100\%$$

But since $W = Q_{hot} - Q_{cold}$, then

$$\epsilon = \left(\frac{Q_{hot} \quad Q_{cold}}{Q_{hot}}\right) \times 100\%$$

$$= \left(1 - \frac{Q_{cold}}{Q_{hot}}\right) \times 100\% \tag{2}$$

This expression is perfectly general and applies to all heat engines.

There is a prescribed sequence of steps in the operation of the cycle of a heat engine. There are many physical systems in which a given sequence of events can be reversed. We have already encountered one such system in the form of chemical reactions. For example, the formation of the oxygen molecule O_2 from two oxygen atoms is

$$O + O \rightarrow O_2$$

The reverse of this would be

$$O + O \leftarrow O_2$$

The different direction of the arrow denotes the reversal. The energetics of these two processes are, of course, different. One requires that energy be added; the other requires that energy be released. The arrows on a heat engine can also be reversed. This means that thermal energy is taken out of the *cold* source, work is done *on* the system, and thermal energy is released to the *hot* reservoir. This is just what a household refrigerator does and an engine operating in reverse is appropriately called a refrigerator. The electric motor on a household refrigerator provides the energy source for *doing* work.

The Carnot Engine

A young French engineer in 1824 devised a theoretical model of a heat engine that established a maximum value for the efficiency of a heat engine operating between a hot reservoir at a temperature designated as T_{hot} and a cold reservoir designated as T_{cold}. This so-called Carnot engine drives a piston in the ordinary sense, but the model requires that there be no friction between the piston and the cylinder walls.

During parts of the cycle, the walls of the cylinder are required to be perfectly insulated from the surroundings so that no heat can either come into or leave the working gas in the cylinder. A process like this that does not exchange heat with its surroundings is called an adiabatic process. The working gas used to drive the piston has an ideal or perfect behavior. Mathematically, the pressure (in N/m^2), volume (in m^3), and temperature (on the Kelvin scale) of an ideal gas are related by

$$PV = kNT$$

where N is the number of molecules in the gas and k, called Boltzmann's constant, is a number equal to 1.38×10^{-23} joule/K·molecule. Interestingly, many gases do perform according to this relation, and indeed it explains many everyday observations about the behavior of gases. For example, the equation says that pressure P increases if temperature T increases. This is consistent with the rise in pressure in an automobile tire after it has been warmed on a hot day or after the car has been run at high speed.* The equation says that if the volume V of a gas goes down, the pressure P goes up. This is consistent with the behavior of a toy balloon when it is squeezed to make its volume smaller. Finally, the equation says that if the number N of molecules is reduced, the pressure P goes down. This is consistent with the behavior of a toy balloon when air is let out of it thereby reducing the number of molecules in the balloon.

   The cycle of the Carnot engine involves four processes. Two of these are adiabatic; two take place at constant temperature. The cycle starts with a certain volume of gas at some pressure and temperature. The gas expands against a piston doing work in the process in such a way that the temperature of the gas does not change. This means that heat must flow into the system to keep the temperature of the expanding gas from changing. The cylinder is then thermally insulated from its surroundings and the gas expands further, doing more work. Since no heat is exchanged, the temperature of the gas decreases. In the third step, the gas is compressed by forcing the piston inward, again in such a way that the temperature of the gas does not change. Thus heat must flow out of the system to keep the temperature of the compressed gas from changing. Finally, the gas is compressed back to its initial volume and pressure and the cycle is complete. This process is done in such a way that no heat is exchanged with the surroundings. The complete Carnot cycle is shown as a pressure versus volume plot in Fig. 9. Knowing how an ideal gas behaves when its pressure and volume are changed (1) at constant temperature and (2) when no heat is exchanged with its surroundings, we can show that the heat rejected divided by the heat extracted is equal to Kelvin temperature of the cold heat sink divided by the Kelvin temperature of the hot heat source. In symbolic form

$$\frac{Q_{cold}}{Q_{hot}} = \frac{T_{cold}}{T_{hot}}$$

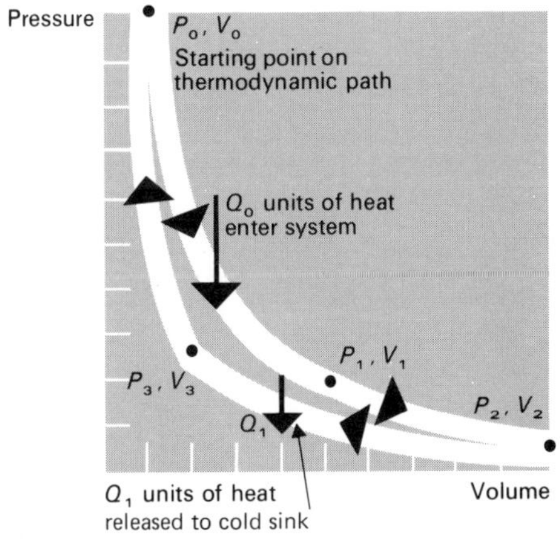

Fig. 9 The complete Carnot cycle as represented on a pressure-volume graph. The arrows denote the sequence of the operations.

---

*The pressure-temperature effect can be dramatically demonstrated by pouring liquid nitrogen on an inflated toy balloon.

Using the generalized efficiency equation for a heat engine (Eq.2) the efficiency of the Carnot engine is

$$\epsilon = \left(1 - \frac{T_{cold}}{T_{hot}}\right) \times 100\% \tag{3}$$

Thus the efficiency of a Carnot engine depends only on the temperatures of the hot heat source and cold heat sink. In a steam turbine, energy is extracted from steam at a temperature typically 825 K. Heat is released to the environment at a temperature of about 300 K. Hence, a Carnot engine operating at these temperatures would have an efficiency of

$$\epsilon = \left(1 - \frac{300}{825}\right) \times 100\% = 64\%$$

Based on typical operating temperatures, the efficiency of a Carnot engine is not particularly impressive. But, interestingly, we can show, using the second law of thermodynamics, that this is the greatest degree of efficiency that can be obtained for these two operating temperatures. We will, in fact, do exactly this in the section on "Maximum Efficiency of a Heat Engine."

The Second Law of Thermodynamcis

A perfect heat engine would convert all the thermal energy that it takes in into mechanical work. A perfect refrigerator would take thermal energy from a relatively cool source and release only that taken in to a source at a higher temperature. Friction is always found in a mechanical system and so we cannot expect to build perfect engines and refrigerators. Even if friction could be eliminated, our experience with thermodynamic processes leads us to the belief that perfect engines and refrigerators still could not be built. In this concept lies the essence of the second law of thermodynamics. There are several ways of stating the law. The following two equivalent ways are sufficient.

1. For a system operating in a *cycle* between heat sources designated as T_{hot} and T_{cold}, it is impossible to extract a net amount of heat from the hot source and convert it entirely into work. Some heat must be rejected to the cold source.

2. For a system operating in a cycle between heat sources designated as T_{hot} and T_{cold}, it is impossible to extract a net amount of heat from the cold source and transform only that amount to the hot source. In other words, heat will not spontaneously flow from a cold source to a hot source. It must be forced to and this results in work being done *on* the system.

These statements cannot be proved, but it is possible to prove that they are equivalent. They would be disproved if an exception were found, but to date, none has been.

Maximum Efficiency of a Heat Engine

If the Carnot engine cannot be constructed, then it seems that the time spent considering it is purely academic. However, the Carnot engine serves an extremely useful function because we can show that within the framework of the second law of thermodynamics no heat engine can be more efficient than a Carnot engine operating between the same two temperatures. Thus if a heat engine is required to operate between two given temperatures, we can immediately determine its maximum efficiency from the simple expression (Eq. 3) for the efficiency of a Carnot engine. We will prove this supposition by taking a real engine, that is, one that could actually be built, and letting it supply the energy for a Carnot refrigerator. Remember a Carnot refrigerator is simply a Carnot engine running backwards.

Fig. 10 Schematic representation of a hypothetical heat engine running a Carnot refrigerator. The output from the engine is coupled to the "driveshaft" of the refrigerator. The numbers are those obtained assuming an efficiency of 60% for the engine and 1 unit of heat absorbed by the engine from the heat source.

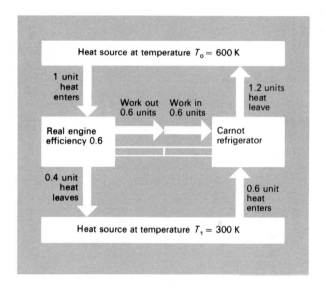

Figure 10 schematically shows a heat engine coupled to a Carnot refrigerator. The *combination* of the engine and refrigerator constitutes a cyclic process, which is necessary for application of the second law of thermodynamics as formulated on p. 100. Note that both the engine and refrigerator use the *same* heat sources. The goal is to show that the engine cannot have an efficiency greater than a Carnot engine operating between the same two temperatures. Let us suppose a specific situation and leave the general proof for a step-by-step exercise. Take $T_{hot} = 600$ K and $T_{cold} = 300$ K. The efficiency of a Carnot engine operating between these two temperatures would be

$$\epsilon_C = \left(1 - \frac{300}{600}\right) \times 100\% = 50\%$$

Assume that the hypothetical engine has an efficiency of 60%; that is, assume that its efficiency is greater than the Carnot efficiency. Now extract 1 unit of thermal

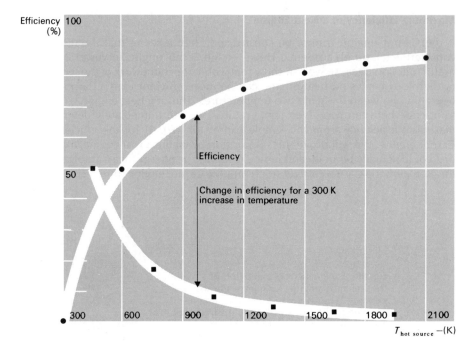

Fig. 11 Efficiency and change in efficiency of a Carnot engine for various temperatures of the hot source assuming a cold sink temperature of 300 K.

energy and see what happens. The engine will convert 60% of this to work and reject the other 40% as thermal energy to the cold reservoir. Thus

work produced = 0.6 unit
heat rejected = 0.4 unit

This 0.6 unit of work is used to run the Carnot refrigerator; hence the net work done by the engine-refrigerator combination is 0. To proceed, the work used by the refrigerator pulls 0.6 units of thermal energy from the cold reservoir and rejects 1.2 units to the hot reservoir. (Note that this is consistent with a Carnot engine which extracts 1.2 units of thermal energy, does 0.6 units of work, and rejects 0.6 units of thermal energy.) In making our final calculations then, we find that the engine-refrigerator combination has extracted a net amount of $0.6 - 0.4 = 0.2$ units from the cold reservoir and rejected $1.2 - 1.0 = 0.2$ units to the hot reservoir and has accomplished no *net* work in the process. Thus the second version of the second law of thermodynamics is violated. The conclusion is that the hypothetical engine cannot have an efficiency of 60% which exceeds the 50% efficiency of a Carnot engine at the specified temperatures. Exercise 16 illustrates that the second law is being followed if the efficiency of the hypothetical engine is less than that of a Carnot engine operating at the same temperatures. You are encouraged to work through this exercise to clarify these ideas.

It seems that there is something very special about the Carnot engine and there is, but only in the sense that the calculations are relatively simple if an ideal gas is used for the working substance. However, the fact is that any reversible

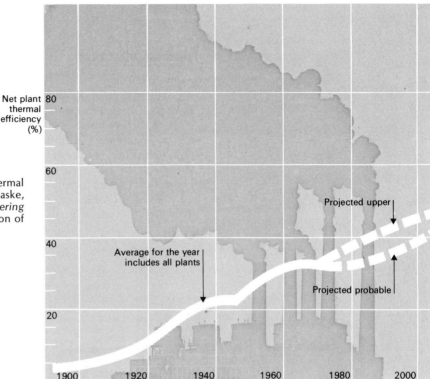

Fig. 12 Past and future trends of the overall thermal efficiency for an electric power plant. (From R. T. Jaske, J. F. Fletcher, and K. R. Wise, *Chemical Engineering Progress* **66**, 11, 17 (1970). Reprinted by permission of R. T. Jaske.)

engine can theoretically have the same efficiency as a Carnot engine if operated between the same two temperatures, leading us to the inevitable conclusion that there is nothing fundamentally special about the Carnot engine.

PROSPECTS FOR IMPROVED EFFICIENCY OF STEAM TURBINES

The Carnot engine, in addition to giving us a base for understanding the efficiency of a practical steam turbine, illustrates the significance of the temperatures of the hot source and cold sink. Increasing the temperature of the hot source or decreasing the temperature of the cold sink improves the efficiency (Exercise 18). Since the temperature of the cold sink is usually that of the environment (300K), over which we have little or no control, it is tempting to suggest a significant increase in the temperature of the hot source. There are, however, many nontrivial engineering aspects to this. For example, typical metal like copper melts at about 1350 K. Despite these technical difficulties, let us examine the Carnot efficiency for several temperatures of the hot source, assuming that heat is rejected at 300K. The results shown in Fig. 11 illustrate significant improvement in efficiency for *changes* between 300 and 1500K, but a change in temperature from 1200 to 1500K produces only a 5% increase in efficiency. The practical aspects of making such a temperature increase are forbidding, and the reward in efficiency small. What might be expected, then, in the way of increased efficiency of practical turbines? The projections in Fig. 12 are not encouraging, in fact, they indicate the imperativeness of seeking completely new technologies for generating electricity.

THERMAL POLLUTION

A pollutant is normally a biologically hazardous or otherwise undesirable particle or chemical compound in the environment. Several pollutants, though, add no mass to the environment. Thermal pollution (bad for fish), light pollution (bad for astronomers), and noise pollution (bad for people) are examples. Thermal pollution is the addition of unacceptable quantities of thermal energy to the environment. Knowing the many uses of heat in our everyday lives, we find it hard to think of rejected thermal energy as useless. It is useless only because as yet it is not economically practical to capitalize on it. Environmental considerations may in time alter this state of affairs, and thermal pollution may change to thermal enrichment.

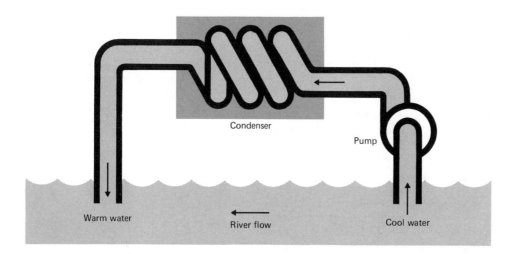

Condenser

Pump

Warm water

River flow

Cool water

Fig. 13 Schematic diagram of the condenser cooling system in a fossil fuel electric power plant.

How does thermal pollution come about? After steam passes through a turbine, it must be condensed back to the liquid state to complete the thermal cycle (Fig. 13). Thermal energy liberated in this process must be absorbed by some other system. This other system is the environment, and the transfer is effected by a continuous flow of water which is in thermal contact with the condenser associated with the turbine. Water for this heat-transfer system comes from a river, lake, or some man-made reservoir (Fig. 3-13). The water may return to its source some 10–20°F warmer. If the heat-dissipating abilities (the natural flow of a river, for example) of the source cannot cope with this heat load, the temperature of the source may rise. For example, the temperature of the entire Yellowstone River below Billings, Montana rose 1–1.5°F as a result of a 180-megawatt plant near Billings.*

Increased water temperature has several observable effects on a body of water.

1. As the temperature of the water increases, the capacity to hold dissolved oxygen decreases. At the same time, oxygen utilization by aquatic life increases.

2. Addition of heat can cause temperature stratification of a body of water because of the lower density at elevated temperatures.

3. Chemical reactions are accelerated at elevated temperatures.

The anticipated effects of elevated temperatures on aquatic life include the following.

1. Decreased spawning success of fish.

2. Decreased survival probability of young fish.

3. Limitation of migration patterns.

4. Changes in processes that depend on biological rhythms.

5. Increased susceptibility to disease.

6. Enhanced eutrophication.

Many laboratory studies have shown that some of the anticipated effects are very real. For example, 50% of a sample of brook trout will die in 133 hours in water at 77.0°F if the fish have been acclimated to a temperature of 68.0°F.† Thermal effects in actual streams or lakes are less well known, but research is proceeding. Preliminary reports of studies still going on in the estuaries of Maryland make it clear that the addition of heat to an aquatic environment could cause changes in its biota.‡ *Science News*, in the August 1, 1970 issue, reports that "there appears to be little doubt that effluents from the existing plants have done severe damage in Biscayne Bay (Florida)." Thermal effects are especially destructive in bodies of water that are already at temperatures near a critical level.

The concern over thermal pollution is not so much about present problems as about what may happen if precautions are not taken. For example, about 10% of the total energy presently consumed in the United States takes the form of waste heat from electric power plants (Exercise 20). The annual discharge in 1969 was

---

*Science News, **98**, # 5, (August 1, 1970).

†"Selected Materials on Environmental Effects of Producing Electric Power" p. 305.

‡Report of the Thermal Research Advisory Committee to the Maryland Department of Water Resources, December 1969.

about 6×10^{15} Btu, and it is expected to reach 20×10^{15} Btu in 1990.* By the year 2000 about 20% of the nation's average runoff of water would have to be used to cool power-plant condensers if present trends in power consumption continue and technology remains unchanged.† Fears of thermal pollution are real and the utilities industry is being challenged. Construction of a power plant on Cayuga Lake in New York was halted partly because of potential thermal effects on the lake. However, the thermal effects from power plants already in existence should be carefully studied, because the deleterious effects that have been anticipated may not in fact actually occur.‡ There is some evidence that heated water actually resides in a fairly thin layer on top of the main body of water and does not affect aquatic life in the main body.

Efforts toward making thermal enrichment a reality are in progress. A major problem with using heat from present-day plants is that it is low grade, meaning low temperature ($\sim 90°$F), and consequently its applications are limited. The Scots find that, in a fishery warmed by waters from their Hunterston nuclear power plant, sole mature in two years rather than the normal four. Similar uses of warm water are made in Japan. Experiments are under way in this country to improve the growth rates of fish, shellfish, and crustaceans with warmed water. Although still in the development stage, research indicates that the heat could be used in large-scale (300–500 acre) greenhouses. The temperature of the warm water can be increased by sacrificing the thermal efficiency of the turbine. Water at higher temperatures is more useful, especially for heating purposes. So the lower efficiency may be tolerable if greater use is made of the rejected thermal energy. The feasibility of this approach is being studied by the Department of Housing and Urban Development.

THERMAL STANDARDS

A single cooling process applicable to all power plants would be ideal from a practical standpoint. Yet even before environmental concerns were an issue, this was not possible. The common practice of using water from a river or lake is simply not feasible in many areas.

There is also no single thermal standard which is appropriate for all states. The temperatures of streams and lakes vary widely throughout the country, and the standards required to protect salmon in the Northwest may be quite different from those required to protect catfish in the Ohio River. The Water Quality Act of 1965 allows individual states to devise their own standards subject to approval by the Department of the Interior. If not approved, the Secretary may decide to set

---

*Problems in Disposal of Waste Heat From Steam-Electric Plants, Federal Power Commission, Bureau of Power (1969).

†Selected Materials on Environmental Effects of Producing Electric Power, p. 30 (Data from Federal Power Commission).

‡"The Calefaction of a River," by Daniel Merriman, Scientific American, 222, 42 (May 1970).

the standards himself. All 50 states have proposed standards and most have been approved. These standards are available from the Department of the Interior and can be found in a number of publications.* The general guidelines for setting the standards are:

1. Streams with cold-water fisheries (salmon, bass, trout, etc.)—68°F maximum temperature with a maximum change of 0 to 5°F above that expected for a given month.
2. Streams with warm-water fisheries (catfish, shad, bluegills, etc.)—83–93°F temperature with a maximum change of 4 to 5°F above that expected for a given month.

It should be pointed out, though, that at the time of this writing a senate sub-committee on air and water pollution headed by Senator Edmund Muskie is proposing national water quality standards to take effect January 1, 1975. The details are as yet unavailable.

Our next step is to see what methods are being used to dispose of waste heat within the framework of the existing standards and to study the physics of the processes used.

HEAT TRANSFER TO THE ENVIRONMENT

Microscopic Description of a Gas and a Liquid

The rejected heat from a power plant ultimately ends up in the atmosphere. The problem is to transfer it from the heated water in the condenser to the atmosphere in an acceptable way. Heat can be transferred by three basic mechanisms—conduction, convection, and radiation. Heat is also removed from a liquid in the process of vaporization. All these processes are currently employed to some extent by industry. Study of them at the microscopic (or atomic) level provides us with an ideal opportunity to understand more fully the concept of heat and the second law of thermodynamics.

Microscopic models are intrinsically more difficult to deduce than macroscopic models. The reason is simple. The forces between atoms and molecules must either be known or assumed. Even if known, the physics and mathematics are usually difficult. We clearly cannot do a rigorous analysis, but some of the results of such calculations might be made plausible and acceptable here.

The most elementary approach to a microscopic model of a gas seeks an explanation of the ideal gas equation, $PV = kNT$. As a start, we should agree that the gas is a collection of a very large number of molecules. The molecules are in motion and collisions between the molecules and the walls of the container give rise to pressure on the walls (Fig. 14). Volume is the same as that of the container.

Fig. 14 Schematic representation of molecules in a container exerting a force on the walls.

---

*See, for example, *Selected Materials on Environmental Effects of Producing Electric Power* (1969), available from the U.S. Government Printing Office.

But what about temperature? As yet we have no obvious microscopic interpretation. Therefore, the approach is to calculate the product of pressure and volume and equate the results to kNT to interpret T. The assumptions made are:

1. The size of a molecule is so small that it can be treated as a point. In other words, the volume of all the molecules is negligible compared to the volume of the container (see Exercise 26).

2. No force is exerted on a molecule except during collision with a wall or with another molecule.

3. The kinetic energy of a molecule is unchanged after a collision.

The result is that the average kinetic energy of a molecule is proportional to the temperature (in Kelvins) of the gas. Expressed as an equation, this concept becomes

$$\text{average kinetic energy} = 3/2\, kT \qquad\qquad (4)$$

We now have a microscopic interpretation of temperature. Let us see if this interpretation agrees with intuition and with experimental results.

A molecule colliding with a wall is much like a ball striking a wall, and surely the force on the wall increases as the energy of the ball increases. Since pressure is proportional to temperature (from the ideal gas equation), it is entirely reasonable to infer, first, that molecular energy is related to temperature (Fig. 15).

Second, if work is done on a gas by compressing it adiabatically, its pressure

and temperature increase. Third, if work is done on a mechanical system, its kinetic energy, or speed, increases. This is the principle employed in using a steam turbine to drive an electric generator. Likewise the work done on a gas might result in an increase in kinetic energy, or speed, of the molecules.

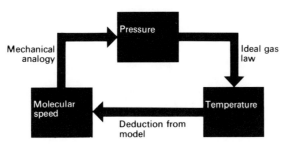

Fig. 15 Interrelation of pressure, temperature, and molecular speed. By analogy with a ball striking a wall, molecular speed is related to pressure. Pressure is related to temperature through the ideal gas law, for example. Therefore, it is reasonable that temperature is related to molecular speed.

The fact that our conclusions involve only the average speed may bother some readers. However, a collection of molecules in a container is somewhat like a collection of bees in a hive. Neither the molecules nor the bees are in such an orderly state that all have the same speed. Clearly, however, they would have an average speed and, perhaps, some most probable speed. We also know from our study of mechanics that direction of motion is just as important as speed. A little thought should convince you that there is no preferred direction of motion for the molecules in a gas (Question 4); the energy associated with molecular motion is completely random. This random energy must be extracted from the gas to drive the blades in a steam turbine.

When an object at temperature T_{hot} is placed in contact with an object at a lower temperature T_{cold}, the hotter object cools down and the cooler object heats up until an equilibrium temperature is achieved. Macroscopically we would say that *heat* has flowed (or been conducted) from the hotter to the cooler object. Microscopically we would say that the average kinetic energy of the molecules in the hotter object has decreased and the average kinetic energy of the molecules in the cooler object has increased. At equilibrium, the average kinetic energy is the same (Fig. 16). The transfer of energy has taken place by molecular collisions in much the same way that energy is transferred by colliding billiard balls. Heat is appropriately defined, then, as energy transferred by molecular collisions. This is why we say that heat is energy in transit.

It is important to remember that the concept of heat is meaningful only in terms of a *transfer* of energy. The energy *content*, that is, the internal energy, is just the kinetic and potential energy of the molecules. In an ideal gas, the energy content is all linear kinetic energy but many nonideal substances have intermolecular forces that give rise to a variety of types of potential energy.

A numerical calculation illustrates the magnitude of energy available in fresh water. Let us use water at 300K (room temperature) and assume that each molecule has average energy

$$E_{av} = 3/2\, kT, \quad \text{with } k = 1.38 \times 10^{-23} \text{ joule/K-molecule.}$$

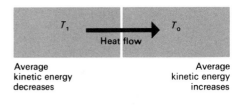

Average kinetic energy decreases   Average kinetic energy increases

Average kinetic energy and temperature are the same

Fig. 16 Macro- and micro-interpretation of heat transfer.

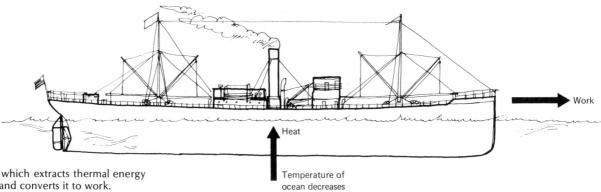

Fig. 17 A ship which extracts thermal energy from the ocean and converts it to work.

Work

Heat

Temperature of ocean decreases

Fig. 18 Manner in which circulating air current is produced in a room.

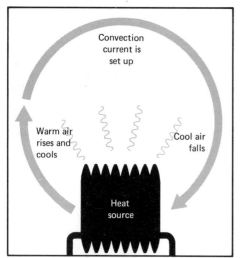

Convection current is set up

Warm air rises and cools

Cool air falls

Heat source

Then

$$E_{av} = 3/2 \, (1.38 \times 10^{-23}) \, 300$$
$$= 6.21 \times 10^{-21} \text{ joule/molecule}$$

The molecular weight of water is 18, and it is known that a mass (expressed in grams) equal to the molecular weight contains Avogadro's number, $N_A = 6.023 \times 10^{23}$, of molecules. Since the density of water is 1 g/cm³, the volume of these 18 g is 18 cm³ (18 cm³ is about the volume of a small bottle of aspirin). So the total energy is

$$E_{total} = N_A E_{av}$$
$$= 6.023 \times 10^{23} \, (6.21 \times 10^{-21})$$
$$= 3720 \text{ joules} = 3.7 \text{ Btu}$$

Hence in the oceans and the atmosphere there is an enormous amount of internal energy. Then why can't a heat engine be built to extract some of this energy and convert it to useful work, with the only environmental effect being a cooling of the source because energy has been extracted? (See Fig. 17). The answer is that although energy is available, it is random. To do useful work, the random motion must be converted to organized motion. Doing this requires a heat sink at a lower temperature so that energy can be transferred from a region of higher average kinetic energy to a region of lower average kinetic energy.

Conduction, Convection, and Radiation

Transfer of heat by conduction is the transfer of energy by molecular collisions. Note that there is no transfer of mass.

It is also possible to transfer energy by literally moving material. If a mass of air with a given average molecular energy moves into a region with lower average molecular energy, then energy will clearly be transferred. How might this be accomplished in practice? Suppose that a container of air, a room for example, is heated from the bottom (Fig. 18). The air in the vicinity of the source warms, the average kinetic energy increases, and the density of the air decreases (Exercise 27).

The less dense air rises in the room and is replaced by cooler (higher density) air coming down from the top. As a result, a current, or mass flow, of air is set up. This is the process of convection; it is the main mechanism by which heat is circulated in houses. Convection currents also give rise to wind. Since convection also occurs in liquids, it is partially responsible for ocean currents.

Radiation is the transfer of heat energy by electromagnetic waves. This mechanism is an important one but it is not primarily involved in any man-made system for disposing of thermal energy from a steam turbine.

Evaporation

To conclude our discussion of thermal-energy processes, let us consider evaporation. Evaporation is literally the escape of molecules from a liquid. As an analogy, imagine a box of bees with one end covered by a thin membrane which some bees are able to penetrate (Fig. 19). Their motion is random, with a distribution of energies and speeds. All bees with sufficient energy can escape, but those energetic ones near the surface are more likely to escape. The same is true of molecules in a liquid. Surface tension, the result of the attraction of the molecules in the liquid for each other, tends to act like a membrane in keeping the molecules inside the container. Only the more energetic molecules, especially those near the surface, are able to escape. A standard experiment illustrating the existence of surface tension is to lay a clean, dry needle on the surface of water (Fig. 20).

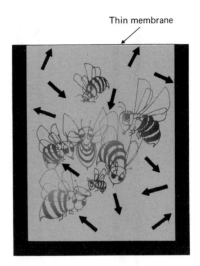

Fig. 19 A container of liquid is analogous to a hive of bees.

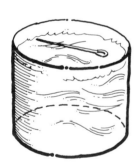

Fig. 20 Surface tension at the surface of water is sufficient to support a clean, dry sewing needle.

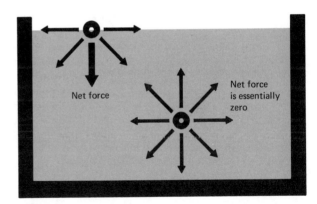

Fig. 21 Schematic illustration of forces on a molecule in the interior and on the surface of a liquid.

The origin of the molecular force that causes surface tension is understandable. Molecules experience attractive intermolecular forces from their neighbors. A molecule in the interior of a liquid (Fig. 21) is closely surrounded on all sides by similar molecules (see Exercise 26). Therefore, it experiences forces from all directions. However, a molecule at the surface has few or no neighbors beyond the surface. As a result, it is subject only to forces which tend to pull it back into the liquid.

If these forces did not exist, then every molecule would readily escape. Since only the most energetic molecules escape, the average kinetic energy of the liquid

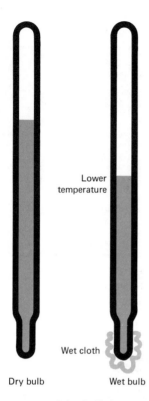

Lower temperature

Wet cloth

Dry bulb   Wet bulb

Fig. 22 A wet- and dry-bulb thermometer. The relative humidity is obtained from the difference in the readings of the two thermometers.

decreases and temperature goes down. Evaporation is therefore a cooling process. Also, since evaporation is a surface phenomenon, it follows that the larger the surface area, the greater the cooling effect.

The reverse process is also possible. A molecule in the vapor above a liquid can be captured by the liquid. This process is called condensation; in any real situation, both evaporation and condensation are taking place simultaneously. If evaporation predominates, there will be a loss in the volume of the liquid and a cooling of it. The rate of evaporation depends strongly on the amount of vapor above the liquid. The smaller the amount of vapor, the greater the evaporation rate. The amount of vapor that can exist in the air before the vapor begins to condense is limited and this amount, as we would expect from everyday experience, depends on the temperature. When air contains the maximum amount of water vapor, it is said to be saturated. The word *humidity* refers to the amount of water vapor in the air. The ratio of the amount of vapor actually in the air to the amount that would saturate the air at that same temperature is called the *relative humidity*. Expressed as a percent

$$\text{relative humidity} = \frac{\text{actual vapor content}}{\text{saturated content}} \times 100\% \qquad (5)$$

The evaporation rate depends on the relative humidity. Relative humidity can be measured by determining the difference in the dry-and-wet-bulb temperatures. The dry-bulb temperature is simply the reading of an ordinary thermometer in the environment in question. If the bulb of a similar thermometer is surrounded by a wick which is kept moist, it is referred to as a wet-bulb thermometer (Fig. 22). The wet-bulb thermometer will have a lower reading because the evaporation of water molecules will lower the temperature. The amount of lowering is a measure of the relative humidity. Normally, a table is used to determine the relative humidity from the temperature readings. The wet-bulb temperature is the lower limit for cooling processes utilizing an evaporative process. The dry-bulb temperature is the lower limit for a cooling process based on a conduction process. This turns out to be quite important for two of the current schemes used to cool the condenser water from a steam turbine.

METHODS OF DISPOSING OF THERMAL ENERGY

The common methods of disposing of the thermal energy within the framework of acceptable environmental conditions are shown in Table 2. We will now proceed to discuss these methods.

Table 2. Environmentally acceptable ways of thermal energy disposal

Method	Physical method	Comments
Once-through cooling	Evaporation, radiaiation, conduction	Utilizes a natural (lake, river) or artificial (cooling pond) watercourse
Wet-type cooling tower	Evaporation	Recirculates water to condenser after evaporative process.
Dry-type cooling tower	Conduction, convection	Heat dissipated to air through heat exchanger (radiator)

Fig. 23 Schematic diagram of the thermal stratification of a lake.

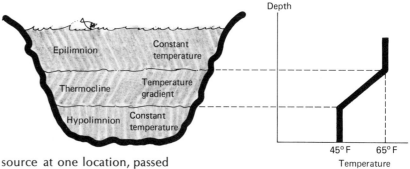

Once-through Cooling

In this process, cooling water is drawn from a source at one location, passed through the condenser, and returned to the source at a different location. Rivers, lakes, reservoirs, ponds, estuaries, and oceans are possible sources. The heated water is dispersed in a variety of ways. In a river, the natural flow is utilized. In all cases, the heat is dissipated from the body of the water to the atmosphere primarily by evaporation. Since evaporation is a surface phenomenon, the idea is to disperse the heated water to as large a surface area as possible. The longer the water stays in the condenser, the more heat it absorbs and the more its temperature increases. Hence, it is reasonable to expect that the rise in temperature of the water is inversely proportional to the rate of flow of the water through the condenser.

$$\Delta T = \text{increase in temperature } \alpha \ \frac{1}{\text{rate of flow of water}} \qquad (6)$$

About 0.01 gal/sec is required for each kilowatt of power for a temperature rise of 15°F. A 1000-MW station would require a water flow of about 10,000 gal/sec. If the temperature increase were to be lowered to 5°F, the water flow would increase to 30,000 gal/sec. Water pumped out of the river must be replenished by the natural river flow at a rate at least equal to the rate of flow through the condenser. Few rivers can meet these requirements and, as a result, rivers are not being considered so frequently as a source for dissipating heat. The sheer magnitude of these flow rates are difficult to appreciate until we realize that such a rate approaches that which would supply the water needs of a city like Los Angeles.

Cooling ponds are artificial bodies of water used to dissipate heat by evaporation. They are sometimes used when a natural supply is not available and land is inexpensive. As we might expect, large areas are required on the order of 1 to 2 acres per megawatt. A 1000-MW plant would require a pond with a surface area of 1000 to 2000 acres. Most cooling ponds in this country are found in the southwest.

Knowing that the U.S. abounds in large and deep lakes, they naturally come to mind as sources of cooling water. As far as predicting the thermal effects, they might seem at first to present an ideal situation. But unlike rivers for which the water flows are fairly well defined, a lake is subject to a variety of flow patterns and conditions which are strongly dependent on the weather. One feature, however, known as thermal stratification, makes a lake a highly attractive source. In winter a lake tends to achieve a nearly uniform temperature throughout. On the other hand, in the summer, when the upper portion is warmed, the lake tends to stratify into two layers of nearly constant temperature which are separated by a layer in which the temperature changes gradually (Fig. 23). The stratification results from

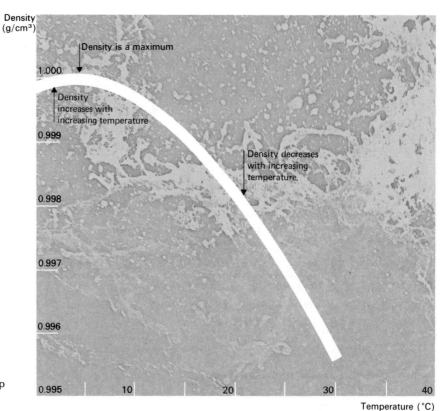

Density (g/cm³)

Density is a maximum

1.000

Density increases with increasing temperature

0.999

Density decreases with increasing temperature

0.998

0.997

0.996

0.995   10   20   30   40

Temperature (°C)

Fig. 24 Density-temperature relationship for water.

another extremely interesting property of water, namely the dependence of its density on temperature.

It is not surprising to discover that density can vary with temperature when we become aware that the size of an object changes with temperature due to a contraction or expansion. What may be surprising to learn is that, in the case of water, the density will either increase or decrease with an increase of temperature depending on the initial temperature (Fig. 24). The condition prerequisite to the floating of something in a fluid is that its density be less than the density of the fluid. A fluid having density less than water will float on the surface of the water. This is why an oil slick floats on water. A lake in winter receives relatively little energy from the sun. As a result, the temperature tends to be low, say about 40°F in central U.S., and nearly constant throughout. In the summer months the lake is warmed to a depth equal to the penetration of the sun's rays and is mixed by the wind. As a result of the warmed layers, which are of lower density, tending to stay on the surface, the lake stratifies. From the standpoint of the cooling condenser of a steam turbine, it is desirable to extract water from the cool hypolimnion layer and disperse the heated water from the condenser over the surface of the warm epilimnion. Done properly, little or no increase in temperature of the surface water results. Although this is an ideal engineering feature, there is an undesirable environmental feature.

Fig. 25 Cooling towers for the Rancho Seco nuclear-electric generating station located near Sacramento, California. Note the relative sizes of the towers, the generating facility, and automobiles. (Photograph courtesy of the Sacramento Municipal Utilities District.)

The dissolved oxygen is lower in the hypolimnion water. When this water, after passing through the condenser, is discharged to the epilimnion, the dissolved oxygen vital for plant and animal life will be reduced. Caution must be exercised to see that this does not in fact happen.

The oceans constitute a nearly inexhaustive source of cooling water and surely they will be tapped. There are unique problems, such as corrosion, which result from the salt in the water and these remain expensive propositions.

Cooling Towers

Wet- or dry-type cooling towers are normally used when a natural or artificial watercourse is unfeasible. The wet and dry nomenclature follows from whether or not water vapor is released to the atmosphere in the cooling process. If a natural source with the proportion of a river or lake is required to achieve adequate cooling, we could well imagine that these tower structures are enormous. Indeed, typical cooling towers (Fig. 25) are some 400 feet high and as much as 400 feet in diameter at the base. Merely constructing these towers presents some challenging engineering and environmental problems.

Mechanical draft tower

Air out

Drift eliminator

Water in →

Packing

Fan

Air in

Water out ←

Wet (evaporative) forced air flow

Fig. 26 Schematic drawing illustrating the principles of the natural draft and forced air flow in wet cooling towers.

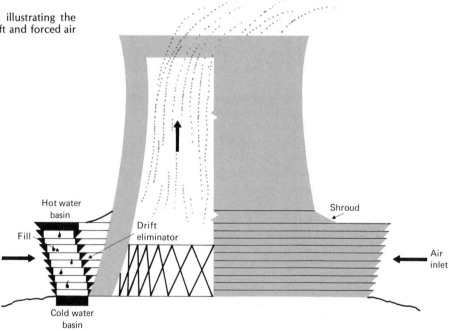

Natural draft tower

Hot water basin

Fill

Drift eliminator

Shroud

Air inlet

Cold water basin

Wet (evaporative) crossflow

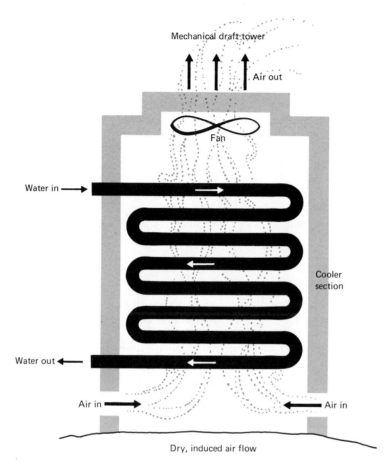

Mechanical draft tower

Air out

Fan

Water in →

Cooler section

Water out ←

Air in → ← Air in

Dry, induced air flow

Fig. 27 Schematic drawing illustrating the principle of the mechanical draft dry-cooling tower.

The wet-type tower utilizes evaporative cooling (Fig. 26). The lowest cooling temperature is therefore the wet-bulb temperature (Fig. 22). As in any evaporative method, the idea is to disperse the water in such a way as to maximize the surface area. This is achieved by either breaking the water into droplets or forming it into thin films. The evaporated water molecules must be removed and this is achieved by either a natural or forced air draft through the tower. The natural draft works by the ordinary chimney effect; the forced air draft functions by using a fan which either pulls the air through from the top or forces it up from the bottom. About 2% of the water is lost through evaporation—a figure which is deceptively small. A 1000-MW facility will require evaporative losses of about 30 ft^3 sec. This is equivalent to a daily rainfall of about 1 inch over an area of 1 square mile (see Exercise 13). Hence, severe fogging and icing could occur in the vicinity of the plant on a cold day.

The dry-type tower is, in principle, a huge radiator of the type found in most automobiles. The water to be cooled is pumped through a heat exchanger (Fig. 27) with a huge surface area which is in contact with air. Heat is transferred from the exchanger to the air by conduction and is carried away by either forced or natural convection currents.

This system seems ideal because it is closed and there are essentially no water losses. However, besides being the most expensive of the systems discussed, it produces an immense heat plume because the energy is transferred directly to the atmosphere. The heat plume is so pronounced that it could produce local alteration of the weather. Although systems of this type have never been used in the United States in conjunction with electric power plants, they have been used on a limited scale in Europe for some time. The operating experience gained there leads some to believe that such systems can be practical in the United States and that they will be a competitive alternative in the near future (see, for example, *Thermal Waste Treatment and Control*, by Frank H. Rainwater, Federal Water Quality Administration, Department of the Interior, presented to the Atomic Industrial Forum, Inc. and Electric Power Council on Environment, Washington, D.C., June 28–30, 1970).

Relative Costs

As a law of ecology, Dr. Barry Commoner proposes that, "There is no such thing as a free lunch."* Somebody has to pay and in the case of cooling condenser water, it will surely be the consumer. The relative costs of the systems discussed are shown in Fig. 28. Any variation in the standard once-through method is more expensive. For the dry-type tower system the added expense is astounding and, although this method is ideal in the sense of conservation of water, it has been economically prohibitive and neither has a large-scale dry-cooling facility been built nor is one planned for this country.

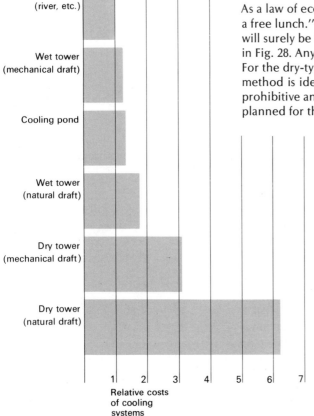

Fig. 28 Relative costs of the various cooling systems for steam turbine condenser water. (Adapted from "Cooling Towers," *Scientific American* **224**, 5, 70 (1971).)

*See, for example, "Our Ecological Crisis," *National Geographic* **138**, No. 6, 772, (1970).

STABLE ATMOSPHERIC CONDITIONS

"The solution to pollution is dilution" is a clever cliche. Nevertheless, much of the disposal of by-products and refuse relies on this principle and assigns nature to be the bearer in the form of wind and ocean currents. A serious situation results if nature rebels and does not provide the necessary atmospheric circulation to disperse the pollutants. There are two common processes that produce stable atmospheric conditions for which we are now prepared to understand the thermodynamic reasons.

Temperature Inversions

Anyone who has gone up a mountain or in a commercial jet airplane knows that the air temperature is lower at the high altitude than it is on the ground. The variation of temperature with altitude is characterized by a quantity called the lapse rate. Typical units for the lapse rate are degrees Celsius per kilometer (°C/km) and degrees Fahrenheit per mile (°F/mi). If the temperature decreases 4°C for each km increase in altitude, we would say the lapse rate is -4°C/km. The $-$ sign reminds us that the temperature decreases. Generally, the temperature decreases up to an altitude of about 40,000 feet (about 8 miles or 13 kilometers). This region is called the troposphere and contains about 80% of the entire mass of the atmosphere (Fig. 29). Note that the temperature decreases about 7°C for each kilometer increase in altitude. Such a condition of decreasing temperature with height is necessary for a mass of warm air to rise.* This is no different than the conditions required for warm air to rise from the floor in a room. The dispersal of pollutants from a smokestack relies on rising warm air to carry the particles into the upper atmosphere. Now suppose that the temperature increases with altitude. This is called a temperature inversion because the condition is inverted from normal. The air from the smokestack does not rise now (Figs. 30 and 31), the pollutants become more concentrated, and serious effects can occur.

Temperature inversions can occur in several ways. The two most common are as follows. In the daylight hours of a normal, sunny day, the temperature will decrease with altitude up to several thousand feet (Fig. 32a.) At night, the ground and air in the first one or two thousand feet cool thereby producing an inversion in the temperature distribution (Fig. 32b). Pollutants released during the night will not rise and will tend to collect in this low "inverted" layer. Normally, when the sun comes up and the ground warms, the situation goes back to normal and the pollutants disperse. But if there is a cloud cover and the ground does not warm up and there is no wind, the pollutants remain and the situation worsens. It is a temperature inversion of this type that prevailed at the time the photograph in Fig. 30 was taken. Temperature inversions of this type are common on cold, clear winter mornings and they can readily be identified simply by observing smoke emanating from a chimney or smokestack.

---

*In the next section we will see that there are conditions under which warm air will not rise even though the temperature decreases with increasing altitude.

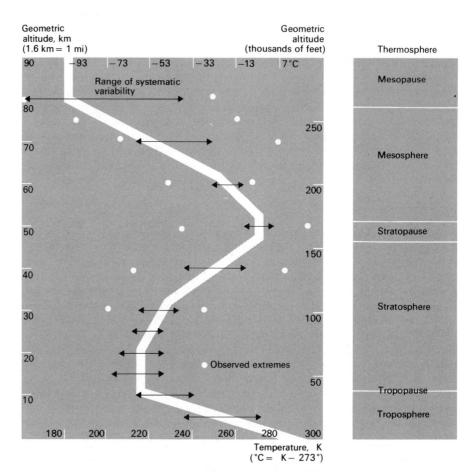

Fig. 29 U. S. standard atmosphere (1962) and range of systematic variability. (From *Climate and Weather*, by John A. Day and Gilbert L. Stearns, Addison-Wesley Publishing Co., Reading, Mass., 1970.)

Fig. 30 An illustration of how smoke fails to rise because of a temperature inversion in the atmosphere. This type of inversion is often seen in the morning on a clear, cold day. After the sun is up for a while, the inversion disappears and the smoke can be seen to rise.

The second common type of temperature inversion results from the phenomenon of *subsidence* (from the word subside, meaning to sink to a lower level). Air pressure, as well as temperature, decreases with increasing altitude. Under certain conditions, cool high level air will sink to a higher pressure level. This happens, for example, when a high level wind loses speed. Once in the high pressure area, the area is compressed and it warms. (Warming of air by compression can be easily observed with an ordinary tire pump or with a device like that shown in

Fig. 31 A device for demonstrating how smoke fails to rise if temperature increases with increasing altitude.

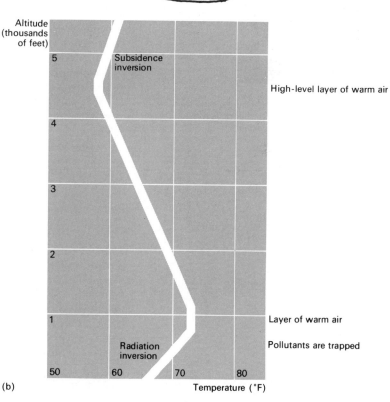

Fig. 32 (a) Typical temperature versus altitude curve on a sunny day. (b) A temperature versus altitude curve illustrating both a radiation and subsidence temperature inversion.

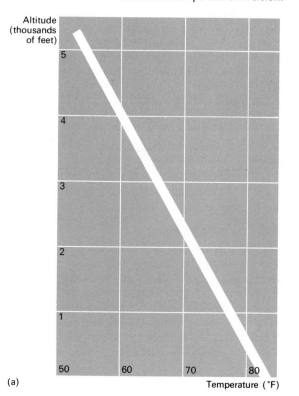

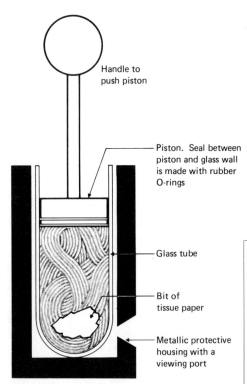

Handle to
push piston

Piston. Seal between
piston and glass wall
is made with rubber
O-rings

Glass tube

Bit of
tissue paper

Metallic protective
housing with a
viewing port

Fig. 33 A device for demonstrating that the temperature of a gas increases when it is compressed adiabatically.

Fig. 34 The first except is an example of a warning of possible conditions for producing a temperature inversion. The second excerpt provides a description of the aftermath of the stagnant air conditions which did occur.

Fig. 33.) As a result, the air temperature is higher than normal in a local area and a temperature inversion results. Inversions of this type tend to form at altitudes of 1000–10,000 feet. Subsidence inversions are common on the West Coast and, because of the enclosing mountainous terrain, produce the stagnant air conditions around Los Angeles. Such inversions are not as common in other parts of the United States but they do happen. The Donora incident mentioned in Chapter 2 resulted from an inversion of this type and conditions there were aggravated by the surrounding mountains. Weather forecasters are often able to sense the onset of conditions which produce a subsidence inversion and are able to forewarn the public. Figure 34 gives an account of such an alert and the polluted conditions which followed. Figure 35 illustrates how sulfur dioxide levels rose in St. Louis in August 1969 as a result of a temperature inversion that affected 22 states east of the Mississippi River.*

From *Cincinnati Enquirer* August 17, 1971

SMOG TRAP PERILS AREA

Weather conditions offer potential for high levels of air pollution in Ohio urban areas, the National Weather Service advised Tuesday.

Smog levels were moderate in Cincinnati Tuesday, but the possibility of increased pollution this morning exists.

The federal agency said a strong inversion could cause the accumulation of pollution until Thursday morning. A Cincinnati spokesman for the agency said the conditions were expected to deteriorate Tuesday night and this morning.

It is during the morning hours that pollution tends to be the heaviest. The concentration of automobile traffic produces a volume of oxides and hydrocarbons that in the sunlight may produce smog.

The weather condition—a large static high pressure system with very light winds—traps the contaminants and holds them close to the ground.

From *Cincinnati Enquirer* August 19, 1971

FRESH WINDS DISPERSE CITY'S DIRTY, STAGNANT AIR

The smog that has accumulated in the city the past couple days and made seeing downtown difficult (even from Mt. Adams) began to disperse Thursday afternoon. The pollution has been trapped by an inversion that kept Cincinnati's air contaminants near their sources. But the approach of a frontal system started breaking up the stagnant air and moving it to the southeast. The change prompted the National Weather Service to lift its pollution advisory at noon Thursday although a spokesman in the Cincinnati office said it would be this afternoon before the smog accumulation in the air is cleared out by the increasing wind activity.

*"Episode 104," Environment **13**, 2 (January/February 1971).

Conditions inclining toward development of temperature inversions are further aggravated if fog is present as often happens in London. The word smog, meaning smoke and fog, was coined for these conditions. More will be said about this in Chapter 6.

A Stable Atmospheric Condition without a Temperature Inversion

Another situation producing a stable atmospheric condition in which pollutants will not disperse can occur *without* a temperature inversion. This happens when the prevailing temperature *decrease* with *increasing* altitude (that is, the lapse rate) is less than a certain reference lapse rate.

If a balloon is seen to accelerate upward, it means that there is some net force acting on it in the direction of the acceleration. We can release a parcel of air and analyze the forces acting on it just as in the case of the balloon. It is somewhat more conceptual because we don't actually see the boundaries of the parcel as we would for the balloon. If the net force on this parcel always tends to restore it to some equilibrium position, then it will not rise or fall and the air will be stable. Analogy can be made to a ball placed inside an upright bowl. A ball at rest on the bottom of a bowl is in equilibrium because the net force on it is zero (Fig. 36a). It is said to be in *stable* equilibrium because it always tends to return to the equilibrium position regardless of the direction in which it is moved (Fig. 36b). If the ball is at rest on top of an inverted bowl it is also in equilibrium but it is unstable because it always tends to move away from the equilibrium position regardless of the direction in which it is moved (Fig. 36c). If we take the parcel of air and change its position in the atmosphere, its internal pressure must adjust to the surrounding pressure. For example, if it is raised to a position of lower pressure, it expands, its internal pressure decreases, and its temperature goes down according to its own internal lapse rate.

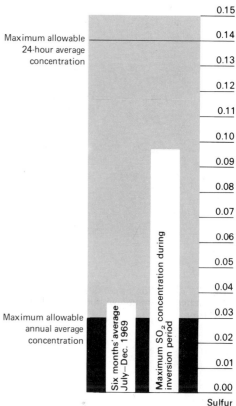

Fig. 35 Illustration of how sulfur dioxide levels rose in St. Louis in August 1969 as a result of a temperature inversion. Particulate concentrations show the same behavior. (From *Environment* **13**, 2 (January/February 1971).)

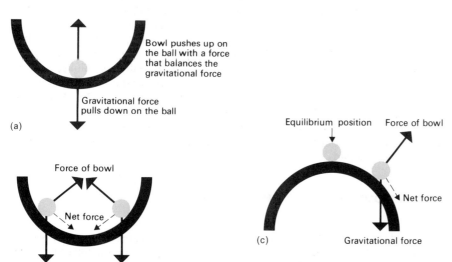

Fig. 36 (a) A ball at rest at the bottom of an upright bowl or at the bottom of an inverted bowl is in equilibrium because the net force on the ball is zero. (b) A ball moved from its equilibrium position at the bottom of an upright bowl will return to the equilibrium position when it is released. This is an example of stable equilibrium. (c) A ball moved from its equilibrium position at the top of an inverted bowl will not return to the equilibrium position when it is released. This is an example of unstable equilibrium.

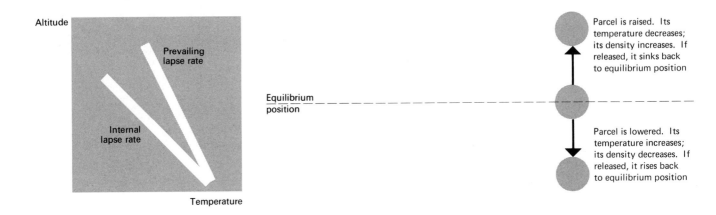

Fig. 37 Illustration of conditions for a stable atmospheric condition without a temperature inversion.

However, air is such a poor heat conductor that it essentially exchanges no heat with the surroundings. Therefore, it is, by definition, an adiabatic process. If there is no condensible water vapor, the lapse rate is about 10°C/km. This is called the dry-adiabatic lapse rate. If the parcel contains condensible water vapor, then when the gas expands and cools the water vapor may condense and release its latent heat of condensation. There is competition between cooling and warming and the internal lapse rate decreases. This is called the wet-adiabatic lapse rate. It depends on pressure, volume, and moisture content and may be as low as 3°C/km. Thus there is a characteristic lapse rate for any parcel of air under consideration and a prevailing lapse rate characteristic of the surroundings. If the prevailing lapse rate is *less* than the internal lapse rate, the air is in a stable condition. For example, if the prevailing lapse rate is −4°C/km and the internal lapse rate is −7°C/km, the air is in a stable condition. The reason for this is that if the parcel is raised, its temperature will be less than its surroundings; therefore its density is greater and it will sink back to its equilibrium position if released (Fig. 37). The converse happens if the parcel is lowered.

CARBON DIOXIDE AND THE "GREENHOUSE" EFFECT

Carbon dioxide (CO_2) is not toxic and does not harm plants or property. Superficially it would seem that there is no concern even though a 1000-MW coal-burning electric generator at full capacity dumps about 29,000 tons of CO_2 into the

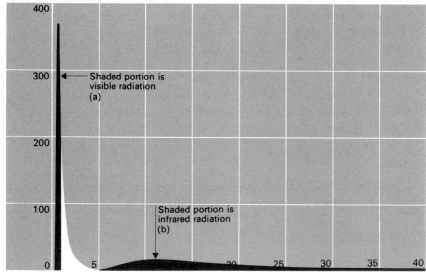

Fig. 38 (a) The approximate radiation spectrum of the sun as seen at the earth's surface. (b) The effective radiation spectrum of the earth.

atmosphere each day. Yet the long-term effects of CO_2 accumulation in the atmosphere may be disastrous. The concern is due to what is popularly called the "greenhouse" effect which *may* cause the mean temperature of the earth to rise just as the temperature inside the glass walls of a greenhouse rises. It has been speculated that the earth could warm to the extent that the polar ice caps would melt and submerge the coastal cities.*

The warming effect can be explained as follows. The amount and type of radiant energy from a heated object depends primarily on the temperature of the object. This is commonly observed in the heating coils of an electric stove. Radiation coming from the sun is mostly the visible type to which our eyes are sensitive. Visible radiation passes through carbon dioxide in the atmosphere with very little attenuation. The radiation, after penetrating the earth's atmosphere, is absorbed and reflected by things on the earth's surface. The earth radiates energy like the sun, but because the temperature of the earth's surface is much lower than the temperature of the surface of the sun the type of radiation is much different. It is mostly of the infrared type. Carbon dioxide strongly absorbs infrared radiation. Thus radiation comes freely through carbon dioxide in the atmosphere, but is prevented from escaping upon reradiation by the earth. The energy is trapped and the temperature of the atmosphere and the earth would obviously rise *if* significant amounts of carbon dioxide were to accumulate.

---

*"Man's Impact on the Global Environment," *Report of the Study of Critical Environmental Problems* (*SCEP*), M.I.T. Press, Cambridge, Mass. (1971).

A more quantitative explanation of this effect can be made using some facts obtained from studies on radiation from heated objects.* There is a simple relation between temperature and wavelength for the greatest amount of radiation which is expressed as

$$\lambda \cdot T = 2900 \text{ micrometers} \cdot \text{Kelvins}$$

wavelengths   source temperature
in micrometers   in Kelvins

The sun radiates at an effective temperature of about 6000K. Hence the most probable wavelength is

$$\lambda = \frac{2900}{6000} = 0.48 \text{ micrometers.}$$

This is in the middle of the visible region of the electromagnetic spectrum (Fig. 38a). The earth radiates at an effective temperature of about 255K. Hence, the most probable wavelength for radiation from the earth is

$$\lambda = \frac{2900}{255} = 11 \text{ micrometers.}$$

This is in the far infrared region of the electromagnetic spectrum (Fig. 30b) and is not perceived by the human eye. Figure 39 shows in a quantitative way how CO_2 selectively absorbs infrared radiation. It is this selective absorption that gives rise to the concerns about a possible temperature increase of the earth.

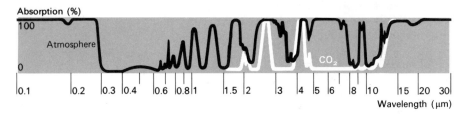

Fig. 39 Absorption characteristics of carbon dioxide and the atmosphere. An absorption of 100% means that all the radiation of that wavelength will be absorbed. Note the selective absorption for certain wavelengths.

It is not particularly difficult to forecast the amounts of CO_2 produced by the combustion of fossil fuels. However, all the CO_2 does not remain in the atmosphere. Significant amounts are taken up by the oceans and some probably stimulates plant growth via photosynthesis. These factors are difficult to estimate. So the only reliable measure is actual monitoring of CO_2 in the atmosphere. Sporadic measurements have been made for over a century. In 1959 systematic monitoring was begun at the Mauna Loa Observatory in Hawaii. These results are shown in Fig. 40. Concentrations have risen from 290 ppm a century ago to 320 ppm in 1969. Twenty percent of this increase has occurred in the last decade, With these em-

---

*A discussion of electromagnetic waves and thermal radiation is given in Appendix 2 for those not familiar with these topics.

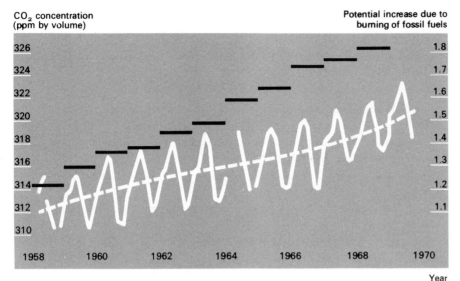

Fig. 40 Long-term variations in the CO_2 content of the atmosphere observed at the Mauna Loa Observatory in Hawaii.

pirical results and the predicted consumption of fossil fuels, it is estimated that the concentrations will rise to about 400 ppm and 500–540 ppm in the years 2000 and 2020, respectively. There have been many predictions about, but little agreement on, the effects of these concentrations. The disagreement is understandable because of the great difficulty in accounting for atmospheric motions. The conclusions reported by the *Study on Critical Environmental Problems (SCEP)** in 1970 were that the projected increase for 2000 *might* increase the surface temperature of the earth by about 0.5°C and a doubling of the CO_2 might increase mean annual surface temperatures by 2°C. A 2°C rise would indeed be significant.

REFERENCES

Basic Thermodynamics

Physics: A Descriptive Analysis, Albert J. Read, Addison-Wesley Publishing Co., Reading, Mass (1970).

Conceptual Physics, Jae R. Ballif and William E. Dibble, John Wiley and Sons, Inc., New York (1971).

Thermal Pollution

Selected Materials on Environmental Effects of Producing Electric Power, U.S. Government Printing Office, pp. 285–335 (1969).

"Thermal Pollution and Aquatic Life," *Scientific American* **220**, 19 (March 1969).

---

*"Man's Impact on the Global Environment," *Report of the Study of Critical Environmental Problems (SCEP)*, the M.I.T. Press, Cambridge, Massachusetts (1971).

"Finding a Place to Put the Heat," *Science News* **98**, No. 5, 98 (Aug. 1, 1970).

"Environmental Side Effects of Energy Production," *Bulletin of Atomic Scientists* **26**, No. 8, 39 (October 1970).

"Thermal Pollution Control," *Science* **168**, 421 (24 April 1970).

"Uses of Waste Heat," *Oak Ridge National Laboratory Review*, **3**, No. 4 (Spring 1970).

Report of the Thermal Research Advisory Committee to the Maryland Department of Water Resources, December 1969.

"Nuclear Power and Thermal Pollution: Zion, Illinois," *Bulletin of the Atomic Scientists* **26**, No. 3, 17 (March 1970).

"The Calefaction of a River," Daniel Merriman, *Scientific American* **222**, 42 (May 1970).

"Heat Rejection Requirements of the Unted States," R. T. Jaske, J. F. Fletcher, and K. R. Wise, *Chemical Engineering Progress* **66**, No. 11, 17 (Nov. 1970).

Cooling Towers

"Cooling Towers," by Riley D. Woodson, *Scientific American* **224**, No. 5, 70 (May 1971).

Weather Phenomena

Climate and Weather, John A. Day and Gilbert L. Sternes, Addison-Wesley Publishing Co., Reading, Mass. (1970).

Weather and Life, William P. Lowry, Academic Press, New York, N.Y. (1969).

Carbon Dioxide Pollution

"The Carbon Cycle," *The Biosphere*, W. H. Freeman and Company, San Francisco (1970) pp. 47–57. Also *Scientific American* **223**, 124 (Sept. 1970).

"Human Energy Production as a process in the Biosphere," *The Biosphere*, W. H. Freeman and Company, San Francisco (1970) pp. 105–115.

"Electric Power Generation and the Environment," *Westinghouse Engineer*, May 1970, pp. 66–80.

"The Atmosphere: A Clouded Horizon," *Environment* **12**, No. 3, 32 (April 1970).

"The Williamstown Study of Critical Environmental Problems," *Bulletin of Atomic Scientists* **26**, No. 8, 24 (Oct. 1970).

Topics and Material for In-depth Studies

The Environmental Effects of Energy Generation on Lake Michigan, Serial 91–96, U.S. Government Printing Office, Washington, D.C. (1970). This is a publication of the proceedings of a U.S. Senate Subcommittee on Energy, National Resources, and the Environment. It gives, in detail, testimony of citizens and groups who are concerned about possible thermal effects on Lake Michigan.

"Episode 104," *Environment* **13**, 2 (1971). This is a detailed study of the effects of an air inversion which touched 22 states during late August, 1969.

Construction of a nuclear electric generator by the New York State Electric and Gas Company on Cayuga Lake near Ithaca, New York was halted because of potential thermal effects on the lake. The impetus for the closure was supplied by citizens groups composed mainly of staff members from Cornell University. It would be very educational to look into calculations and data they used to back up their fears about thermal effects on the lake. An account of this effort can be found in "Nuclear Power and Its Critics; the Cayuga Lake Controversy" by Dorothy Nelkin, Cornell University Press (1971).

QUESTIONS

1. Discuss some possible ways in which the thermal energy that is rejected by a steam turbine could be put to use.
2. A manufacturer of a household dehumidifier will state that it does not provide a cooling of the surroundings. Why is this so? However if you go into a room in which the humidity has been lowered, you feel cooler. Why?
3. An area of skin touched with perfume feels cool. Why?
4. Formulate an argument in support of the following statement: if there were a preferred direction of motion for the molecules of a gas in a container, then the container would experience a net force and could possibly move.
5. A child's swing oscillates back and forth with a frequency that depends only on the length of the swing and the acceleration due to gravity

$$f = \frac{1}{2}\pi\sqrt{\frac{g}{l}}$$

child learns to increase (by pumping) the amplitude of the oscillations by feeding energy into the system at the natural frequency of oscillation. A similar situation occurs for two objects connected by a spring.

The frequency of vibration depends only on the mass of the objects and the strength of the spring. A CO_2 molecule can be thought of as three masses

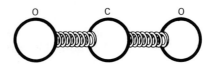

connected by two springs with vibrational frequencies that depend only on the masses and the strength of the spring force. Using this microscopic model, explain why you might expect CO_2 to absorb certain frequencies of electromagnetic radiation which happen to be in the infrared region. These natural vibrations can be easily demonstrated by taking three masses and connecting them with rubber bands.

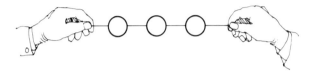

Hold the ends with your hands and set the masses vibrating.

Three masses of string like a CO_2 molecule have some natural frequencies of vibration. This photograph shows one of the natural frequencies for which the two outer masses move in the opposite direction from the center mass. Another natural frequency corresponds to the two outer masses moving in opposite directions with the center mass remaining still.

6. Discuss the idea of defining ecology as "the quantification of the first and second laws of thermodynamics for environmental processes."

7. An animal is, in a sense, a machine. He takes energy in as food and converts it to other forms of energy at a certain efficiency. Suppose that each animal in a carnivorous food chain involving five animals has a conversion efficiency of 0.6. What is the overall conversion efficiency for the five-animal chain? What are the implications of eliminating one animal in the chain?

8. There are many natural cyclic processes in our environment; the carbon dioxide and sulfur dioxide cycles serve as examples. Environmental problems often occur when infringements are made on these cyclic processes. Think of some natural cyclic processes which have had infringements made on them and have consequently developed problems. (A good example is the Aswan Dam project in Egypt.)

9. The benefits versus risks must be weighed in any infringement on a natural cycle. Birth control pills are an infringement on the natural menstrual cycle of a woman. Discuss the benefits versus risks involved in this procedure.

10. The oceans and the atmosphere constitute enormous sources of thermal energy. Why, then, doesn't someone construct a heat engine that uses these energy sources?

11. Cooling towers are the dominant structures in an electric generating station. Think of some architectural problems in trying to make them aesthetically pleasing.

12. The density-temperature relation for water is extremely unusual for a liquid. That for mercury shown in Fig. 41 is more typical. Yet the fact that the density of water decreases from 4°C to the freezing point of 0°C is extremely important as far as life on this planet is concerned. Think a bit about the freezing of a lake or pond and how this might change if water behaved as most liquids. What environmental effects might result?

13. Suppose that an inventor with no knowledge of thermodynamics comes to you with a drawing of a heat engine which does not reject any heat. Since the two of you have no common basis for argument, what might you demand him to provide in the way of evidence supporting his thesis? This is just what would happen if the inventor were to go to the Patent Bureau with his drawings.

14. An irate (husband/wife) unaccustomed to air conditioning threatens to cool the house by opening the refrigerator door. How would you tactfully answer (him/her) as a (wife/husband)? What is the significance of placing the exhaust of a room air conditioner on the outside of a house?

15. An automobile, with the windows up, will warm from the sun's radiation by the "greenhouse" effect. Explain.

16. What effects do you think might contribute to the warming of a greenhouse other than the one described in the text?

17. Suppose that fish were attracted to the warm water effluents of a power plant. Then, for some reason, the power plant had to shut down and the fish died as a result of a cold water kill. What would your reaction be to this type of fish kill?

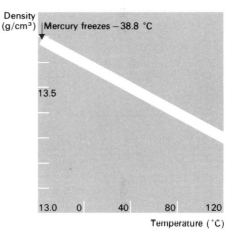

Fig. 41 Density-temperature relationship for mercury.

Exercises

Easy to Reasonable

Exercise 1. A typical temperature in a room is about 80°F. Show that this corresponds to 27°C and 300K.

Exercise 2. There is one temperature for which the Celsius and Fahrenheit values are the same. What is this temperature?

Exercise 3. Suppose that you were not satisfied with any of the three temperature scales mentioned in the text and you chose to assign numbers of 100 and 200 to the ice and steam points. If you characterized your scale by °M, show how you could convert °M to °C.

Exercise 4. What are some physical quantities which change with temperature that could be used for a thermometer?

Exercise 5. Atmospheric pressure at sea level is about 15 lb/in². How big a square on the earth's surface would you need so that the total force exerted on this square would equal the weight of a 135-lb person?

Exercise 6

a) A large tractor exerts considerably less pressure on the ground than a hypodermic needle on the skin of a patient. Why?

b) Suppose that a 10-ton tractor distributes its weight over two tracks each having an area of 20 ft². What is the pressure in lb/ft²?

c) Suppose that the diameter of a hypodermic needle is 1/64 in. How much pressure is exerted on the skin by a 1-lb force supplied by a nurse?

Exercise 7

a) A piston in a cylinder has a diameter of 10 cm. If this piston moves a distance of 4 cm, what is the change of volume (in m³) in the cylinder?

b) If this displacement is caused by a 10,000-N force, how much work is done in the process?

Exercise 8. A heat engine in each cycle extracts 50,000 Btu of thermal energy and releases 20,000 Btu of thermal energy.

a) How much energy is converted to work?

b) What is the efficiency of the engine?

c) What physical principle did you use for part (a)?

Exercise 9. Why is it that a body of water in sunlight warms more slowly than the land which contains it? Land is, of course, not completely composed of iron, but you might give plausibility to your answer by calculating the temperature rise of 1000 g of iron and water if each absorbs 10,000 cal of heat. You might want to use the data in Table 1.

Exercise 10. The pressure exerted by gas molecules on the walls of an automobile tire is greater than the pressure recorded by a "tire gauge" used to measure the tire pressure. On the basis of pressures (or forces) acting on the tire, can you explain why this is so? Can you estimate in lb/in² by how much the tire gauge underestimates the actual pressure in the tire?

Exercise 11. What is the efficiency of a Carnot engine operating between a heat source at 100°C and a heat sink at 0°C?

Exercise 12. Make a pressure-volume sketch for the cycle of a Carnot refrigerator similar to the graph made for a Carnot engine in Fig. 9.

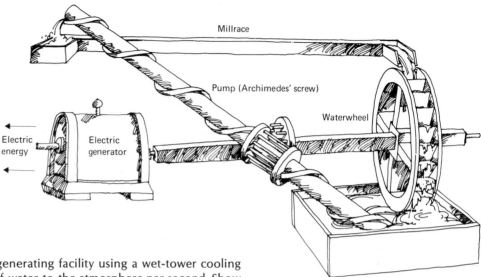

Millrace

Pump (Archimedes' screw)

Waterwheel

Electric
energy

Electric
generator

Exercise 13. A 1000-MW electric generating facility using a wet-tower cooling method will disperse about 30 ft^3 of water to the atmosphere per second. Show that this amounts to a daily rainfall of about 1 inch over an area of 1 square mile.

Fig. 42 A proposed electric generating system.

Exercise 14. Suppose that an inventor comes to you with Fig. 42 and asks you to invest in the construction of this electric energy generating system. Using the first law of thermodynamics as a basis for argument, how would you explain to him why you are not interested? If he is not a believer in the first law, what additional evidence might you demand of him?

Reasonable to Difficult

Exercise 15. Normally an advertisement for an air conditioner will characterize it as having a cooling capacity of so many Btu, for example 24,000 Btu. What important variable has been left out which makes this advertisement deceiving? Explain.

Exercise 16. Following the analysis in the text of a real engine driving a Carnot refrigerator, assume that the temperatures of the hot and cold sources are 600K and 300K and that the efficiency of the real engine is 0.4. Show that the second law is not violated.

Exercise 17. When water boils and changes from a liquid to a vapor, energy must be added. This is called the heat of vaporization and amounts to 540 cal/gram.

a) How does the energy required to vaporize 1 g of water compare with the energy required to raise the temperature of 1 g of water from 0°C to 100°C?

b) When water begins to boil, its temperature does not change. What, then, is the difference between water which is just simmering on the stove and water which is boiling vigorously?

c) Why would you suspect that the air pressure above a liquid affects the temperature at which it boils?

Exercise 18. Assume that a steam turbine in a power plant operates at maximum efficiency. The temperature of the heat source is 900K and heat is rejected at 300K. The efficiency can be improved by either increasing the temperature of the heat source or decreasing the temperature of the heat sink. Suppose that you had a choice of *either* a 25K increase in heat source or 25K decrease in heat sink.

a) What would your choice be?

b) The heat sink for a steam turbine is usually a river or a lake. Do you think it would be significant to try to use water from the lower part of a lake as opposed to that from the upper part? Explain.

Exercise 19

a) Using the basic definition of a calorie, show how it can be used to calculate the heat required to raise the temperature of 100 g of water from 30°C to 80°C.

b) Write down a general expression for the heat required to raise the temperature of m grams of water ΔT degrees Celsius.

From the practical standpoint of cooling the condenser of a steam turbine, it is important to recognize that the specific heat of water is the largest of those listed.

c) Suppose that 1 kg of water and 1 kg of copper each absorb 10,000 cal of heat. Which will experience the greater increase in temperature? Use a numerical calculation to prove your point.

Exercise 20. The total energy consumption in the United States in 1970 amounted to 6.88×10^{16} Btu* whereas the electric energy consumption amounted to 1.5×10^{12} kW-hr.

a) What percent of the total energy is electric?

b) About 83% of this electric energy was generated by fossil fuel plants. What percent of the total energy was generated by fossil fuel plants?

c) If the average conversion efficiency for fossil fuel plants was 35%, what percent of the total energy was used at the input of the power plants?

d) If 45% of this energy is rejected as thermal energy, what fraction of the total energy is rejected as the environment in the form of thermal energy?

Exercise 21. Efficiency is a quantitative measure of what is expected of an engine.

$$\text{efficiency} = \frac{\text{work out}}{\text{energy in}}$$

a) Using the first law of thermodynamics, explain why this quantity is always less than one.

b) In words, what does an air conditioner do?

c) What would you expect a perfect air conditioner to do?

_____

United States Energy—A Summary Review, U.S. Department of Interior, January 1972.

d) Define in your own way a coefficient of performance which would be a quantitative measure of the performance of an air conditioner. Using the first law of thermodynamics, explain why this number is always greater than one.

Exercise 22. Figure 43 schematically shows a real engine providing the energy source for a Carnot refrigerator. It is identical to the example in the text except that it is perfectly general and does not assume a numerical value for the efficiency and thermal energies. The idea is to show, in general, that if the efficiency of the real engine is greater than the efficiency of a Carnot engine, then the second law of thermodynamics is violated. For each question in parts (a) and (b) select one of the following—less than, equal to, or greater than—as the correct answer. The efficiencies for the individual devices are

$$\epsilon_R = \frac{W}{Q_0}, \qquad \epsilon_C = \frac{W}{Q_0'}$$

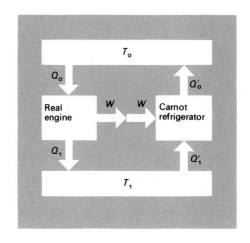

Fig. 43 Real engine running a Carnot refrigerator.

a) If ϵ_R is greater than ϵ_C, then Q_0 is less than, equal to, or greater than Q_0'?

b) Since W is the same for both engines, the first law of thermodynamics tells us that $Q_0 - Q_1$ is less than, equal to, or greater than $Q_0' - Q_1'$?

Using (a) and (b), would you conclude that Q_0' is less than, equal to, or greater than Q_1'. You should be able to show that heat $Q_1' - Q_1$ has been extracted, no work has been done, and the same amount of heat has been rejected at temperature T_0. Hence the second law is violated.

Difficult to Impossible

Exercise 23. An experiment was described in Fig. 31 in which a gas was heated by compression to the extent that a bit of paper within the cylinder was ignited. It is now easy to see why this happens. The compression happens so fast that it is essentially an adiabatic process. Hence we can write

$$P_{before} V_{before}^{\gamma} = P_{after} V_{after}^{\gamma}$$

$$\frac{P_{after}}{P_{before}} = \left(\frac{V_{before}}{V_{after}}\right)^{\gamma}$$

a) Taking $\gamma = 1.5$ and a compressed volume 100 times smaller than the uncompressed volume, show that the pressure has gone up by a factor of 1000.

b) Using the ideal gas equation, we can write

$$\frac{P_{after} V_{after}}{T_{after}} = \frac{P_{before} V_{before}}{T_{before}}$$

Assuming that the temperature before the expansion is 300K, show that the final temperature is 300×10^5 K. Hence it is not surprising that the paper ignites.

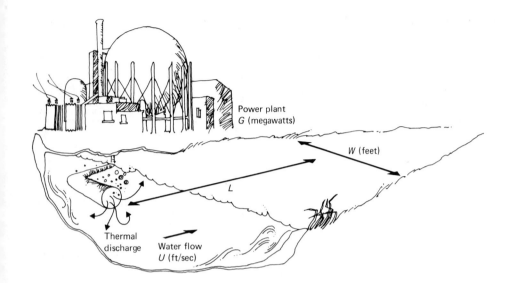

Power plant
G (megawatts)

W (feet)

L

Thermal
discharge

Water flow
U (ft/sec)

Fig. 44 Variables associated with thermal effects in a river.

Exercise 24. The theoretical efficiency of an internal combustion engine is

$$\epsilon = 1 - \frac{1}{r^{\gamma-1}}$$

where γ, the specific heat ratio, is a constant which is typically 1.4 and r is the ratio of the maximum volume of the cylinder to the volume of the gas when it ignites.

a) The efficiency increases as the compression increases? Explain.

b) Has this in fact occurred in automobile engines over the years?

c) What sort of environmental problems do you think might result from high performance engines with large compression ratios?

Exercise 25. Figure 44 shows schematically a generating facility which rejects thermal energy to a stream. Thermal effects are witnessed in the stream for L miles. How would you expect the affected length L to depend on the river width W, water flow U, and generating capacity G? Express your answer in the form of an equation.

Exercise 26. A mass in grams equal to the molecular weight of a substance contains the same number of molecules. This number is called Avogadro's number and is equal to 6.02×10^{23}. This can be used to estimate the separation of molecules in a gas and a liquid. Table 3 gives basic information for nitrogen gas and water. Assume that each molecule occupies a cubic volume. Calculate the length of the side of this volume for the gas and the liquid. Compare these lengths with the size of an atom (10^{-8} cm in diameter). Does it seem reasonable, then, that molecules in a liquid might experience forces that a molecule in a gas would not?

Table 3

	Nitrogen gas	Water
Molecular weight	28.013	18.015
Density	0.00125 g/cm³ at 10°C, 760 mm Hg	1g/cm³
Volume		
Volume shared by each molecule		

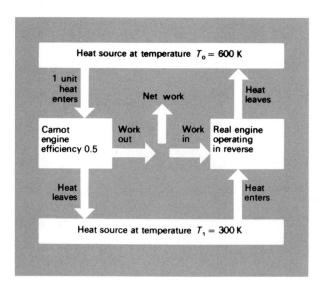

Fig. 45 Carnot engine running a real refrigerator.

Exercise 27. A container of volume V cubic centimeters contains M grams of an ideal gas. The density ρ, is by definition

$$\rho(g/cm^3) = \frac{M(g)}{V(cm^3)}$$

The N in the ideal gas equation $PV = kNT$, stands for the number of molecules in the gas. So if we knew the mass, m, of a single molecule, the number N is

$$N \text{ (molecules)} = \frac{M(g)}{m(g/molecule)}$$

a) Using this result, show that the density of an ideal gas is inversely proportional to absolute temperature.

b) Explain why a heated gas tends to rise.

c) The fact that a heated gas rises can also be explained by pressure differences. Do this assuming an ideal gas.

d) Can you explain why the pressure in an automobile tire increases on a long trip in summer?

Exercise 28. Rather than taking a real engine and running a Carnot refrigerator, suppose that we take a Carnot engine and let it run a real refrigerator (Fig. 45).

Assume that the real engine has an efficiency of 0.6, that is, an efficiency greater than that of the Carnot efficiency. Assuming that 1 unit of heat is extracted, show that a net amount of heat can be extracted from the hot reservoir and converted into work with no net heat rejected at the lower temperature. This would violate the second law of thermodynamics. What would you then conclude?

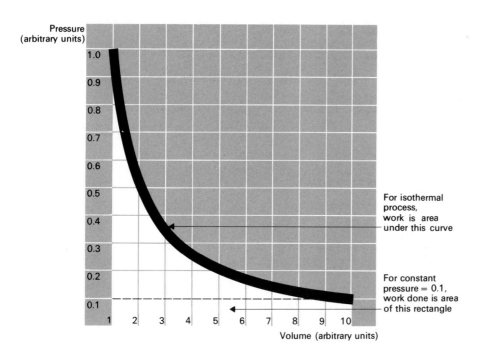

Pressure (arbitrary units)

For isothermal process, work is area under this curve

For constant pressure = 0.1, work done is area of this rectangle

Volume (arbitrary units)

Fig. 46 Isothermal expansion of an ideal gas represented by $P = 1/V$.

Exercise 29. The expression

$$W = \frac{kNT}{0.434} \log_{10} \frac{V_f}{V_i}$$

yields the work done by an isothermal expansion of an ideal gas. The purpose of this exercise is to verify the validity of the expression with a numerical example.

Figure 46 shows a plot of $P = 1/V$. This corresponds to an isothermal expansion of an ideal gas if $kNT = 1$ in the expression $P = NkT/V$. Using the quoted expression for the work done if the gas expands from $V_i = 1$ to $V_f = 10$, we have

$$W = \frac{kNT}{0.434} \log_{10} \frac{V_f}{V_i} = \frac{1}{0.434} \log_{10} \frac{10}{1} = \frac{1}{0.434} = 2.30 \text{ units}$$

Now if an expansion from $V_i = 1$ to $V_f = 10$ took place at a constant pressure of 0.1, the work done would be
$$W = P \Delta V$$
$$= 0.1 \ (9) = 0.9 \text{ units}$$

For both of these processes the work is related to the area under the curve which is, in turn, related to the number of squares. For the constant process there are 225 squares (count them). Estimate the squares under the curve for the isothermal process and by comparison with the constant pressure process verify that the work is 2.30 units.

Exercise 30. Many gases at normal atmospheric pressure (about 10^6 dynes/cm^2) and temperature (about 300K) obey the ideal gas law

$$P V = kNT$$

Pressure Volume   Boltzmann's   Number of   Kelvin Temperature   constant   molecules

If we write this as

$$\frac{N}{V} = \frac{P}{kT}$$

then we obtain an expression for the number of molecules per unit volume. For example, if $V = 1$ cm^3, then N is the number of molecules in this volume of 1 cm^3. It is interesting that this depends only on pressure and temperature and not on the type of molecule in the gas.

a) Using $k = 1.38 \times 10^{-16}$ ergs/K, $P = 10^6$ dynes/cm^2, and $T = 300$K, show that the number of molecules in one cubic *meter* of gas is 2.44×10^{25}. Now if we know the total mass of a certain constituent, say SO_2, then we can calculate how many molecules there are of this constituent.

$$\text{Total mass} = (\text{total number of molecules}) \times (\text{mass per molecule})$$
$$M = N \cdot m$$

We also know that a mass in grams equal to the molecular weight contains Avogadro's number (6.023×10^{23}) of molecules. For example, the molecular weight of SO_2 is 64. So the mass of a single molecule is

$$m = \frac{\text{molecular weight}}{\text{Avogadro's number}} = \frac{A}{N_A} = \frac{64}{6.023 \times 10^{23}}$$

Using this, we find that the number of molecules is $\quad N = \dfrac{M N_A}{A}$

b) Show that there are 7.5×10^{17} SO_2 molecules in a mass of 80 μg of SO_2.

c) If there are 80 μg of SO_2 in 1 m^3 of air, show that this amounts to about 0.03 SO_2 molecules per million total molecules.

Exercise 31. The pressure, volume, and temperature of an ideal gas are related by $PV = kNT$. Thus if an ideal gas expands from pressure P_0 and volume V_0 to pressure P_1 and volume V_1 at constant temperature, we can write $P_0 V_0 = P_1 V_1$. The pressure and volume of an ideal gas during an adiabatic process are related by $PV^\gamma = $ constant. Thus if an ideal gas expands adiabatically from pressure P_1 and volume V_1 to pressure P_2 and volume V_2, we can write $P_1 V_1^\gamma = P_2 V_2^\gamma$. Using these relations and Fig. 9 as a guide, show that

$$\frac{Q_0}{Q_1} = \frac{T_{hot}}{T_{cold}}$$

4

NUCLEAR-FUELED ELECTRIC POWER PLANTS: PHYSICS, POLLUTION, AND OUTLOOK

Photograph courtesy of the Yankee Atomic Power Plant, Rowe, Massachusetts.

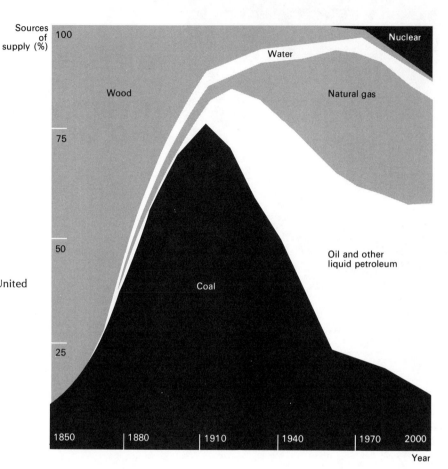

Fig. 1 Various sources of energy in the United States.

INTRODUCTION

The sun constitutes the earth's most abundant energy source and must be considered in any planning for our long-term energy needs. With the few exceptions of energy derived from external sources such as the energy associated with ocean tides, all other sources are locked up in the earth's structure. Interestingly enough many of these earthly sources are derived directly from the sun's energy. The fossil fuels, for example, result from small, but crucial, long-term imbalances in the carbon cycle involved in the process of photosynthesis which is dependent upon the sun. Wood, the prime source of fuel in the United States until 1880 (Fig. 1), is produced by photosynthesis.

Each new discovery of an energy source involves increased difficulty and time delay in successful application. The discovery of coal, oil, and natural gas introduced problems in mining, drilling, and transportation. Economic transportation of natural gas across an ocean requires liquefaction at $-259°F$ and one can well

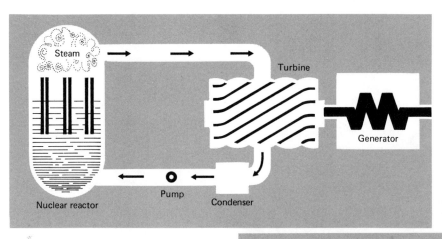

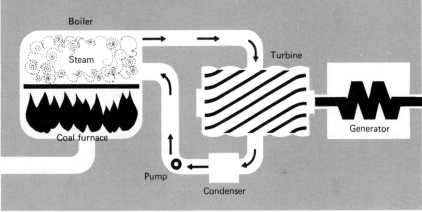

Fig. 2 Schematic illustrations of the nuclear fuel and fossil fuel electric generating systems.

imagine the involved technical and safety problems. The feasibility of nuclear energy as a practical source was demonstrated at the University of Chicago in 1942, yet some 30 years later nuclear energy accounts for only 0.2% of the United States energy budget. Many of the problems are technical, but there has never been a technology more carefully and rightfully evaluated for environmental effects, and this has also contributed to the time delay. We can be assured that any new energy technology will require an equivalent development time. Nuclear fusion is a distinct possibility for a clean, nearly inexhaustible energy source, but the research has not reached a stage of development comparable to that achieved in the work at the University of Chicago in 1942. Energy from nuclear fission offers at least an interim solution to some of the problems of limitation of fossil fuels and pollution. There is real hope that it will provide a long-term solution.

The nuclear and fossil fuel electric generating systems differ only in the nature of the heat source (Fig. 2). The thermal pollution problem exists in both and is, in fact, worse at present for nuclear systems because the thermal efficiency is lower.

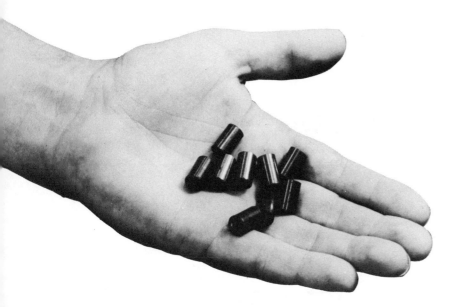

Pellets of uranium dioxide that will be fabricated into fuel assemblies. When these assemblies are placed into the core of a nuclear reactor, each pellet can produce as much as, or more than, a ton of coal. (Photograph courtesy of the United States Atomic Energy Commission.)

The particulate, carbon dioxide, nitrogen oxide, and sulfur oxide pollution problems are eliminated, but a new one—radioactivity pollution—is introduced. Nuclear fuel offers the possibility of much cheaper electricity, but for a variety of reasons this has yet to be demonstrated.

The general public may not understand the physics of nuclear energy, but they are aware that the mechanism is the same as that in the atomic bomb. As a result, the element of fear has been introduced and the public is rightfully demanding assurance that a holocaust will not result from an explosion in a nuclear power plant.

This chapter is devoted to the physics of the nuclear energy process, and to an examination of the origin and magnitude of radioactivity from a nuclear power plant, the environmental and biological effects of radiation, and the public reaction to the idea of nuclear power. The treatment is necessarily brief; however, the topics are extremely important in our present society. There is considerable material available which can be used to extend the coverage presented here.*

MOTIVATION

The intrinsic energy associated with mass is absolutely intriguing. To illustrate, let us calculate the rest mass energy of 1 gram of matter. (One gram of water could be put in an eye dropper.)

---

*One such appropriate book in paperback which covers these topics in a detailed and readable way is *Nuclear Energy—Its Physics and Its Social Challenge* by David R. Inglis, Addison-Wesley Publishing Co., Reading, Mass., 1973.

Atomic nuclei are a vast source of energy. A single truck-load of nuclear fuel can supply the total electrical power needs of a city of 200,000 people for a year. (Photograph courtesy of the United States Atomic Energy Commission.)

1 gram $= 0.001$ kg

$c = 3 \times 10^8$ m/sec

$E = mc^2 = (0.001)\,(3 \times 10^8)$

$\quad\quad = 9 \times 10^{13}$ J

The thermal energy released in the combustion of one ton of coal is about 2.7×10^{10} J. Thus, there is about 1000 times more intrinsic energy in one gram of matter than that derived from the combustion of one ton of coal. In fact, the bombs exploded on Hiroshima and Nagasaki in World War II released energy equivalent to about 1 gram of matter. The question is, "How do you convert even a portion of this energy to a useful form?" The answer, is "Through a nuclear reaction."

RULES FOR THE NUCLEAR DOMAIN

Our discussions thus far have required an atomic model consisting only of a positively-charged nucleus surrounded by negatively-charged electrons. This model is adequate because those things discussed microscopically (molecules, chemical reactions, etc.) involved only interactions of the electrons. Nuclear reactions involve interactions of the nuclei of atoms and this requires that the model be improved. However, many of the concepts used in the atomic model are also useful in the refinement. For example, both the atom and the nucleus are *bound states*. The atom is bound together by attractive *electric forces* between the positively-charged nucleus and negatively-charged electrons. The nucleus is bound together by "strong" *nuclear forces* between its constituents which are protons and neutrons. A proton has a charge equal in size to the electron but it is positive. Its mass is 1.6725×10^{-27} kg. The neutron is electrically neutral and its mass is slightly larger, 1.6748×10^{-27} kg. The proton and neutron are, respectively, 1836 and 1839 times as

Fig. 3 Schematic illustration of atomic and nuclear forces. Note the differences in energy, distances, and form of the potential energy curves. The potential energy of a proton would be different from that for the neutron because of the additional energy due to its electric interaction with the other proton.

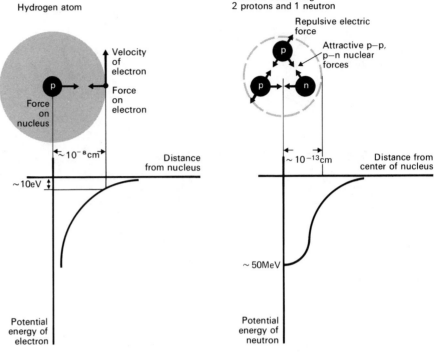

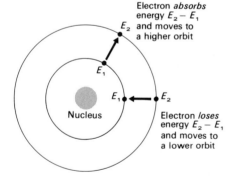

Fig. 4 An atom in an energy state E_1 absorbs energy and the atom "moves" to a state with larger energy E_2. In the Bohr model of the atom this corresponds to an electron moving to an orbit of larger diameter. An atom in an energy state E_2 loses energy and the atom "moves" to a state with less energy E_1. In the Bohr model of the atom this corresponds to an electron moving to an orbit of smaller diameter.

massive as the electron. Since the atom is electrically neutral and only protons in the nucleus are charged, this means that an atom has equal numbers of electrons and protons. The nuclear force has characteristlcs radically different from the gravitational and electric forces. First, it is enormously stronger. For example, the total energy, kinetic plus potential, of an electron about a nucleus is about 10 electron volts whereas the energy of a proton in the nucleus is about 50,000,000 electron volts.* Secondly, the nuclear force acts only over a very short distance, something like 10^{-13} cm which is about the "size" of a nucleus (Fig. 3). The mathematical form of the electric force between two charges is quite simple and found to be proportional to the product of the two charges and inversely proportional to the square of the distance between the charges.

$$F \propto \frac{Q_1 Q_2}{r^2}$$

*An electron volt, symbolized eV, is the energy acquired by an electron which moves through a potential difference of one volt. This energy, in general, is $W = QV$ (see Chapter 2). The charge of an electron is 1.602×10^{-19} coulombs (C). Hence, in basic MKS units an electron volt is

$$W = QV$$
$$= (1.602 \times 10^{-19} \text{ C}) \cdot (1 \text{ V})$$
$$= 1.602 \times 10^{-19} \text{ J}$$

It is this simplicity which allows fairly elementary calculations of the properties of atoms. The Bohr model of the atom, for example, evolves from such a calculation and gives rise to the idea that electrons revolve about the nucleus in orbits much as planets revolve about the sun in our own solar system. The electrons exist in certain well-defined orbits each characterized by a well-defined (quantized) energy. An electron can move to a higher orbit by absorbing just the right amount of energy. Conversely, an electron can descend to a lower orbit by releasing a well-defined amount of energy. (See Fig. 4.) It is *not* the concept of orbits that is particularly important. In fact, this concept does not even arise in more sophisticated calculations. However, the idea of well-defined energies is a characteristic of all atomic models. Although it is known that the nuclear force is very strong and acts over a very short distance, its precise mathematical form is still not completely understood. The search for its precise nature constitutes one of the challenges and major efforts of current research. Consequently, there is no such thing as an *elementary* calculation for a model of the nucleus. However, like electrons around the nucleus, protons and neutrons (or nucleons as they are termed collectively) have kinetic energy due to their motion and potential energy associated with forces operating within the nucleus. Therefore, it is probably not surprising that the total energy of a nucleus, like the atom, is quantized. The language for absorption and emission of energy still applies, but the energy scale is characteristic of the nuclear force (Fig. 5).

A nucleus is identified by the number of protons and neutrons that it possesses. The same chemical symbol is used to denote an atom or its nucleus. When referring to the nucleus, it is customary to also affix the proton, neutron, and

Fig. 5 (a) Schematic energy level diagram for a system in which the energies take on only discrete values. (b) Energy level diagram for the nucleus of the nitrogen atom.

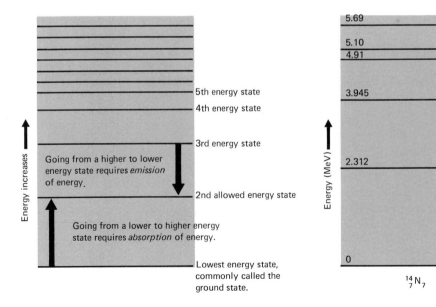

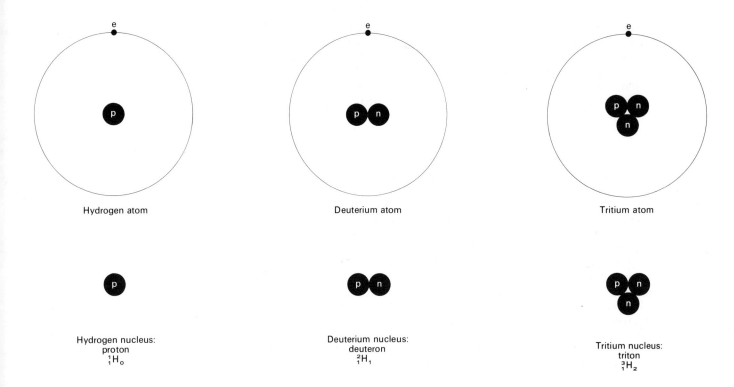

Fig. 6 Schematic illustration of the three isotopes of hydrogen.

nucleon number to the chemical symbol in the manner $^A_Z X_N$, where X denotes the chemical symbol, and Z, N, and A are the proton, neutron, and nucleon numbers, respectively. The nucleon number is simply the sum of the proton and neutron numbers, $A = N + Z$. Hence the nucleus of the hydrogen atom, for example, would be denoted $^1_1 H_0$. This complete description is necessary because although most properties of an atom are independent of the number of neutrons, the nuclear properties are not. Those atoms with nuclei having the same number of protons but a different number of neutrons are called isotopes. The nuclei of the hydrogen, deuterium, and tritium atoms all have one proton but 0, 1, and 2 neutrons, respectively (Fig. 6). The most abundant isotope of carbon has 6 protons and 6 neutrons. It is this isotope which is assigned a mass of exactly 12 on the atomic mass scale (amu). All other nuclear and atomic masses are then measured relative to this isotope of carbon. The most abundant isotope of oxygen, $^{16}_8 O_8$, has a mass of 15.994915 on this scale.

The notion of stability arose in our discussion of dispersal of pollutants by rising air masses. For example, a parcel of air is stable when the *net* force on it is zero. This idea of stability also enters in the energetics of nuclei. Figure 7 shows some of the known stable and unstable nuclei. Those nuclei in Fig. 7 that are stable have nearly equal numbers of neutrons and protons. For example, $^2_1 H_1$, $^3_2 He_1$, $^4_2 He_2$, $^6_3 Li_3$, $^9_4 Be_5$, $^{10}_5 B_5$, $^{12}_6 C_6$, $^{16}_8 O_8$, and $^{17}_8 O_9$ are all stable. As the number of nucleons in a nucleus increases, the neutron number gets progressively larger than the proton number. For example, the one stable isotope of cesium has 55 protons and 78 neutrons.

A plot of N versus Z for the stable nuclei defines a "region" of stability as shown in Fig. 8. Most nuclei are, however, unstable and may release energy and descend to a lower energy state (Fig. 9). Ultimately the nucleus proceeds to a stable state in which no more spontaneous energy transformations are possible. These energy transitions are accompanied by emission of radiation which is either particle or electromagnetic in nature. A nucleus which is capable of emission of energy is said to be radioactive and the actual process of energy emission is called radioactivity. These terms evolved from early studies and really say little about the actual mechanism. Any radiation which is emitted is of great environmental concern because of its possible biological effects on man.

Types of Nuclear Radiation

Electromagnetic radiation (refer to the second section of Appendix 2) is released from an atom when an electron moves from a higher energy state designated E_{higher} to a lower energy state designated E_{lower}. The frequency of this radiation is given by

$$f = \frac{E_{higher} - E_{lower}}{h}$$

where h is Planck's constant. Exactly the same mechanism can take place in a nucleus. However, the energies involved are characteristic of nuclear processes and therefore the frequency of the radiation is some 1 million times greater than for an atomic process. This nuclear radiation is called gamma radiation. It is in the form of photons with energies greater than about 100 keV. X-rays which are nothing more than photons with energies 1 to 100 keV are often emitted from nuclei by the same mechanism.

Frequency, wavelength, and speed of electromagnetic radiation are related by $v = f\lambda$. The speed in vacuum is independent of wavelength and frequency. So if the frequency goes up, the wavelength goes down. As the wavelength goes down, the ability to penetrate matter increases. Although the wavelength of visible light (about 5×10^{-5} cm) is comparatively short, it is not short enough to enable the light to penetrate even a few sheets of paper or the surface layer of skin. However, the wavelength of an X-ray (about 10^{-8} cm) or a gamma ray (about 10^{-11} cm) is short enough that either can easily penetrate matter. This explains why an X-ray can be used to photograph the bone structure of a person and also why it is difficult to protect oneself against the biologically damaging effects of such radiations.

A nucleus (somewhat like a person) in a state of excess energy often has different avenues open for release of the energy. For example, a nucleus may release its energy as electromagnetic radiation or it may give the energy up to one of the electrons that surround the nucleus. The electron is then released from the atom with kinetic energy equal to the excess nuclear energy minus the energy required to eject it from the atom (Fig. 10). This process is called internal conver-

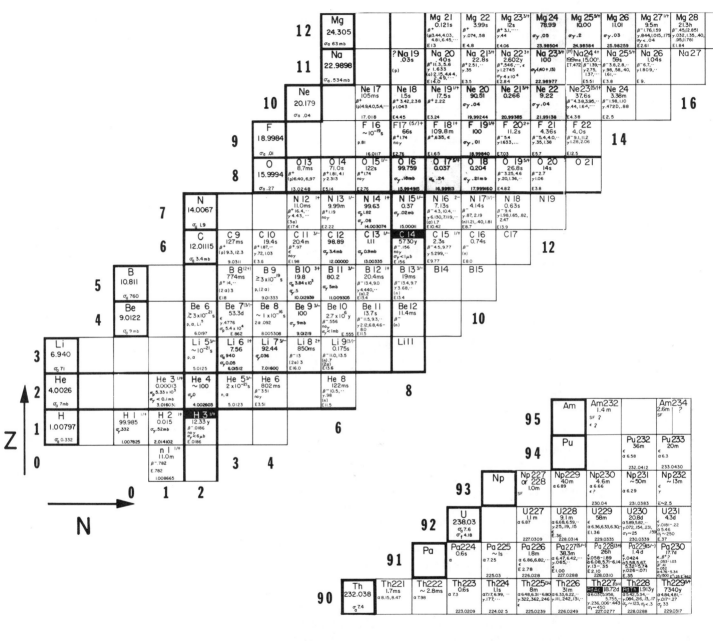

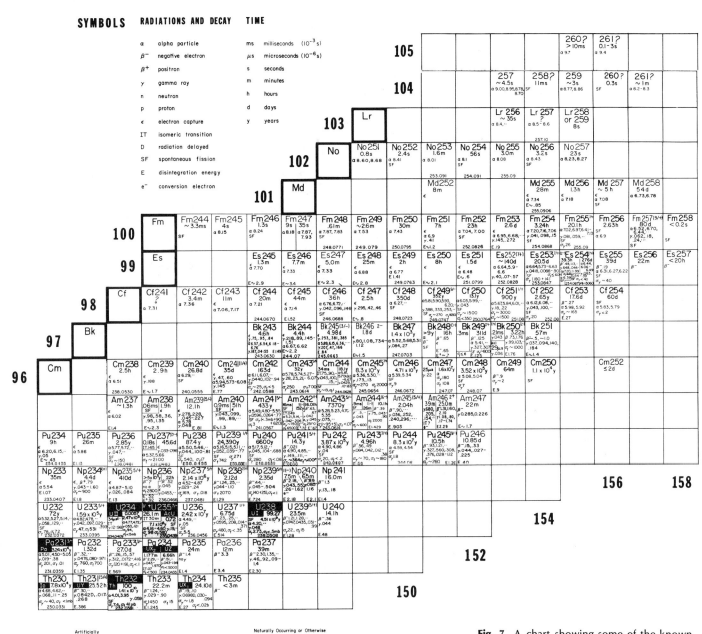

Fig. 7 A chart showing some of the known stable and unstable nuclei and their emissions. (From *Introduction to Atomic Physics*, by H. A. Enge, M. R. Wehr, and J. A. Richards, Addison-Wesley, Reading, Mass., 1972, p. 380.)

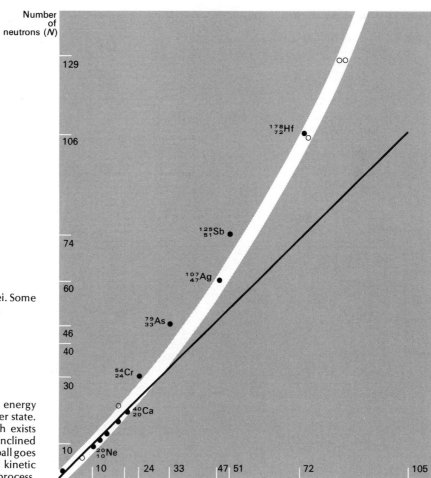

Fig. 8 The stability "region" for nuclei. Some stable nuclei are shown for reference.

Fig. 9 A nucleus in a state of excess energy may give off energy and drop to a lower state. It is analogous to the situation which exists when a ball is located on top of an inclined plane. By rolling down the plane, the ball goes to a lower potential energy state, but kinetic energy is imparted to the ball in the process.

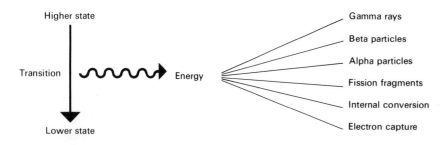

sion. Internal conversion is not as common a process as gamma emission. However, it always competes with gamma emission to some extent.

Beta particles emanate from the nucleus and come in two types which are identical except that one has negative charge, (β^-), the other positive, (β^+). In fact, β^- differs in no way from an electron. A beta particle does not exist as an entity in the nucleus for this would contradict the proton-neutron model which provides for only two discrete particles. Rather, a beta particle is created at the time of emission by virtue of the "weak" nuclear force. Symbolically, the processes are:

$$n \rightarrow p + \beta^- + \bar{v}$$

neutron is *transformed* into a proton, a β^-, and an antineutrino

$$p \rightarrow n + \beta^+ + v$$

proton is *transformed* into a neutron, a β^+, and a neutrino

With "the stroke of the pen," a neutrino and antineutrino are introduced. Although not important as an environmental danger, they are extremely important for the physics of the process. Like photons (see Appendix 2), they possess no charge and no mass. Their interaction with matter is extremely weak. For example, if the earth were bombarded with neutrinos, with high probability, they would whiz unscathed through the earth. But, importantly, they have energy and, if for no other reason, are necessary to conserve energy in the transformation.

It is important to note that charge and number of nucleons are conserved in the process of beta decay. However, since the newly formed nucleon remains in the nucleus, the net charge of the nucleus will always change, thereby changing the chemical species. To illustrate, consider the beta decay of the triton and $^{11}_{6}C_5$ (see Fig. 11).

β^- emission occurs in nuclei that have more than enough neutrons for stability and β^+ emission occurs in nuclei that have more than enough protons for stability. Thus β^- and β^+ decay would occur for nuclei to the left and right, respectively, of the stability region shown in Fig. 8.

Electron capture is a process that competes with β^+ decay. A proton in the nucleus captures an orbital electron. Symbolically $p + e^- \rightarrow n + v$. Thus the charge of the nucleus goes down one unit just as in β^+ decay.

Given a nucleus containing Z protons and N neutrons it is not hard to envision that some of these nucleons might get together and form their own nuclear state and split off from the main nucleus. Alpha particle emission is just such a process. An alpha particle is a bound nucleus of 2 protons and 2 neutrons. In other words, it is identical to the nucleus of a helium atom. The nuclei for which alpha particle emission is observed are usually very heavy. The alpha particle decay of $^{238}_{92}U_{146}$ is an example,

$$^{238}_{92}U_{146} \rightarrow ^{234}_{90}Th_{144} + ^{4}_{2}He_2$$

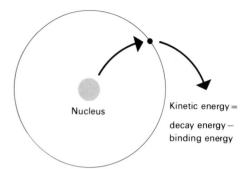

Fig. 10 Schematic description of the process of internal conversion. The excess energy of the nucleus of the atom is released internally to one of the atom's electrons.

Nucleus

Kinetic energy = decay energy − binding energy

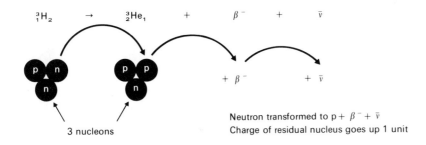

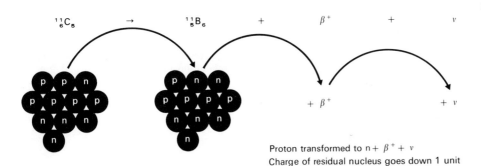

Neutron transformed to p + β^- + $\bar{\nu}$
Charge of residual nucleus goes up 1 unit

Fig. 11 Schematic illustration of beta decay. The charge of the residual nucleus always changes by one unit in beta decay. The charge goes down in β^+ decay; up in β^- decay.

Proton transformed to n + β^+ + ν
Charge of residual nucleus goes down 1 unit

Note that the process is simply the removal of 2 protons and 2 neutrons from the original nucleus. One might suspect that there is really nothing special about a helium nucleus and therefore one should see nuclei spontaneously emitting all sorts of other nuclei. However, most of these processes are prohibited when energy balances are considered.

Nuclear fission is a process, in principle, like alpha emission. However, it constitutes a gross fracturing or fissuring of a nucleus. A nucleus will be split into two or more (but usually two) fragments of comparable mass plus some lighter fragments which are usually neutrons. Spontaneous fission is very rare. It competes with alpha emission in the very heavy nuclei, but is typically 10^7 to 10^8 times less probable. However, nuclear fission can be induced by an appropriate projectile and it is this process which is exploited as a source of energy. We will have much more to say about this later on.

A nucleus which emits a particle or a photon is said to disintegrate. From the standpoint of protection against nuclear radiations, it is important to know the rate at which a sample of nuclei is disintegrating. The curie (Ci) is used as a measuring unit for disintegration rate.

1 curie = 3.70×10^{10} disintegrations per second

Most radioactive sources encountered in environmental situations have disintegration rates much less than 1 Ci. Therefore, these rates are often expressed as millicuries (mCi), a thousandth (10^{-3}) of a curie; microcuries (μCi), a millionth (10^{-6}) of a curie; or picocuries (pCi), a trillionth (10^{-12}) of a curie.

Half-life of Radioactive Nuclei

Suppose that you have samples of tritium and ^{11}C each having the same number of nuclei at some given time. Will the decay rate of these two samples at this given time be the same? Interestingly, they will not. The ^{11}C decay rate will initially be more than 100,000 times larger than the tritium decay rate. This may seem surprising at first. However, in principle, it is much like taking two equal groups of people having average ages of 65 and 25 and asking if the death rates are the same. Both of these groups are people but their physical characteristics are quite different. Tritium and ^{11}C are both radioactive but they are bound together quite differently and, therefore, their decay characteristics are different. Given a collection of people, we can say with absolute certainty that everyone in the collection will eventually die. If the collection is large enough and there is enough information about the previous history of the types of people involved, we can determine with great accuracy how many will die in a certain time period. (Life insurance companies survive on this principle.) But we can never say when a given individual in the collection will die. Similarly, given a large collection of tritons, we can determine with a certain accuracy how many will decay in a certain time interval and what is the probability that a given triton will decay. This type of information is extremely important from the standpoint of environmental effects.

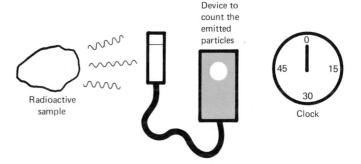

Fig. 12 Schematic diagram of a nuclear particle counting apparatus.

A purist would prefer to calculate this probability from first principles which would mean starting with the nuclear force. But like all nuclear calculations this is extremely difficult. Rather, let us look at what happens experimentally and then deduce a model for explanation. We use the language of a nuclear process. But we always keep in mind that the analysis may apply to other systems with different characters. We start with a sample of radioactive nuclei, a clock, and some device for counting the number of products emitted (Fig. 12). The measurement involves the rate at which nuclei decay. This simply means measuring the number of particles detected in a certain time interval.

$$\text{Decay rate} = \frac{\text{number of particles detected}}{\text{time interval}}$$

$$R = \frac{N}{t}$$

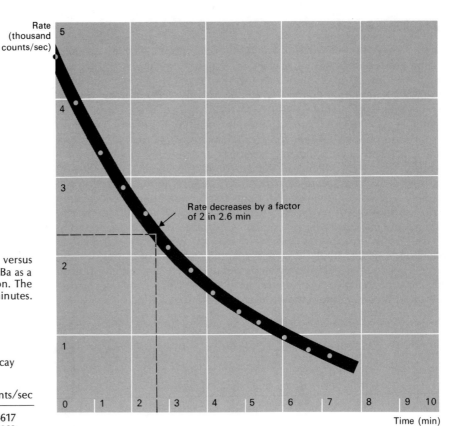

Fig. 13 Counting rate versus time using radioactive ^{137}Ba as a source of gamma radiation. The measured half-life is 2.6 minutes.

Table 1. Data for the decay rate of ^{137}Ba.

Time (min)	Counts/sec
0	4617
0.6	3963
1.2	3332
1.8	2904
2.4	2542
3.0	2114
3.6	1820
4.2	1541
4.8	1279
5.4	1151
6.0	960
6.6	789
7.2	728
7.8	603

R could be expressed in counts/second which could be related to the curie unit mentioned earlier. This measurement is taken at regular intervals of time. Table 1 shows data from such an experiment using ^{137}Ba which is a gamma ray emitter.* Figure 13 shows a plot of the data in Table 1. Note that the counting rate decreases by a factor of two at regular intervals of time. This is just what was found for electric energy production in the U.S. except that the progression represented an increase, indicating growth, instead of a decrease. It is appropriate, then, to talk about a time required for the sample to decrease by a factor of two. This is called the half-life. If you start with a sample of N_0 radioactive nuclei, then

---

*Several companies sell a small device which separates ^{137}Ba from a solution. Using this and a simple counter such as a survey meter which produces an audible sound when a gamma ray is detected, we can easily demonstrate the radioactive decay process. There are also radioactivity analogs which use many 20-sided "dice" to simulate nuclei. One face is marked to designate the emission of a radioactive particle. The "dice" are rolled and the number of decays recorded. Both of these devices are available, for example, from The Sargent-Welch Scientific Company.

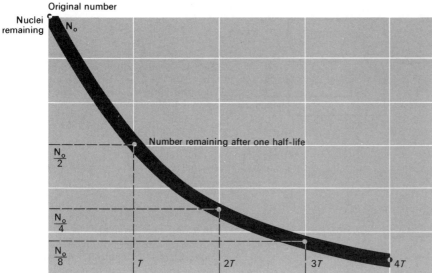

Original number

N_o

Number remaining after one half-life

$\frac{N_o}{2}$

$\frac{N_o}{4}$

$\frac{N_o}{8}$

T $2T$ $3T$ $4T$

Time in units of half-life

Fig. 14 A generalized plot for radioactive decay.

there will remain $N_0/2$ nuclei after a time T has elapsed. After a time $2T$, $3T$, and $4T$ has elapsed, there will *remain $N_0/4$, $N_0/8$, and $N_0/16$. This is shown as a general-ized plot in Fig. 14.

The range of half-lives of radioactive nuclei is impressive—less than 10^{-20} seconds to more than 10^{15} years. The half-lives of some nuclei of en-vironmental interests are shown in Table 2.

Exoergic Nuclear Reactions

The microscopic origin of energy from either fossil or nuclear fuel is from an exoergic reaction which might be symbolized:

$$A + B \quad \rightarrow C + D + Energy$$

reacting	reaction
particles	products

The reacting particles in a chemical reaction are atoms and molecules. In a nuclear reaction, they are the nuclei of atoms. The principles of analysis are very much the same, but some of the bookkeeping language is different.

In a nuclear reaction the number and types of nuclei may be different after the reaction. (Recall that the number and types of atoms are unchanged in a chemical reaction.) But the number of nucleons and the total charge are unchanged. The nucleons may simply be bound together to form different species. To illustrate,

Table 2. Half-lives of some nuclei of environmental concern

Nuclide	Half-life
$^{3}_{1}H_2$ (tritium)	12.26 years
^{90}Sr (strontium)	28.8 years
^{137}Cs (cesium)	30.2 years
^{131}I (iodine)	8.05 days
^{85}Kr (krypton)	10.76 years
^{133}Xe (xenon)	5.27 days

The half-lives of radioactive nuclei can be found in nearly all nuclear physics texts, for example, *Introduction to Nuclear Physics* by H. A. Enge, Addison-Wesley Publishing Co., Reading (1966). They are also given on the General Electric Chart of the Nuclides, a portion of which is shown in Fig. 7.

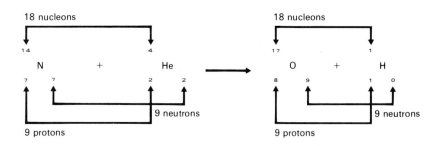

Fig. 15 Symbolization of the interaction of $^{14}_{7}N$ with $^{4}_{2}He_2$. Note that the total number of neutrons and protons is the same before and after the interaction.

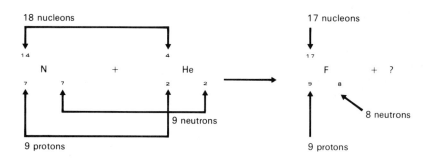

Fig. 16 The nuclear reaction shown in Fig. 15 is only one possibility when $^{14}_{7}N_7$ interacts with $^{4}_{2}He_2$. Another possibility is the production of $^{17}_{9}F_8$ and a neutron. Any reaction consistent with the laws of physics is possible.

consider the interaction of $^{14}_{7}N_7$ with $^{4}_{2}He_2$ to form $^{17}_{8}O_9$ and $^{1}_{1}H_0$ (see Fig. 15).

In the same interaction suppose that two nuclei are formed, one of which is $^{14}_{9}F_8$. What is the other nucleus? (See Fig. 16.) The second particle has 0 protons and 1 neutron; that is, it is a neutron

$$^{14}_{7}N_7 + {}^{4}_{2}He_2 \rightarrow {}^{17}_{9}F_8 + {}^{1}_{0}n_1$$

The analysis of the energetics of a nuclear reaction is precisely the same as for a chemical reaction. However, the magnitude of the energies involved are usually more than a million times larger. For comparison, the energy released in an exothermic chemical reaction is of the order of electron volts while in an exoergic nuclear reaction the energy released is of the order of millions of electron volts.

The inclusion of mass energy is crucial in a nuclear reaction. To illustrate, consider the reaction in which the neutron was discovered by James Chadwick.

$$^{9}_{4}Be_5 + {}^{4}_{2}He_2 \rightarrow {}^{12}_{6}C_6 + {}^{1}_{0}n_1$$

$$\begin{aligned} \text{mass of reacting nuclei} &= m(^{9}_{4}Be_5) + m(^{4}_{2}He_2) \\ &= 9.012186 + 4.002603 \\ &= 13.014789 \text{ amu} \end{aligned}$$

$$\begin{aligned} \text{mass of reaction products} &= m(^{12}_{6}C_6) + m(^{1}_{0}n_1) \\ &= 12.000000 + 1.008665 \\ &= 13.008665 \text{ amu} \end{aligned}$$

The mass of reacting nuclei is greater than the mass of reaction products:

mass difference = 13.014789 − 13.008665

= 0.006124 amu

In terms of energy equivalence of mass, 1 amu = 931 MeV.

Hence we have

mass energy difference = 0.006124 amu × $931\dfrac{\text{MeV}}{\text{amu}}$

= 5.70 MeV

This difference in mass energy will appear as energy of the reaction products. An exoergic nuclear reaction is one for which the mass of the reacting nuclei exceeds the mass of the reaction products. If the mass of the reaction products is greater than the mass of the reacting nuclei, the reaction is endoergic, and the mass energy difference must be made up by adding kinetic energy to the reacting nuclei (Exercise 30).

Some further insight into nuclear properties can be obtained from consideration of the deuteron. Since the deuteron is the simplest bound nuclear system with more than one nucleon, it plays much the same role in nuclear physics as the hydrogen atom in atomic physics. Let us compare the mass of the deuteron with the mass of its constituents.

deuteron
$m_d = 2.014102$

separated constituents
$m_n = 1.008665$; $m_p = 1.007825$
$m_n + m_p = 2.016490$

Thus, m_d is less than $m_n + m_p$

$(m_n + m_p) - m_d = 0.002388$ amu

The mass energy difference is 2.22 MeV. In the "assembly" process, this 2.22 MeV of energy is carried away by a gamma ray.

$n + p \rightarrow d + \gamma$

The total energy of the deuteron is thus 2.22 MeV less than the sum of the total energies of the proton and neutron (Fig. 17). To separate the deuteron into a proton and neutron, 2.22 MeV of energy must be added. This is referred to as the binding energy. To separate a hydrogen atom into an electron and proton, 13.6 eV of energy must be added. This illustrates the comparative magnitude of energies associated with nuclear and electric forces.

The amount of energy added to any nucleus to separate it into its constituent nucleons is called the binding energy, E_B. The greater the number of nucleons, the greater the binding energy (Exercise 27). A useful concept for seeing how energy is derived from a reaction is the binding energy per nucleon, E_B/A where A

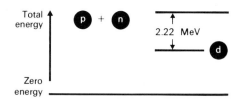

Fig. 17 The total mass energy of a deuteron is 2.22 MeV less than the sum of the mass energies of a proton and a neutron. This amount of energy would have to be added to a deuteron in order to separate it into a proton and a neutron.

represents the number of nucleons. This is a measure of the average energy to remove a nucleon from the nucleus. To illustrate, $E_B = 2.224$ MeV for the deuteron. Hence, $E_B/A = 1.112$ MeV/nucleon. Table 3 lists E_B/A for some representative nuclei and a plot of E_B/A versus A which is representative of all nuclei is shown in Fig. 18.

Recall that the rule for deriving energy from a reaction is to end up with less mass. Another way of saying the same thing is, the binding energy per nucleon of the reaction products must be greater than that of the reacting nuclei. Two ways of doing this can be inferred from Fig. 18.

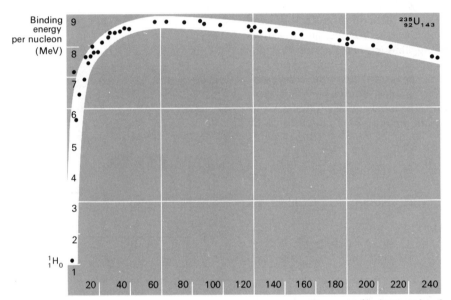

Table 3. Binding energy and binding energy per nucleon for selected nuclei.

Nucleus	E_B (MeV)	E_B/A (MeV/nucleon)
$^{2}_{1}\text{H}_1$	2.224	1.112
$^{27}_{13}\text{Al}_{14}$	224.95	8.331
$^{56}_{26}\text{Fe}_{30}$	492.26	8.790
$^{98}_{42}\text{Mo}_{56}$	846.25	8.635
$^{127}_{53}\text{I}_{74}$	1072.59	8.446
$^{152}_{62}\text{Sm}_{90}$	1253.09	8.244
$^{235}_{92}\text{U}_{143}$	1783.85	7.591

Fig. 18 Binding energy per nucleon as a function of nucleon number A. (From *Introduction to Atomic Physics*, by H. A. Enge, M. R. Wehr, J. A. Richards, Addison-Wesley, Reading, Mass., 1972, p. 403.)

1. If a nucleus with A near 200 can be split into two nearly equal fragments with, for example, $A = 100$ and $A = 80$, the binding energy per nucleon has increased in the reaction products. Hence energy will be released. This is the process of fission. The word evolves from the notion that the nucleus is "fissured" or split into two or more fragments.

2. If two light nuclei, say around $A = 2$, are combined to form a single nucleus with $A = 4$, the binding energy per nucleon of the reaction product has increased. Hence energy is released. This is the process of fusion. The word evolves from the notion that two nuclei are "fused" together.

Fission probability
(arbitrary units)

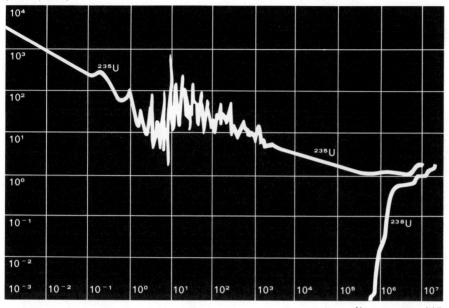

Neutron energy (eV)

Fig. 19 Relative probability of fission for ^{235}U and ^{238}U. Note that both the horizontal and vertical scales are logarithmic. (From *Introduction to Nuclear Physics*, by H. A. Enge, Addison-Wesley, Reading, Mass., 1966, p. 444.

The Nuclear Fission Reaction

A neutron is a good nuclear "axe" for splitting a nucleus. If $^{235}_{92}U_{143}$ reacts with a neutron, then the following reaction *may* occur with a release of about 200 MeV energy (Exercise 29).

$$^{235}_{92}U_{143} + ^{1}_{0}n_1 \rightarrow \underbrace{^{141}_{56}Ba_{85} + ^{92}_{36}Kr_{56}} + 3\,^{1}_{0}n_1$$
$$\text{Two nearly equally massive fission fragments}$$

The "may occur" is emphasized for two reasons.

1. There are a variety of possible pairs of fission fragments. The only restriction is that the reaction be consistent with conservation laws for nucleons.

2. Even though the conservation laws are satisfied, there is still no guarantee that the reaction will proceed. The conservation laws are a necessary condition but not a sufficient condition. There is only some probability that the reaction will proceed and this depends on the nature of the nuclear force between projectile and target and the structure of the interacting nuclei. The $^{238}_{92}U_{146}$ is an isotope of uranium with only 3 more neutrons than $^{235}_{92}U_{143}$. Yet the probability of it fissioning when bombarded with neutrons less than 1 MeV in energy is nearly zero (Fig. 19).

The energy aspect of the fission reaction is impressive, but an even more intriguing prospect is the use of the neutrons emitted in a reaction to instigate other reactions (Fig. 20). This is the mechanism of a chain reaction which gives rise to self-sustaining, energy-producing nuclear reactions. It is this process which is exploited in nuclear reactors.

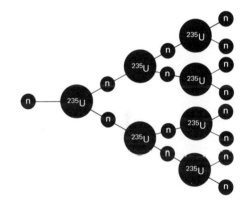

Fig. 20 Schematic illustration of a nuclear chain reaction. The neutrons released when ^{235}U fissions as a result of interacting with a neutron are used to instigate other fission reactions and keep the process going.

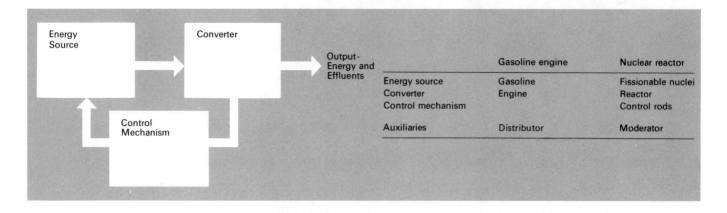

	Gasoline engine	Nuclear reactor
Energy source	Gasoline	Fissionable nuclei
Converter	Engine	Reactor
Control mechanism		Control rods
Auxiliaries	Distributor	Moderator

Fig. 21 The basic elements of a gasoline engine and a nuclear reactor.

THE NUCLEAR REACTOR

Principle

Knowing that a gasoline vapor-air mixture will explode when ignited is sufficient to recognize its feasibility as an energy converting device. However, the transition from feasibility to practicality involves many technological hurdles. The same is true with nuclear energy. But the basic ideas are the same even to the environmental price to be paid. This is illustrated in Fig. 21.

The layout of a typical nuclear reactor as shown in Fig. 22 provides a basis for our subsequent discussion of the function of each of the components.*

Fuel

A cursory examination of the basic energy release reaction

$$\underbrace{\text{H. N.}}_{\text{heavy nucleus}} + n \rightarrow \underbrace{X + Y}_{\text{fission fragments}} + \text{neutrons} + \text{energy}$$

suggests taking any heavy nucleus, bombarding it with a neutron, then using the neutrons produced to instigate other reactions. However, as Fig. 19 illustrates, the probability of fission depends strongly on the energy of the bombarding neutron. Unless the neutrons produced in the reaction have the proper energy, they will not produce fission reactions. It turns out that the neutrons are emitted with several MeV of energy. Furthermore, from Fig. 19 it would appear that either ^{235}U or ^{238}U would suffice as a fuel. However, there are many processes other than fission which compete for the neutron. The neutron may be absorbed, scattered,

*"Principles of Thermal, Fast and Breeder Reactors" is a good 10-minute color film which describes reactor principles. It is an Atomic Energy Commission film which is available, for example, from Inspiration Film Service, P.O. Box 682, 220 South Bluff Avenue, LaGrange, Illinois, 60525.

A view into the core of a nuclear reactor. The glow near the bottom of the reactor is called Cerenkov radiation (see Chapter 8). It occurs when charged particles pass through a transparent medium at a velocity in excess of the speed of light in that medium. (Photograph courtesy of the United States Atomic Energy Commission.)

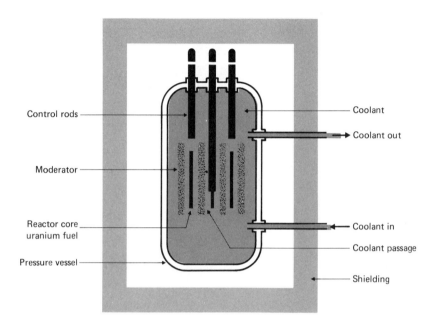

or involved in another type of reaction. It happens that those neutrons which are not removed are rapidly degraded in energy, especially by collisions with hydrogen nuclei in the coolant, to the point that the probability of their inducing fission of ^{238}U is significantly reduced. Hence, given ^{235}U and ^{238}U as possible sources, only ^{235}U has a chance. Unfortunately, natural uranium contains only 0.7% ^{235}U. To obtain a source sufficiently enriched to use as a fuel thus requires a very elaborate and complex separation scheme. Huge plants at Oak Ridge, Tennessee, Portsmouth, Ohio and Paducah, Kentucky achieve this separation by the gaseous diffusion process. Other sources suitable as fuel are ^{239}Pu and ^{233}U, but they are made artificially by bombarding ^{238}U and ^{232}Th, respectively, with neutrons (Chapter 5 and Exercise 28).

The energy release process becomes self-sustaining when more neutrons are generated than are lost in processes that compete with the fission. In this sense, it is somewhat like a profit-oriented business that becomes self-sustaining when more dollars are returned than invested. This condition is achieved only if the resources of the business are properly managed. There are many things such as taxes, overhead, labor, etc. that compete for the invested dollar and often the best management can make a profit only if there is a critical size for the business. Many small stores, for example, fail to survive because they cannot compete with

Fig. 22 Schematic cut-out view of a nuclear reactor. (From *Research Reactors*, United States Atomic Energy Commission Understanding the Atom Series.)

the giants of the industry. An analogous situation occurs in a nuclear reactor. The nuclear chain reaction is self-sustaining (that is, profitable) only when there is a critical size or critical mass. It is fairly easy to see why this should be. Figure 23 show schematically a neutron released somewhere within a spherical volume. This volume contains fissionable material as well as material which simply removes neutrons. The neutron will bounce around, but eventually it enters into a nuclear reaction or escapes from the reactor. Clearly, if the size of the reactor is too small, the probability of escape is large and the probability of fission is small. Therefore, a critical size and arrangement is required which depends on such things as:

1. the geometry of the fuel elements,

2. the type of moderator,

3. the purity of the fissionable material,

4. the neutron energy used for the fission process, and

5. the type of material used to reflect neutrons back into the system.

When the fission process is self-sustaining, the reactor is said to be critical. This concept of criticality is extremely important both in the making of a successful nuclear reactor and a nuclear bomb.

Moderator

Since the fission probability of ^{235}U is higher for low-energy neutrons, an effort is made to degrade the fission neutrons as rapidly as possible. This is the purpose of the moderator. The moderator is made of a material which will degrade the neutron energy by collisions with its atomic constituents. A specific particle will lose the maximum amount of energy in a collision with another particle of the same mass (Exercise 34). Hence the most efficient moderators are made of very light materials. The original reactor built by Fermi, et al., used graphite, which is pure carbon. Present day reactors use water which is primarily hydrogen. The water serves as both the moderator and coolant.

Control Mechanism

The chain reaction is something like a series of collisions which sometimes occurs on a crowded highway. That both of these processes can get out of control is obvious. The nuclear chain reaction is controlled by introducing a material which is a more efficient absorber of neutrons than the fissionable material. Cadmium and boron function well in this capacity.

At first thought, it might appear that the chain reaction happens so fast that any mechanical method of control would be ineffective. However, all neutrons from the fission processes are not emitted instantaneously. About 0.7% of the neutrons are delayed in emission by times of the order of minutes. This delay in emission is enough to allow mechanical devices to operate.

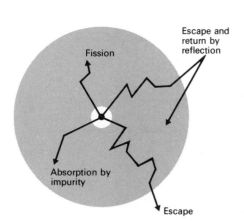

Fig. 23 There are several fates of a neutron in a nuclear reactor. This figure schematically illustrates a few of the important ones.

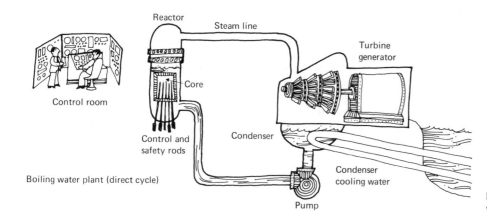

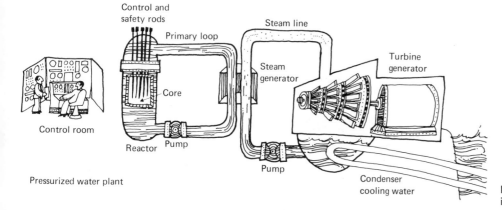

Fig. 24 The water-steam circuit in a boiling water reactor.

Fig. 25 The water steam circuit in a pressurized water reactor.

Converter

The energy of the fission products ultimately produces thermal energy in the reactor core. This thermal energy is usually extracted by water, but gases such as helium and carbon dioxide are sometimes used. There are two basic types of water cooled reactors which are called boiling water reactors (BWR) and pressurized water reactors (PWR). These are shown schematically in Figs. 24 and 25. In the PWR system, the water is kept under pressure so that it doesn't come to a boil. By doing this the temperature of the water can be increased. For example, the pressure and temperature in a BWR is about 1000 pounds per square inch and 545°F whereas, in a PWR, it is about 2250 pounds per square inch and 600°F. The steam that ultimately drives the turbine in a PWR comes from water brought to boil in a heat transfer system. Water is in intimate contact with the reactor cores in both systems, although the contact with the fuel is not direct because the fuel is clad in a tight container. Nonetheless there is always the possibility of leaks. For this reason, the coolant is always contained in a closed cycle system.

Output Energy and Effluents

Table 4 gives a quantitative comparison for electric-power stations using fossil fuel and those using nuclear fuel. The lower efficiency for the nuclear plant is a disadvantage because the thermal energy disposal problem is compounded. However, its efficiency (based on present reactor technology) is expected to surpass that for fossil fuel plants in the next generation of reactors (see Chapter 5). Table 4 also shows that the emissions from fossil fuel plants which have been demonstrated to be hazardous are completely eliminated in the nuclear systems. However, a different type of effluent emerges from the nuclear fuel plants, namely radioactive nuclei, and it is appropriate now to discuss the consequences. It is interesting to note that a certain amount of radioactivity is released in the fossil fuel system. This radioactivity is not a product of the basic energy conversion mechanism. Radioactive atoms exist as unwanted by-products in the fuel and are simply released when the fuel is burned. Although the quantities appear small,

Table 4. Effluents from a 1000 megawatt electric power station.*

	Type of Fuel			
	Coal	Oil	Gas	Nuclear
Annual fuel consumption	2.3×10^6 tons	460×10^6 barrels	6800×10^6 ft^3	2500 lb
Annual release of pollutants (millions of pounds)				
Oxides of sulfur	306	116	0.03	0
Oxides of nitrogen	46	48	27	0
Carbon monoxide	1.15	0.02		0
Hydrocarbons	0.46	1.47		0
Aldehydes	0.12	0.26	0.07	0
Fly ash (97.5% removed)	9.9	1.6	1.0	0
Annual release of nuclides (Ci)				
1620-year ^{226}Ra	0.0172	0.00015		0
5.7-year ^{228}Ra	0.0108	0.00035		0
10.8 year ^{85}Kr + 5.3-day ^{135}Xe	0	0	0	
Radioactive noble gases				
PWR				600
BWR				1.11×10^6
Overall Efficiency		38%		32%

*From "Radiation in Perspective: Some Comparisons of the Environmental Risks from Nuclear- and Fossil-Fueled Power Plants" by Andrew P. Hull in *Nuclear Safety*, Vol. 12, No. 3 (May–June 1971). The data for fossil-fueled plants are published by J. G. Terrill, E. D. Harward, and I. P. Leggett in *Ind. Med. Surg.*, **36**, 412 (1967). The data for nuclear-fueled plants are from "Hearings Before the Joint Committee on Atomic Energy, Congress of the United States, Ninety-First Congress, Second Session on Environmental Effects of Producing Electric Power, January 27, 28, 29, 30; February 24, 25, and 26, 1970, Part 2 (Vol. II), Superintendent of Documents, U.S. Government Printing Office, Washington, D.C. 1970. The data for radioactive noble gases are from "Radioactivity in the Atmospheric Effluents of Power Plants That Use Fossil Fuels" by M. Eisenbud and H. Petrow, *Science*, **144**, 288 (1964).

Table 5. Principal fission product radioisotopes in radioactive wastes, IT denotes isomeric transition which means that the nuclear state lasts for a comparatively long time. β, γ, and e^- denote beta emission, gamma emission, and internal conversion, respectively.

Radioisotope	Atomic number	Half-life	Radiation emitted	Radioisotope	Atomic number	Half-life	Radiation emitted
Krypton-85	36	4.4 hr (IT) $\to$ 9.4 yr	β, γ, $e^- \to \beta$, γ	Iodine-129	53	1.7×10^7 yr	β, γ
Strontium-89	38	54 days	β	Iodine-131	53	8 days	β, γ
Strontium-90	38	25 yr	β	Xenon-133	54	2.3 days (IT) $\to$ 5.3 days	e^-, $\beta \to \beta$, γ
Zirconium-95	40	65 days	β, γ				
Niobium-95	41	90 hr (IT) $\to$ 35 days	$e^- \to \beta$, γ	Cesium-137	55	33 yr	β, γ
				Barium-140	56	12.8 days	β, γ
				Lanthanum-140	57	40 hr	β, γ
Technetium-99	43	5.9 hr (IT) $\to 5 \times 10^5$ yr	e^-, $\gamma \to \beta$	Cerium-141	58	32.5 days	β, γ
				Cerium-144	58	590 days	β, γ
Ruthenium-103	44	39.8 days	β, γ	Praseodymium-143	59	13.8 days	β, γ
Rhodium-103	45	57 min	e^-				
Ruthenium-106	44	1 yr	β	Praseodymium-144	59	17 min	β
Rhodium-106	45	30 sec	β, γ				
Tellurium-129	52	34 days (IT) $\to$ 72 min	e^-, $\beta \to \beta$, γ	Promethium-147	61	2.26 yr	β

they should not be overlooked since some believe them to be as problematic as those from nuclear plants.*

Radioactive wastes can be broadly categorized as low and high activity. The first comes from routine operation of the reactor and fuel processing plants. Contributions come from such things as leaks in the cladding of the fuel cells and irradiation of the coolant and air by neutrons. This radioactive material is collected by various filtering systems and released to the environment under controlled conditions. Gaseous radioactive wastes are normally vented to the atmosphere. Solid and liquid radioactive wastes are buried. In either case, the disposal is done within a framework of regulations administered by the Atomic Energy Commission (A.E.C.). Disposal of the wastes is considered a low hazard process because, to be sure, those released to the environment are well below the federal standards. *But* many people are questioning the standards. We will have more to say about this later.

The high-level activity comes from the spent fuel in the fission process. The fission fragments are highly radioactive and emit a variety of gamma rays and beta particles. As one would expect (Exercise 6), these fission fragments are primarily β^- emitters because they contain an excessive number of neutrons (Table 5). This waste is analogous to the hot ashes that are left over in a coal burning stove or furnace. Hot ashes can be cooled by dousing with water. However, the radioactive "hot ashes" cannot be cooled by any external process. They rid themselves of the excess energy on a time scale characterized by the half-lives of the nuclear constituents. Some of these nuclei have half-lives greater than 10 years (Table 5). The levels of radioactivity involved are so large that often more than ten half-lives must

---

*"Radioactivity in the Atmospheric Effluents of Power Plants," by M. Eisenbud and H. Petrow, *Science* **144**, 288 (1964).

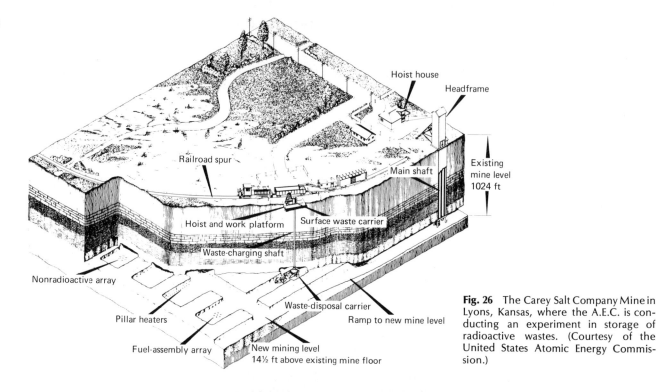

Hoist house

Headframe

Railroad spur

Main shaft

Existing
mine level
1024 ft

Hoist and work platform

Surface waste carrier

Waste-charging shaft

Nonradioactive array

Pillar heaters

Waste-disposal carrier

Ramp to new mine level

Fuel-assembly array

New mining level
14½ ft above existing mine floor

Fig. 26 The Carey Salt Company Mine in Lyons, Kansas, where the A.E.C. is conducting an experiment in storage of radioactive wastes. (Courtesy of the United States Atomic Energy Commission.)

Table 6. A.E.C. projections for reprocessing requirements and high-level and alpha waste from the nuclear fuel cycle industry

	Calendar Year Ending			
	1970	1980	1990	2000
Installed nuclear electric capacity, MW	6,000	150,000	450,000	940,000
Fuel reprocessed, metric tons/year	55	3,000	9,000	19,000
Solidified High-Level Waste				
Annual volume, thousands of ft^3	0.17	9.7	33	58
Accumulated volume, thousands of ft^3	0.17	44	290	770
Accumulated radioactivity, megacuries	200	19,000	110,000	270,000
Alpha Waste				
Annual volume, thousands of ft^3*	1,000	1,000	1,000	1,000
Accumulated volume, thousands of ft^3	1,000	11,000	21,000	31,000
Accumulated radioactivity, megacuries	0.05	0.6	1.1	1.6

*Based on initial design capacity of the facility.

elapse before they reach a safe level. Thus the only solution is to place the spent fuel in a stable area remote from the general public.

There is contained in the spent fuel cell some valuable unspent fuel. Once the fuel cell is removed from the reactor it is sent to a reprocessing center where the unspent fuel is recovered and the waste converted to a liquid. The A.E.C. presently has 93 million gallons stored. However, the technology of this process is being refined and by 1980 there is expected to be about 5 million gallons of this waste in storage even though the number of operational reactors has increased. Much research is being done to convert this liquid waste to a solid which would reduce the volume by a factor of ten. Still, finding a safe place and method for storage is no mean task. The waste container must obviously be permanent. Developing such a container is complicated by structural damage that results both from the radioactivity and heat. The storage area must be protected from the weather and from water sources; be located, ideally, in an area of no seismic activity; be structurally strong; and be shielded for nuclear radiation. One possible storage area is in the salt deposits which underlie many areas of the U.S. Salt formations are normally dry and sufficiently strong that storage areas formed by removal of the salt would not collapse. The requirements on area for storage would seem to be completely unreasonable. Yet despite these requirements the amount of radioactive waste expected to be stored by 2000 would occupy only 1% of the available 400,000 square miles of salt formation in the United States. The A.E.C. has operated a pilot project at Lyons, Kansas (Fig. 26).* However, the plan for using the Lyons site as a major storage area is opposed by

*This project is described in a 28-minute color film "Safety in Salt: The Transportation, Handling and Disposal of Radioactive Wastes." Available on loan from Audio Visual Branch, Department of Public Information, U.S.A.E.C., Washington, D.C. 20545.

state officials on the basis that it is not completely structurally safe. So although the use of salt formations appears to be the best scheme for storing high-level nuclear wastes, it is not clear at the time of this writing what salt formation will be used. Table 6 dramatically illustrates the magnitude of the storage problem and the need to find immediately a satisfactory storage area or alternate method. An alternate method that has been proposed is to launch the radioactive wastes into space with a rocket.

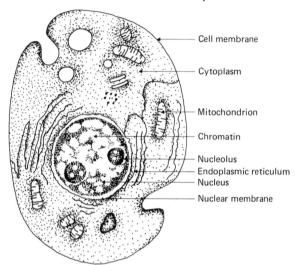

Fig. 27 Schematic depiction of a cell showing its main components.

Cell membrane

Cytoplasm

Mitochondrion

Chromatin

Nucleolus
Endoplasmic reticulum
Nucleus

Nuclear membrane

BIOLOGICAL DAMAGE DUE TO RADIATION

Microscopic Mechanisms for Biological Damage

Many of our discussions have used the notion of bound systems built from some basic constituent(s). Living tissue can be viewed within this same framework with cells as the basic units (Fig. 27). Cells are responsible for both the metabolic processes (movement, respiration, growth, and reaction to environmental changes) and reproduction. Those which control metabolism and reproduction are called somatic and genetic cells, respectively. Control of these cell functions is exercised by the nucleus of the cell. Cells reproduce in tissue through a process of division called mitosis. The hereditary aspects of life are passed on from the parent cell to the divided ones (daughters) in chromosomes which are rich in a substance called deoxyribonucleic acid (DNA). The growth of a cell and its reproduction is governed by biochemical reaction involving many complex molecules. Damage to the molecules involved in these processes *may* alter the function of a cell to the extent that it:

1. dies outright and is removed from life processes,

2. is damaged so that it cannot reproduce and therefore it eventually dies, or

3. is able to divide but the function of the new cells is altered (mutations).

Some mutations may be inconsequential. However, some can produce somatic effects such as uncontrolled cell growth which results in cancerous growths. Mutations in genetic cells can be passed on to succeeding generations producing defects in the newborn.

Any mechanism which can supply the energy necessary to break a chemical bond, ionize an atom, or alter the chemistry of cells is capable of producing biological damage. Roughly it takes 10–30 eV to damage a molecule. The alpha, beta, gamma, and neutron radiations from radioactive nuclei typically have energies in the MeV range. Thus they are capable of damaging hundreds of thousands of cells. The actual damage produced depends on the physics of the interaction which, in turn, depends on the type and energy of the particle involved.

A particle loses energy by some appropriate force. In the process, its energy is transformed or converted. For example, the brakes of an automobile exert a force on the wheels which decelerates the automobile. The kinetic energy of the car is converted into heat. A charged particle like a beta—or an alpha—particle loses its energy primarily through the electric interaction with the charged particles associated with atoms. Atoms may absorb energy and move to higher energy states, or, if enough energy is transferred, electrons may be completely removed from the atoms. The amount of energy lost depends on the energy (E), charge (Z), and mass (M) of the radioactive particle. As a rule, the energy lost in a thin piece of material of thickness ΔX is

$$\Delta E \propto \frac{Z^2 M}{E} \Delta X.$$

An alpha particle being about 2000 times as massive and having twice the charge of an electron loses energy at a much greater rate. For example, it only takes 76 micrometers (0.0076 cm) of silicon to stop a 10 MeV alpha particle but it takes 2.2×10^4 micrometers (2.2 cm) of silicon to stop a 10 MeV beta particle. Thus, an alpha particle can produce much more local damage than a beta particle of the same energy.

A neutron, by virtue of having no charge, is very difficult to stop. It loses energy by collisions with nuclei the same way billiard balls lose energy in collisions. However, if a neutron does make a collision and loses its energy it can do considerable damage. Since a neutron does not lose energy continuously as a charged particle does, it is not meaningful to talk about a thickness required to stop a neutron of given energy. Rather one talks about a thickness required to reduce the intensity of a neutron source by some given amount. Concrete is often used as a material to attenuate neutrons. Ten inches of concrete will reduce the intensity of a 10 MeV neutron source by 90%.

A gamma ray has no mass and no charge. It does have electromagnetic energy and does experience electromagnetic forces when it traverses matter. A gamma ray loses energy by three mechanisms:

1. It may scatter from an electron and impart some energy to it. This is called the Compton effect.
2. It may be absorbed by an atom whereupon energy is relinquished to one of the electrons surrounding the nucleus. This is called the photoeffect.

3. It may interact with an atom and transform its energy into an electron (β^-) and positron (β^+). This is called pair production. Pair production requires a minimum of 1.022 MeV of energy to compensate for the masses of the electron and positron.

Like neutrons, gamma rays do not lose energy continuously and again one talks about a thickness required to reduce the intensity of a beam of gamma rays (Exercise 23). Lead is often used to attenuate a source of gamma rays. It takes about 4 cm of lead to reduce the intensity of a beam of 10-MeV gamma rays by 90%.

Radiation Units

In terms of energy, a 9-ounce baseball and a 9-ounce chunk of glass moving 60 miles per hour are identical. To a physician who has to repair the damage inflicted if a person has the misfortune to intercept these particular objects, they produce quite different effects. The situation for nuclear radiation is quite similar, with different effects produced by the various types of radiation.

In some cases it makes sense to talk only about the energy deposited in some type of matter. This is referred to as a *dose*. The earliest unit for quantifying this concept was the roentgen (abbreviated R). It is a useful measure for X-rays and gamma rays but is not a unit appropriate for radiations in general. A more appropriate unit is the Rad, meaning radiation absorbed dose. One Rad is defined as 100 ergs of energy absorbed per gram of substance. When the substance is tissue, the Rad and roentgen are essentially the same so that they are often used interchangeably. The Rem, meaning roentgen equivalent man, is a biological unit which accounts for the biological damage produced. It is much less precise than a physical unit because of the many factors entering into biological damage. The energy of the particle is a major consideration. For example, visible light and gamma rays are both electromagnetic radiation. However, a photon of visible light is not biologically dangerous because its energy is too small. Intense local tissue damage can be more serious than diffuse damage. For this reason, an energetic fission fragment which travels a very short distance in tissue is more dangerous than 250-keV

Table 7. Some representative RBE's

Radiation-type examples	Biological effect	Recommended RBE
X, gamma, and beta rays (photons and electrons) of all energies above 50 keV	Whole-body irradiation, hematopoietic system critical	1
Photons and electrons 10 − 50 keV	Whole-body irradiation, hematopoietic system critical	2
Photons and electrons below 10 keV, low-energy neutrons and protons	Whole-body irradiation, outer surface critical	5
Fast neutrons and protons, 0.5 − 10 MeV	Whole-body irradiation, cataracts critical	10
Natural alpha particles	Cancer induction	10
Heavy nuclei, fission particles	Cataract formation	20

X-rays which distribute their energy over a much longer distance. Also, some body organs are much more susceptible to radiation damage than others. The amount of radiation damage is, however, directly proportional to the dose received so that the Rem and Rad are related to each other. But, since some radiation is more effective than others in producing biological damage, this proportionality factor, called the relative biological effectiveness (RBE), depends on the type and energy of the particle. Stated formally

dose equivalent in Rems = RBE times dose in Rads
$$DE = RBE \cdot D$$

It is customary to use the damage done by a whole body irradiation of 250-keV X-rays as the norm and assign an RBE of unity to this radiation. The RBE for any other particle is then measured relative to this standard. Table 7 lists some representative RBE's for various radioactive particles of interest.

Background Radiation

Knowing that radiation can produce very undesirable biological damage, it is only natural to seek protection by avoiding radiation exposure. It is, however, impossible to avoid all radiation because everyone is routinely exposed to a variety of natural and unnatural radiation sources. As inhabitants of the earth we are continually bombarded with nuclear radiations coming from radionuclides in the earth (Table 8) and from radiations produced by the interaction of cosmic rays with elements of the atmosphere. Cosmic rays are primarily very high-energy protons and gamma rays of extraterrestrial origin. There is considerable variation in the intensities of these radiations depending on geographic location and altitude. The cosmic background radiation at Boulder, Colorado may be a factor of two larger than in an eastern city which is at a substantially lower elevation. The water from wells in Maine has some 3000 times more radium content than water from the Potomac River. On the average, a person in the U.S. receives an annual exposure of 50 mrem* each from cosmic radiation and from the earth and building materials. An additional 25 mrem is received internally from inhalation of air (5 mrem) and

Table 8. Primary radionuclides from the earth that contribute to background radiation

Isotope	Half-life (years)	Type of radiation
Radium-226 (^{226}Ra)	1622	alpha, gamma
Uranium-238 (^{238}U)	4.5×10^9	alpha
Thorium-232 (^{232}Th)	1.4×10^{10}	alpha, gamma
Potassium-40 (^{40}K)	1.3×10^9	beta, gamma
Tritium- (^{3}H)	12.3	beta
Carbon-14 (^{14}C)	5730	beta

*mrem is the symbol for the millirem which is one-thousandth of a rem.

isotopes formed naturally in human tissue (20 mrem). From diagnostic X-rays and radiotherapy, a person will receive about 60 mrem and from the nuclear industry, television, radioactive fallout, etc., 5 mrem. Thus the grand total comes to about 190 mrem/year.*

Radiation Doses for Various Somatic Effects

It is clear that human beings will always be exposed to some radiation over which we have no control and that some risk will always be taken when there is exposure to additional amounts. It is only natural to want to know the extent of this risk. This means that we want to know how much radiation is required to produce a given effect. This is a very difficult question to answer because

1. controlled experiments cannot be done on human subjects, and
2. damage to genetic cells only shows up in succeeding generations.

So information must be obtained from controlled experiments on animals, accidental (and unfortunate) exposures to human subjects, and scientific judgment. The most severe radiation exposures to large numbers of people occurred in the World War II nuclear bombing of Japan and in the Marshall Islands following the hydrogen bomb testing at Bikini Atoll in 1954. Studies of these and other victims have yielded considerable information on somatic effects. Some of these results are shown in Table 9. Most of the effects given in Table 9 are for one time, whole-body exposures. Larger exposures are required to produce the same effects if the net dose is accumulated over a period of time. This is because somatic cell damage is repairable to some extent. A sunburn, for example, produces considerable cell damage which is normally repaired by metabolic processes.

Table 9. X-ray and gamma ray doses required to produce various somatic effects

Dose (Rads)	Effect
0.3 weekly	Probably no observable effect
60 (whole body)	Reduction of lymphocytes (white blood cells formed in lymphoid tissues as in the lymph nodes, spleen, thymus, and tonsils)
100 (whole body)	Nausea, vomiting, fatigue
200 (whole body)	Reduction of all blood elements
400 (whole body)	50% of an exposed group will probably die
500 (gonads)	Sterilization
1000 (skin)	Erythema (reddening of the skin)

Radiation Standards

From our knowledge that risks are involved in exposure to nuclear radiations and that a certain amount of radiation will be released to the environment in the

---

*See for example, "Ionizing-Radiation Standards for Population Exposure," by Joseph A. Lieberman, *Physics Today*, **24**, 11, 32 (November 1971).

routine operation of a nuclear reactor, the question arises of "Who will decide the maximum allowable exposure for the general public?" In the U.S. this job is given to the Environmental Protection Agency (E.P.A.). It is the function of the Atomic Energy Commission (A.E.C.) to set emission standards consistent with the guidelines of the E.P.A. These standards have changed throughout the years but since the mid-1950's the maximum allowable *average* exposure for the general population has been set at 170 mrem/year above background. This figure is based on the recommendation that a person should receive no more than 5 rems of radiation in a 30-year period. The recommendation is based on studies of the genetic effects of radiation by a committee established by the U.S. National Academy of Sciences–National Research Council and the U.K. Medical Research Council in the mid-1950's. If we add this value of 170 mrem/year to that of the approximate 190 mrem/year that a person normally receives from natural and manmade sources, we find that his total could mount to about 360 mrem/year.

Now that we have a guideline, the question becomes, "Will the emissions from a nuclear reactor fall within these standards?" Table 10 gives data for a 1000-megawatt nuclear plant. Although these data are for a single plant, it is clear that the emissions are orders of magnitude lower than the maximum E.P.A. levels and, therefore, a high level of confidence should be placed in the ability of nuclear power plants operating routinely to stay within the established radiation levels. There is, however, a widespread controversy on whether or not the 170 mrem/year exposure for the general population is acceptable. This controversy is led by Doctors John Gofman and Arthur Tamplin of the Lawrence Radiation Laboratory in California. It is their opinion that some 16,000 additional cancer cases will result each year in the U.S. if the general population is exposed to an additional 170 mrem/year of radiation. Briefly, their reasoning is this. That radiation induces human cell damage and forms of cancer such as leukemia is unquestioned. In fact, for exposures above 100 rads, experimental evidence indicates a direct linear relationship between incidences of cancer and radiation exposure (Fig. 28). Reliable data for exposures below 100 rads is lacking. Data for these low exposure levels are of extreme importance because it applies to levels of radiation in the realm of background exposures and those which could be achieved under present E.P.A. guidelines. Any estimate of cancer incidences at low dosages must be based on an extrapolation of this curve. The feeling is that if the incidence rate increased as the exposure decreased, then it would be observed; it has not been. There are two other alternatives, commonly offered. The first is that no effect occurs until some level of exposure, say around 100 mrad, is obtained. Then the cancer incidence rate increases with exposure as observed. This is referred to as the threshold hypothesis. If this were true, it would be the best alternative. The second proposal is that the observed linear relationship continues down to zero exposure. This is the linear hypothesis. In the absence of conclusive contrary evidence, the linear hypothesis is the safest assumption. Gofman and Tamplin make this assumption. Based on what they believe to be an appropriate evaluation of published data on cancer and radiation exposures, they arrive at a 1% increase in incidence rate per year per rad. Since the incidence rate is about 280 cases per 100,000 people per year, this would mean 14,000 additional cases per year for persons over 30 years of age if the entire population received 170 mrad/year for

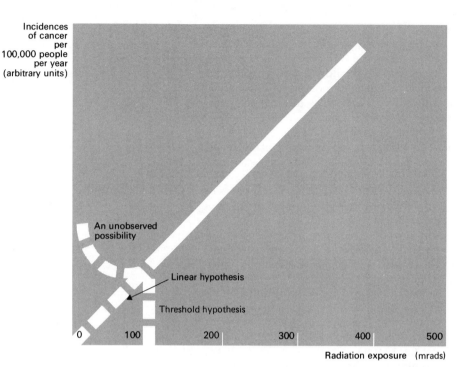

Fig. 28 Three possible ways of extrapolating the cancer incidence–radiation curve for low-dose exposures.

Table 10.   Anticipated radiation exposures of a person located 5 and 20 miles from a nuclear power plant. Those labeled A are based on a 1% fuel leak. The actual fuel leak is expected to be somewhat less than 1% which would make the exposures lower. The actual expected exposures are labeled B. (From "Electric Power Generation and the Environment," by James H. Wright, *Westinghouse Engineer*, May 1970. Reprinted by permission.)

		Low population zone 5 miles from site	General population zone 20 miles from site	
A.	Air	0.1040	0.0156	Reactor designed for 1% fuel leak
	Water	0.0103	0.0093	
		0.1143 mrem/year	0.0249 mrem/year	
B.	Air	0.0001	0.0000	Reactor with less than 1% fuel leak
	Water	0.0022	0.0020	
		0.0023 mrem/year	0.0020 mrem/year	

30 years (Exercise 25). Also they assume an additional 2000 cases/year for the population under 30 years of age giving a grand total of 16,000 additional cancer cases per year. Their estimates are probably realistic *if* the entire population did, in fact, get 170 mrem/year. The A.E.C. maintains that by the year 2000 the radiation from nuclear facilities will be only about 3 mrem/year. Nevertheless, whether the E.P.A. is responding to the pressures of Gofman and Tamplin or whether they are confident that nuclear facilities can operate under much more stringent standards, they have made the following changes in the radiation levels to be achieved by light-water-cooled nuclear power reactors (it is only this type which is widely used at present).

Proposed Rule Making*

1. For radioactive material above background in liquid effluents to be released to unrestricted areas by *all* light-water-cooled nuclear power reactors *at a site*, the proposed higher quantities or concentrations will not result in *annual* exposures to the whole body or any organ of an *individual in excess of 5 millirems*; and

2. For radioactive noble gases and iodines and radioactive material in particulate form above background in gaseous effluents to be released to unrestricted areas by *all* light-water-cooled nuclear power reactors *at a site*, the proposed higher quantities and concentrations will not result in *annual* exposures to the whole body or any organ of an *individual in excess of 5 millirems.*

This is 100 times lower than the previous standard and is 10 times lower than the reduction suggested by Gofman and Tamplin.

Much has been written about the radiation effects controversy. A list of references is included at the end of the chapter. You are strongly encouraged to delve further into this matter.

POSSIBILITY OF THE REACTOR BEHAVING LIKE A BOMB

When a person driving a car spots a stalled vehicle in his lane, information is fed to his brain which causes him to respond by applying brakes to the car. This is an example of the notion of *feedback* which is widely used in many disciplines, but especially in physics and engineering (Question 13). Basically, it means that something from the output of a system is fed back to its input (Fig. 29). The "something" can be information, as in the example cited, or such things as energy and electrical signals (voltage or current). For instance, some of the electric energy from a power plant is fed back to the plant to run the electrostatic precipitators. Now the feedback may either tend to increase or decrease the output and thus is referred to as being either positive or negative feedback, respectively. The feedback in the car example is negative because it results in a decrease in output, namely the speed

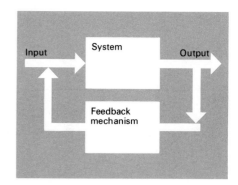

Fig. 29 Schematic representation of a system with a feedback mechanism.

---

*From *Federal Register*, Volume 36, Number 111, Wednesday, June 9, 1971.

of the car. In an atomic bomb the idea is to have a positive feedback mechanism so that the fission reaction proceeds uncontrolled. This is achieved by

1. using no control mechanisms such as cadmium rods,
2. using nearly 100% pure ^{235}U or other fissionable material, and
3. forcing and holding together two pieces of the fissionable material to form a critical mass.

Since a nuclear reactor employs the fission process, the question naturally arises, "Is it possible that these conditions could be accidentally achieved?" And the answer is no for the slow neutron reactors considered here. The reasons are as follows. First, although the ^{235}U is enriched above its natural 0.7% abundance, it is still only about 3% of the uranium used. Thus it is some 30 times *less* concentrated than the fuel needed for bombs. Secondly, every design precaution is taken to see that a critical mass cannot develop. If an uncontrolled reaction does start, the intense heat developed will melt the uranium containers and the uranium will separate, tending to decrease the fission reactions. Finally, there are a multitude of fast-acting, negative feedback mechanisms which control the fission process. The control rods are the only moving parts in the reactor. If an "excursion" from normal is sensed, these rods are slammed into a position in which the reactor cannot possibly operate. Again, the time involved in delayed neutron emission is crucial for success of the control operation. Irregular excursions are indicated by such things as abnormal variations in the pressure and temperature of the reactor and in the flow rate of the coolant water. If the water supply fails or is purposely shut off, the neutron moderating mechanism is removed which drastically reduces the reaction probability for fission.

POSSIBILITY OF RELEASE OF LARGE AMOUNTS OF RADIATION

Of greater concern is the possibility of accidental release of large amounts of radioactivity from the fuel elements. This could occur from a mechanical failure in the cladding surrounding the fuel element or from a sudden failure in the cooling system which would produce a meltdown of the fuel element. To say that the reactor can be built in such a way to ensure against such an occurrence is like saying that a fail-safe dam can be constructed. There are no guarantees against accidental radioactive release, but through careful design, engineering, and control the possibility can be minimized. Such procedures are painstakingly followed in the construction of modern reactors. The reactor is designed to contain the radioactivity resulting from what is termed as the "maximum credible accident." This is an accident that could occur if any combination of major safeguards fail. There are two containment procedures currently used. One employs a large, spherical or cylindrical steel vessel which essentially surrounds the entire reactor. This is the function of the spherical structure seen on many reactor installations (Fig. 30). In the second type, the reactor vessel is located in a steel containment tank which is surrounded by thick layers of high-density concrete (Fig. 31). This structure is also located, at least partially, underground.

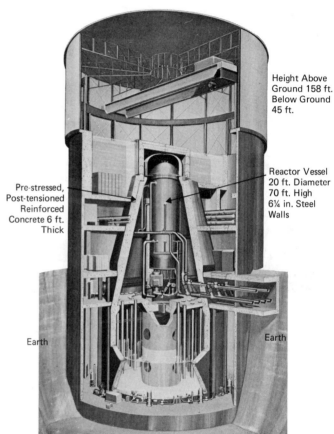

Height Above Ground 158 ft. Below Ground 45 ft.

Reactor Vessel 20 ft. Diameter 70 ft. High 6¼ in. Steel Walls

Pre-stressed, Post-tensioned Reinforced Concrete 6 ft. Thick

Earth

Earth

Fig. 30 Photograph of the San Onofre nuclear generating station near San Clemente, California which uses the spherical containment structure. (Photograph courtesy of the United States Atomic Energy Commission.)

All good intentions and designs are of no avail unless they are strictly followed. The A.E.C. is the agency to which the responsibilities of control and enforcement have been assigned. No reactor may be built or operated without a license from the A.E.C. It is probably fair to say, and it is extremely gratifying to realize, that no industry has been more carefully evaluated and controlled for safety than the nuclear industry. The A.E.C. has spent some $250,000,000 on safety research.

Fig. 31 Drawing of a boiling water nuclear reactor using steel, concrete, and earth barriers to contain radiation in the unlikely event of an accident. (Courtesy of the General Electric Company.)

'WHAT WOULDEST THOU HAVE? MAKE UP YOUR MIND!'

Cartoon by L.D. Warren, *Cincinnati Enquirer*. Reproduced by permission.

LIFETIME OF URANIUM FUEL SUPPLIES

When discussing energy sources of the magnitude of the reserves of a country, it is appropriate to define a heat unit equal to 10^{18} Btu. This is symbolized by Q which has a dual purpose of denoting quintillion and the normal symbol for heat or thermal energy. The total thermal energy derivable from fossil fuels in the United States amounts to about 6–8 Q* (Exercise 21). Depending on projected estimates of consumption rates (Chapter 1), this supply may last for 100–150 years. The estimated fission energy available from thorium and uranium in the United States is about 900Q. So it would appear that if nuclear energy becomes an acceptable energy source, it would extend our energy resources by hundreds of years. And it can, but not with light-water reactors. The reason is that although this 900Q is available it includes ^{238}U and ^{232}Th which will not work in a light-water reactor because of the low fission probability for slow neutrons. Since the useful material,

---

*This estimate and those which follow are taken from *Energy and Power*, A Scientific American Publication, W. H. Freeman and Co., San Francisco (1971), p. 9. They are based on quantities available at no more than twice present costs.

^{235}U, constitutes only about 0.7% of natural uranium, the amount of energy available using nuclear reactor technology like that discussed in this chapter would be comparable to that available from fossil fuels. The ^{235}U resources could last, perhaps, 25 years. These facts create somewhat of a misunderstanding between the public and the promoters of nuclear energy who lead us to believe that the energy available from nuclear sources is nearly limitless. What may not have been explicitly stated is that although the source is nearly limitless, the present technology is not available to extract it. There are, indeed, genuine prospects, but they are possibly 10 to 15 years from fruition. It is this new technology which is our order of business in Chapter 5.

REFERENCES

Radiation Effects Controversy

"Radiation Risk: A Scientific Problem?" Robert W. Holcomb, *Science* **167**, 853 (6 February 1970).

"Tamplin-Gofman, Pauling and the AEC," *Bulletin of the Atomic Scientists* (September 1970).

Dr. Ernest J. Sternglass, Department of Radiology and Division of Radiation Health, University of Pittsburgh has tried to associate infant mortality and nuclear tests. This is also a subject of controversy. "Infant Mortality and Nuclear Tests", Ernest J. Sternglass, *Bulletin of the Atomic Scientists* (April 1969).

"Infant Mortality Controversy," Leonard A. Sagan and "A Reply" by Ernest J. Sternglass, *Bulletin of the Atomic Scientists* (October 1969).

"Fetal and Infant Mortality and the Environment," Arthur R. Tamplin, *Bulletin of the Atomic Scientists* (December 1969).

Radioactive Waste Disposal

"The Kansas Geologists and the AEC," *Science News* (March 6, 1971).

"Nuclear Waste: Kansans Riled by AEC Plans for Atom Dump," *Science* **172**, 249 (16 April 1971).

"The Radioactive Salt Mine," Richard S. Lewis, *Bulletin of the Atomic Scientists* (June 1971).

The Atomic Energy Commission publishes a series of booklets called "Understanding the Atom Series." Several of these are particularly useful for the material covered in Chapter 4. These are:

Atomic Fuel	Nuclear Reactors	The Natural Radiation
Atomic Power Safety	Radioactive Wastes	Environment
Genetic Effects of Radiation	Sources of Nuclear Fuel	Your Body and Radiation
Nuclear Power Plants	The First Reactor	The Breeder Reactor

Any three of these can be obtained free by writing U.S.A.E.C., P.O. Box 62, Oak Ridge, Tennessee.

The following popularly written article, "A Citizen's Guide to Nuclear Power," by Ralph E. Lapp, covers many of the aspects of Chapter 4 such as:

1. Where Will We Get the Energy?
2. How Safe are Nuclear Power Plants?
3. Power and Hot Water.
4. Radiation Risks.

Available from: The New Republic, 1244 19th Street, N.W., Washington, D.C. 20036. The price, including postage and handling is:

1–10   85¢ per copy
10–50   75¢ per copy
Over 50   65¢ per copy

QUESTIONS

1. Neutrons are sometimes referred to as the "glue" needed to hold the nucleus together. Explain the origin of this expression using Fig. 7 as a guide.

2. The intensity of light or radioactivity decreases as one moves away from the source. Explain the reason for this and comment on the effectiveness of distance for protection against the hazards of nuclear radiation.

3. The two types of charges are called + and − because the superposition of charges obeys the laws of algebra. If a β^+ and β^- are brought together, their net charge would be zero. Furthermore, the β^+ and β^- would actually annihilate each other. What do you suppose is the disposition of the two particles?

4. What are some other natural processes involving large amounts of energy which are difficult to harness?

5. It is conceivable that neutrons of the proper energy to instigate fission of ^{235}U could be in, or be created by, cosmic radiation. Why doesn't one of these neutrons instigate a self-sustaining chain reaction with the ^{235}U that exists in the earth?

6. Is it possible that a nuclear power plant could reject water to the environment that has less radioactivity than the water it extracted? Explain.

7. Discuss the benefits versus risk of getting a chest X-ray.

8. What are the hazards of accumulation of radioactive substances in lakes and rivers?

9. Suppose that a credit card when new was impregnated with a small known amount of a radioactive material with a known half-life. Explain how you could determine how long the card was in use.

10. A forest fire can get out of control much like a nuclear fission reaction. Discuss some of the positive and negative feedback mechanisms in a forest fire.

11. If you were going to establish a home downstream from a large dam, what assurances would you desire that the dam would not break? If you were to build a home near a nuclear power plant, what assurances would you desire that a large release of radioactivity would not occur? In your opinion, are these assurances generally fulfilled?

12. Suppose that someone persisted in drinking a radioactive solution for some unknown reason. Even though you know the biological hazards of nuclear radiation you might argue that it is his business and therefore let him go on doing it. But suppose he were a sloppy drinker and some of the solution spilled off on you. Would your attitude change? Discuss the analogy between this and smoking.

13. The electric power system in the United States can be thought of as a system which is being driven by positive feedback. The desires of the public for more electric energy have been fed back to the electric power industry which has responded accordingly. The electric power industry has also been involved through encouragement to use more electric energy. This driving of the system has led to some forms of pollution and concern for depletion of fuel supplies. What sort of negative feedback do you think is being or could be used to effect more control on the system?

14. Some people argue that the only safe level of radioactivity emissions is zero. On this basis, what would be a safe speed for an automobile? How many deaths by automobiles are recorded in the United States on a three-day holiday weekend by deviating from this "safe" speed? (Actually the number of deaths on a three-day holiday weekend is not a lot different from any other three-day period.)

15. What is your reaction to the idea of using terminal cancer patients for nuclear radiation studies on humans?

EXERCISES

Easy to Reasonable

Exercise 1. If it takes 10 eV of energy to break an important biological molecule, how many of these molecules is a 1-MeV gamma ray capable of disrupting?

Exercise 2. Suppose that a radioactive material has a half-life of 10 years and that at some given time it had a strength of 640 Ci. If it can be safely disposed of when its strength has diminished to 10 Ci, how many years will you have to wait?

Exercise 3. If $^{235}_{92}U_{143}$ spontaneously fissions into two nuclei one of which is $^{130}_{54}Xe_{76}$, what are the proton, neutron, and nucleon numbers of the other nucleus?

Exercise 4. $^8_4\text{Be}_4$ is unstable and decays by emitting two identical particles. What are the particles?

Exercise 5. ^{239}Pu releases an average of 2.7 neutrons per fission, compared to 2.5 for ^{235}U. Which of these two elements do you suppose has the smaller critical mass? Why?

Exercise 6. Suppose that $^{235}_{92}\text{U}_{143}$ splits into $^{112}_{41}\text{Nb}_{71}$ and $^{113}_{41}\text{Nb}_{72}$. Using Fig. 7b, show that these two nuclei are unstable and should be β^- emitters.

Exercise 7. A kilowatt-year (kW-yr) is sometimes used when discussing energy resources.

   a) Using the basic ideas of work and energy, show that a kW-yr is a unit of energy.

   b) Show that a kW-yr is equivalent to 30×10^{15} Btu.

   c) Show that a kW-yr is equivalent to 0.03 Q.

Exercise 8. Using data in the text for the masses (in amu) of a proton and neutron, and following the same procedure used in the case of making such determinations for the deuteron, determine the binding energy and binding energy per nucleon of $^{12}_6\text{C}_6$.

Exercise 9. If a deuteron is "assembled" from a proton and a neutron, a 2.22 MeV gamma ray is released. If a hydrogen atom is "assembled" from a proton and electron, what is the analogous development?

Exercise 10. If a nucleus of an atom were the size of an orange (about 3 in. in diameter), about how many city blocks away would an electron be? Assume that a city block is 1000 ft wide.

Exercise 11. The half-life of a certain radioactive nucleus is 6 hr.

   a) What fraction of a given sample will remain after a 12-hr period has elapsed?

   b) What fraction of a given sample will have decayed after a 12-hr period has elapsed?

   c) How long will it take for $\frac{7}{8}$ of a given sample to decay?

Exercise 12. ^{90}Sr and ^{91}Sr are radioactive isotopes of strontium. Both are produced in nuclear reactors and nuclear bombs. However, only ^{90}Sr is of much environmental concern. Why?

Exercise 13. Fission fragments and X-rays have relative biological effectivenesses of 20 and 1, respectively. How many rads of fission fragments would be biologically equivalent to 5 rads of X-rays?

Exercise 14. What features of a nuclear reactor make it particularly useful as a source of power for a submarine?

Exercise 15.   Among the isotopes of oxygen are $^{15}_{8}O_7$, $^{16}_{8}O_8$, and $^{19}_{8}O_{11}$.

  a) Given that one of these is stable, which one would you suspect?

  b) The other two are beta-emitters. One is a β^+ emitter; the other, a β^- emitter. Identify which is which. You might want to check your answer by referring to Fig. 7.

Exercise 16.   Many environmentally-important molecules such as water (H_2O) contain hydrogen. Why might you suspect, then, that there is concern over tritium ($^3_1H_2$) in our environment?

Exercise 17. $^{14}_{6}C_8$ is a β^- emitter with a half-life of 5730 years. It is commonly used to radioactively date historic objects in our environment. Write down the reaction for the β^- decay of $^{14}_{6}C_8$.

Exercise 18.   Sun tanning is produced by the ultraviolet radiation in sunlight.

  a) Why is it impossible to get a suntan in a room in which sunlight has to pass through a window?

  b) Suntanning actually produces cell damage in the skin. Why is ultraviolet radiation capable of producing this damage but visible radiation is not?

Exercise 19.   The complete fission of 1 kg of ^{235}U will produce about 10^{14} J of energy. How many Btu per pound weight is available from the fission of ^{235}U? (1 kg mass is equivalent to 2.2. lb weight.) How does this compare with the thermal energy available from a pound of coal?

Reasonable to Difficult

Exercise 20. $^{131}_{53}I_{78}$ has a half-life of 8 days. It decays by emitting a β^- particle.

  a) What are the proton and neutron numbers of the nucleus that results from the β^- decay of ^{131}I?

  b) What is the strength of a 1-Curie source of ^{131}I after 1 month?

Exercise 21.   Most of the fossil fuel energy resources are in the form of coal. An estimate of the amount of coal remaining in the United States was given in Chapter 1. Also, the heat of combustion for a type of coal, which was probably of fairly good quality, was given in the development of the fossil fuel electric power plant. Using these data, show that the amount of energy available from coal amounts to about 30Q.

Exercise 22.   Consider the following game. A circle is laid out on a flat piece of ground. You are placed within the circle and your movement is confined to the area of the circle. Another person, preferably of the opposite sex, runs through the circle and you try to capture her (him). Give an argument as to why your probability of success is inversely proportional to the speed of the person running through the circle. In what ways is this similar to the capture of a neutron by a nucleus?

Exercise 23. A piece of paper, say 1 mm thick, will completely stop the radiation from an ordinary light bulb. The same piece of paper would have little effect on X- or gamma-radiation from a nuclear source. However, any thickness of material will always remove some of the photons from a beam and therefore reduce its intensity. The reduction in intensity depends on the energy of the photon and the material (Fig. 4.32). The data below are for a beam of 2-MeV gamma rays incident on various thicknesses of lead.

Thickness (cm)	Reduced intensity (arbitrary units)
1	61
2	37
3	22
4	14
5	8
6	5

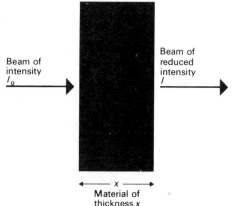

Beam of intensity I_0

Beam of reduced intensity I

← x →

Material of thickness x

Figure 32

a) Show that the intensity decreases exponentially with thickness. You might, for example, plot intensity versus thickness on either linear or semilog paper.

b) What is the intensity with no absorber?

c) What thickness is required to reduce the intensity by 50%? 100%? Comment on some of the problems of totally shielding a source of gamma radiation.

d) Which would you expect to be a better absorber of gamma rays—lead or aluminum? Why?

Exercise 24. The nuclear fission chain reaction requires that more neutrons be produced than are absorbed. In the fission of ^{235}U there are, on the average, 2.5 neutrons produced per fission. Physicists for many years have used (n, 2n) reactions for studies of nuclear structure. For example

$$n + {}^{9}_{4}Be_5 \rightarrow {}^{8}_{4}Be_4 + 2n$$

However, a self-sustaining chain reaction cannot be produced using this reaction. Try to figure out why.

Exercise 25. Using the data in the text, show how Doctors Gofman and Tamplin calculated 14,000 additional cancer cases per year if it is assumed that the entire population is exposed to 170 mrem/year.

Exercise 26. A free neutron is unstable and decays into a proton, a beta particle, and an antineutrino.

$$n \rightarrow p + \beta^- + \bar{\nu}$$

It might be suspected then, if for no other reason than symmetry, that a free proton would also decay.

$p \rightarrow n + \beta^+ + v$

This is precisely the mechanism used to explain positron emission from a nucleus. However, the free proton is stable and will not decay.

a) What physical principle forbids this decay?

b) If the proton were unstable, what environmental changes might you expect for life on earth.

Exercise 27. A nucleus is held together by forces between neutrons and neutrons (n–n), protons and protons (p–p), and protons and neutrons (p–n). There is a contribution to the binding energy from each of these forces. Thus the total binding energy would be expected to increase with increasing nucleon number. Knowing that the binding energy per nucleon is approximately constant, how will the total binding energy depend on nucleon number A?

Exercise 28. Natural uranium is about 99.3% ^{238}U which is not a useful fuel source in a nuclear reactor. Plutonium 239 (^{239}Pu) is fissionable, but does not occur naturally. It can, however, be made from ^{238}U by first allowing ^{238}U to capture a neutron. The product thus formed then ultimately decays to ^{239}Pu. Try to figure out the reactions which ultimately lead to ^{239}Pu.

Exercise 29. Table 1 shows that the binding energy per nucleon for ^{235}U and ^{127}I is 7.6 MeV and 8.4 MeV, respectively. Using these data, explain why about 200 MeV of energy is released in the fission of ^{235}U.

Exercise 30. The nuclear "pickup" reaction is extremely useful for determining the structure of nuclei. Figure 33 shows a nuclear reaction in which a nucleus is bombarded with a proton. A deuteron is formed as a result of the interaction.

a) Suppose that the target nucleus is $^{13}_{6}$C$_7$. What would the residual nucleus be?

b) Determine whether the reaction is endoergic of exoergic. Masses for the nuclei involved can be found in Fig. 6.

Difficult to Impossible

Exercise 31. Assuming that 180 MeV of energy is released in the fission of a single ^{235}U nucleus, calculate the fission energy (in kW-hr) in 1 kg of pure ^{235}U. Take the atomic weight to be 235 and remember that a mass in grams equal to the atomic weight contains Avogadro's number of atoms.

Exercise 32. Neutrons in a reactor are produced in cycles. When the number in some cycle just equals the number produced in the previous cycle, the reaction is just self-sustaining. If the number produced in some cycle is only slightly larger than the number produced in the previous cycle, the neutron population increases rapidly. We can talk about the number of cycles it takes for the population to double. If the production ratio is 1.1, how many cycles are required for the population to double?

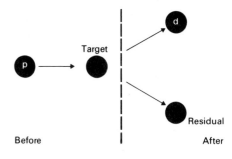

Fig. 33 Schematic illustration of a nuclear (p, d) reaction. It is called a "pickup" reaction because the projectile picks up a neutron from the target to form a deuteron.

Exercise 33.   The rate of decay of a radioactive sample can be written as

$$\frac{\Delta N}{\Delta t} = -\frac{0.693}{T} \cdot N$$

where T is the half life and N is the number present at any given time. Suppose that nuclei are also being produced at some constant rate R. Then you have both production and loss through decay. The net rate is then

$$\frac{\Delta N}{\Delta t} \quad\quad R \quad - \quad \frac{0.693}{T} N$$
$$ \uparrow \uparrow$$
$$ \text{production} \quad \text{decay}$$

Show that there will be no change in the net amount when the number N reaches

$$N = \frac{RT}{0.693}$$

Exercise 34.   Consider a head-on collision of two equal masses. Assume that one is initially at rest. Let V be the speed of the projectile before the collision and let U and W be the speeds of the projectile and target after the collision. Conservation of linear momentum demands that

$$mV = mU + mW$$

Dividing by m, we obtain

$$V = U + W$$

If the only energies involved are kinetic, then

$$\frac{mV^2}{2} = \frac{mU^2}{2} + \frac{mW^2}{2}$$

Dividing by $m/2$, we obtain

$$V^2 = U^2 + W^2$$

So we have

$$V = U + W$$
$$V^2 = U^2 + W^2$$

Show that the only solution to this is $U = 0$ and $W = V$, which means that the mass initially moving has stopped and the mass initially at rest proceeds with the speed of the projectile. Where in bowling and billiards do you see this demonstrated? It is this feature which makes hydrogen a good "stopper" of neutrons.

Exercise 35.   Aluminum has only one stable isotope and it contains 13 protons and 14 neutrons in its nucleus. The mass of this nucleus is 26.981539 amu.

a) Calculate the mass of this nucleus in grams.

b) If the nucleons are contained in a solid sphere, the radius of the sphere would be about 3.6×10^{-13} cm. Calculate the volume in a cubic centimeter of this nucleus.

c) Density is defined as the ratio of the mass of an object divided by its volume. Calculate the density of this nucleus.

d) One pound is equivalent to 2200 g and there are 2000 lb in a ton. Calculate the density in tons per cubic centimeter. Does this give you any indication of the amount of nuclear matter that could be contained in a teaspoon?

5

ELECTRIC ENERGY TECHNOLOGY FOR THE FUTURE

Photograph courtesy of Plasma Physics Laboratory, Princeton University.

INTRODUCTION

Our discussions thus far have stressed the urgency for procuring an environmentally acceptable source of energy that can be readily converted to electric energy. There are sources such as rivers (hydroelectricity) and tidal energy which present no air pollution problems but which cannot provide the quantities demanded. There are sources such as coal which are reasonably abundant, but are still limited, and which have environmental drawbacks. In degree of difficulty, the development of a clean, abundant energy source is not a lot different from putting a man on the moon. And like the latter, it will require massive funding and direction at the federal level. Several ventures are underway.* There is an effort to make current technology acceptable through development of sulfur oxide control technology and coal gasification. The new technologies being actively pursued are the nuclear breeder reactor, fusion energy, magnetohydrodynamics, and solar energy. However, the major thrust is toward the development of the nuclear breeder reactor.† One main reason for this is that operational prototypes have already been built and the feasibility has been demonstrated. These systems are also expected to be 7–10% more efficient than conventional reactors which would alleviate the thermal pollution problem. It is the goal of this chapter to elucidate the principles and assets of these future energy sources now under development.

THE BREEDER REACTOR

Principle

To review, ^{233}U, ^{235}U, and ^{239}Pu are the only three nuclei considered to be useful nuclear reactor fuels. Of these, only ^{235}U occurs naturally and it comprises only 0.7% of natural uranium which is nearly all ^{238}U. ^{233}U and ^{239}Pu must be produced through nuclear reactions and transmutations. A breeder reactor starts with a certain amount of fissionable material for the initial fuel. Then, besides deriving energy in much the same way as in a normal reactor, fissionable material like ^{233}U or ^{239}Pu is produced. It is possible to produce more fuel than is actually "burned" in the energy producing process.

The principle is as follows. When a nucleus is fissioned with a neutron, there is no way to predict beforehand just what reaction products will be produced. Some events will be more probable than others. Some will produce no neutrons and others will produce many. Furthermore, the number will depend on the neutron energy. Depending on the nucleus, the average number produced per fission reaction is between 2 and 3. One of these neutrons is required to sustain

---

*"Nixon Outlines His Energy Plan," *Science* **172**, 11 June 1971.

†"Nuclear Power in the United States" is a comprehensive 28-minute, color movie which describes the fast breeder reactor program of the United States. It is available on loan from Audio-Visual Branch, Division of Public Information, U.S.A.E.C., Washington, D.C., 20545.

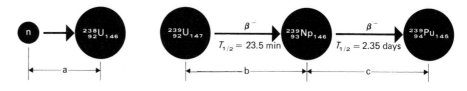

Fig. 1 Steps involved in producing ^{239}Pu by bombardment of ^{238}U with neutrons. (a) $^{238}_{92}$U$_{146}$ captures a neutron forming $^{239}_{92}$U$_{147}$. (b) $^{239}_{92}$U$_{147}$ decays by β^- emission to $^{239}_{93}$Np$_{146}$ (neptunium) with a half-life of 24 minutes. (c) $^{239}_{93}$Np$_{146}$ decays by β^- to $^{239}_{94}$Pu$_{145}$ with a half-life of 2.3 days.

a) $^{238}_{92}$U$_{146}$ captures a neutron, forming $^{239}_{92}$U$_{147}$.

b) $^{239}_{92}$U$_{147}$ decays by β^- emission to $^{239}_{93}$Np$_{146}$ (neptunium), with a half-life of 23.5 minutes

c) $^{239}_{93}$Np$_{146}$ decays by β^- emission to $^{239}_{94}$Pu$_{145}$, with a half-life of 2.35 days

the fission process. But the remainder can be used for any desired purpose. The purpose in a breeder reactor is to produce a fissionable material by allowing the neutron to be captured by an appropriate nucleus, fittingly called a fertile nucleus. Because of its abundance, ^{238}U is a convenient, fertile nucleus. The process is shown in Fig. 1. A similar process can be devised using ^{232}Th as the fertile isotope (Exercise 17).

Either ^{233}U or ^{239}Pu could serve as a fuel in a breeder reactor. However, the mechanics of operation would be quite different. The use of ^{233}U as a fuel requires slow neutrons (energies less than about 100 eV) to instigate fission whereas the use of ^{239}Pu requires fast neutrons (energies greater than about 1 MeV). The fast breeder reactor, meaning fast neutrons and ^{239}Pu fuel, is the type receiving primary emphasis.

In a breeder reactor, more fissionable material is ultimately produced than is consumed. Like many other systems we have encountered, it is meaningful to speak of a doubling time. Here it is the time required to produce as much fuel as was present at the beginning. This is estimated as being from 5 to 20 years depending on the type of reactor used. If the breeder reactor is successful, our energy resources will be increased by 100 to 1000 times the energy available from fossil fuels.

Safety Considerations

Unlike the light water reactor which requires a moderator to degrade the energy of the fission neutrons, the fast breeder must be built with materials that do not moderate the neutrons. This means, then, that water, which functions both as a moderator and heat transferrant, cannot be used. The alternatives are not numerous and the one that has emerged makes use of either metallic sodium or a sodium compound such as salt (NaCl).

Most emphasis is being placed on the former, which involves a type of reactor termed a liquid metal fast breeder reactor and is designated LMFBR. Sodium melts at 210°F, but does not boil until 1640°F. Unlike water at high temperature, its vapor pressure is quite low—a desirable feature because of the minimization of pressure containment problems. Sodium is not a good "capturer" or "moderator" of neutrons and it has excellent heat transfer properties. It will also react with many of the fission products which may leak into the coolant and tie them up

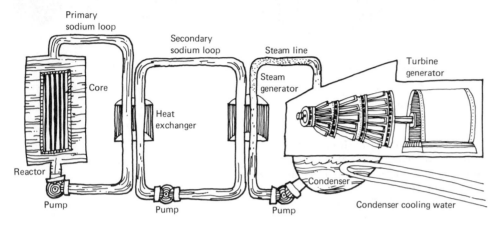

Fig. 2 The heat transfer circuits in a fast breeder reactor.

chemically so they cannot escape to the environment. Although the LMFBR seems like a reactor designer's dream, there are disadvantages. Liquid sodium will oxidize and burn if exposed to the air and will react violently producing hydrogen if exposed to water. Continual bombardment of the sodium coolant produces the radionuclides ^{22}Na and ^{24}Na (Exercise 7) which can accumulate to the extent of being hazardous. So, despite the fact that few problems are encountered in the pressure considerations involved with containing liquid sodium, extensive precautions still must be taken to ensure that the sodium does not escape.

The thermal energy is extracted from the reactor core by the liquid sodium, but the turbine which drives the electric generator still utilizes steam. This is accomplished by heat transfer from the sodium cooling circuit to a water cooling circuit. As a precaution against radioactivity leaking into the water circuit, an intermediate cooling circuit also utilizing liquid sodium is used (Fig. 2).

A second feature of the fast breeder reactor which requires safety considerations not found in the light water reactor is the so-called *secondary criticality*. Criticality in a light water reactor is attained after the fission neutrons have been thermalized, or, in other words, sufficiently reduced in energy. If the coolant which also serves as the moderator is accidentally removed, then the mechanism for achieving criticality is removed and the chain reaction tends to stop (negative feedback). However, the sodium functions only as the coolant and if it is suddenly stopped, the reactor is still critical until other sensors detect that something is wrong. If a sudden meltdown of the core occurs, there is the possibility of the fuel assuming a different geometric arrangement and developing a critical mass which can produce an uncontrolled chain reaction. This is the secondary criticality. Because of this, the design considerations of fast breeder reactors are much more stringent. Two fast breeder reactors have experienced partial core meltdown but neither resulted in secondary criticality.

Status of Breeder Reactor Programs

The possibility of developing a breeder reactor was recognized as early as 1945. Several experimental types used for generating electric energy have already been built and the race for producing a commercial system is worldwide.* Table 1 gives a listing of the American breeder reactor facilities and their characteristics.

Table 1. American breeder reactors and their characteristics.

Name	Location	Thermal output (MW)	Electrical output (MW)	Initial operation
EBR-II	Idaho Falls, Idaho	62.5	16.5	1963
Fermi	Monroe County, Michigan	200	67	1963
SEFOR	Washington County, Arkansas	20		1969
FFTF	Richland, Washington	400		1974
Demo No. 1	Tennessee	750–1250	300–500	1978–1980
Demo No. 2	Undetermined	750–1250	300–500	?

The Enrico Fermi nuclear power station located near Monroe, Michigan. This plant was the world's first large breeder nuclear power plant. Considerable effort is being made toward the further development of this type of facility. Photograph courtesy of United States Atomic Energy Commission.

*See, for example, "Energy from Breeder Reactors," by Floyd L. Culler, Jr. and William O. Harms, *Physics Today*, Vol. 25, No. 5, 28 (May 1972).

The Enrico Fermi reactor located at Lagoona Beach on Lake Erie near Monroe, Michigan about 35 miles southwest of Detroit is of special interest because: (1) it is the first to produce a significant electric power output; and (2) it suffered a partial meltdown of its core on October 5, 1966 due to a blockage in the sodium coolant lines. This second item of interest has been the subject of controversy. Some contend that this clearly demonstrates that accidents can occur which are worse than the maximum credible accident and thus for which no provisions have been made.* In any event, the safety considerations for the fast breeder reactor are much more stringent than for the conventional light water reactor. But there is no reason why they cannot be made safe. It will require money for research, as the President has requested, and time.

Plans for construction of the first major American breeder reactor electric generating station were announced in February 1972. It is to be built somewhere in eastern Tennessee by a joint effort involving the Commonwealth Edison Company of Chicago and the Tennessee Valley Authority (TVA). The plant will provide 300–500 megawatts of electric power for the TVA network. Its cost of about $500 million will be shared by the AEC ($150 million), TVA ($100 million), and private firms ($240 million). It is expected to be completed before 1980 which is the goal set by President Nixon's Energy Plan.

Although the decision has been made to expedite the LMFBR program, there are other alternatives using the breeder concept which have merit. The thermal breeder using fuel bred from ^{232}Th and the gas cooled fast breeder reactor (GCFBR) are examples. The A.E.C.'s Oak Ridge, Tennessee laboratory is investigating, on a limited scale, a thermal breeder using molten salt as the heat transferrant. This system has continuous, on-site fuel reprocessing facilities which could be a tremendous environmental and economic advantage. In addition, the fuel doubling time is expected to be much shorter than for the LMFBR. The GCFBR utilizes gaseous helium as the heat transferrant. This scheme is already used commercially in a type of ^{235}U-burning reactors. Helium has the advantage of being chemically inert. It is also very difficult to form radioactive products through neutron bombardment of helium. The main disadvantage is that the helium would have to be compressed to pressures about 100 times atmospheric pressure, a process which would entail extremely reliable pressure vessels and auxiliary equipment. How-

---

*The incident at the Enrico Fermi nuclear reactor which caused a partial meltdown of the reactor core was caused by a blockage in one of the cooling lines. The operation of the reactor was terminated routinely. The plant officials and the A.E.C. believe that the reactor design was adequate for this incident. Others do not. For an engineering account of the accident see "Fuel-Melting Incident at the Fermi Reactor on Oct. 5, 1966," by R. L. Scott, Jr. *Nuclear Safety*, Vol. 12–2, March/April 1971. *Nuclear Safety* is a journal dedicated to the problems of safety in the nuclear industry and is available from the Superintendent of Documents, U.S. Government Printing Office, Washington, D.C., 20402. A viewpoint counter to the A.E.C. position is presented in *The Careless Atom* by Sheldon Novick, Houghton-Mifflin Publishing Co. Reviews of this book and some very interesting commentary can be found in *Selected Materials on Environmental Effects of Producing Electric Power*, U.S. Government Printing Office, Washington, D.C., 20402, August 1969 ($2.50).

Controlled fusion research device in toroidal geometry, located at Princeton University Plasma Physics Laboratory and sponsored by the United States Atomic Energy Commission.

ever, experience with helium in conventional reactors leads many to believe that these problems are not insurmountable.

NUCLEAR FUSION

The Nuclear Fusion Process

It has already been established that the sun is the most abundant energy source for the earth. Nothing was said, though about how energy is produced in the sun. The fact that the sun emits light and is obviously very hot suggests a burning process; current models reinforce this supposition. But it is a nuclear burning rather than the atomic (or chemical) burning characteristic of an ordinary fire. Interestingly, the principles are very similar; only the energy scales are different. Nuclear burning proceeds by "fusing" nuclei into more tightly bound nuclei with a subsequent release of energy. Atomic burning proceeds by exothermic chemical reactions in which atoms are combined into more stable molecules. Both require some ignition temperature and rely on the released energy to sustain the burning.

The sun consists primarily of protons, electrons, and nuclei having nucleon numbers less than or equal to four. It is believed that these nuclei are formed by the so-called proton-proton cycle.

$$2\,{}^{1}_{1}H_{0} + 2\,{}^{1}_{1}H_{0} \quad \rightarrow \quad 2\,{}^{2}_{1}H_{1} \; + \; 2\beta^{+} \; + \; 2v \; + \; \text{Energy}$$

| 4 protons are fused together | to form | 2 deuterons | 2 positrons | 2 neutrinos |

Then the process continues using the deuterons produced.

$$2 {}_1^1H_0 + 2 {}_1^2H_1 \quad \rightarrow \quad 2 {}_2^3He_1 \quad + \quad \gamma$$

$\underbrace{\phantom{2 {}_1^1H_0 + 2 {}_1^2H_1}}$ \quad $\underbrace{\phantom{2 {}_2^3He_1}}$

2 protons and 2 deuterons   2 helium nuclei gamma
are fused together   of mass 3   photons

The helium-3 nuclei yield the following reaction.

$$ {}_2^3He_1 + {}_2^3He_1 \quad \rightarrow \quad {}_2^4He_2 + 2 {}_1^1H_0$$

$\underbrace{\phantom{ {}_2^3He_1 + {}_2^3He_1}}$ \quad $\underbrace{\phantom{ {}_2^4He_2}}$ \quad $\underbrace{\phantom{2 {}_1^1H_0}}$

2 helium-3 nuclei   alpha   2 protons
are fused together   particle

These three reactions release a total of 26.73 MeV of energy (Exercise 14) most of which initially goes into kinetic energy of the reaction products and is eventually converted to electromagnetic radiation appearing as visible light.

It is simple to write down the equation for the proton-proton fusion reaction and determine the energy released. However, as we might expect, it is no easy task to achieve this result in practice. One reason is that the interacting protons possess charge. Therefore, a repulsive electric force exists which tends to separate them. Energy must be supplied to overcome this electric repulsion, and this energy must be sufficient to get the particles to within the range of the nuclear force ($\sim 10^{-13}$ cm). Even when this is achieved, the fusion process is only one possible nuclear event. For example, the protons could simply scatter from each other like two billiard balls. In fact, the fusion process is very improbable. In the laboratory the necessary energy can be supplied by a proton accelerator such as a cyclotron. In the sun and any practical fusion energy source this energy is supplied thermally which requires a temperature of millions of degrees (Exercise 15). This necessary temperature and the high density of protons for large numbers of collisions are available only in the deep interior core of the sun. Because of the low reaction probability and large ignition temperature, the proton-proton reaction as a practical energy source is extremely remote.

There are several other fusion reactions which have the same exoergic character as the proton-proton reaction but which have greater reaction probability. In particular

$$ {}_1^2H_1 + {}_1^3H_2 \rightarrow {}_2^4He_2 + {}_0^1n_1 + 17.6 \text{ MeV}$$

$\underbrace{\phantom{ {}_1^2H_1 + {}_1^3H_2}}$ \quad $\underbrace{\phantom{ {}_2^4He_2}}$ \quad $\underbrace{\phantom{ {}_0^1n_1}}$

deuteron–triton   alpha   neutron
                    particle

$$ {}_1^2H_1 + {}_1^2H_1 \quad \rightarrow \quad {}_2^3He_1 \quad + {}_0^1n_1 + 3.3 \text{ MeV}$$

$\underbrace{\phantom{ {}_1^2H_1 + {}_1^2H_1}}$ \quad $\underbrace{\phantom{ {}_2^3He_1}}$ \quad $\underbrace{\phantom{ {}_0^1n_1}}$

deuteron–deuteron   helium-3 nucleus neutron

$$ {}_1^2H_1 + {}_1^2H_1 \quad \rightarrow \quad {}_1^3H_2 + {}_1^1H_0 + 4.0 \text{ MeV}$$

$\underbrace{\phantom{ {}_1^2H_1 + {}_1^2H_1}}$ \quad $\underbrace{\phantom{ {}_1^3H_2}}$ \quad $\underbrace{\phantom{ {}_1^1H_0}}$

deuteron–deuteron   triton proton

The deuteron-deuteron reactions are extremely attractive as an energy source because deuterium is stable and occurs naturally. Although it is only 1 part in 7000 of natural hydrogen, there is essentially an unlimited supply in the oceans because of the amount of hydrogen in the water (Exercise 21). However, even though tritium does not occur naturally and is unstable, the deuteron-triton reaction is more appealing because: (1) the energy release is substantially larger, and (2) the ignition temperature is lower (about 40,000,000°C compared with about 400,000,000°C). Still the thought of producing an ignition temperature of millions of degrees for a sufficient length of time and of containing the "nuclear fire" seems technologically inconceivable. Nevertheless, the practical application of fusion energy is within the grasp of man and it offers a nearly unlimited source of energy without the bulk of radioactivity problems of nuclear fission.

Prospects for Practical Fusion Energy Sources*

An ordinary flame burns at a temperature of about 1000°C. The thermal energy associated with this temperature is about 0.1 eV. This is sufficient to excite some atoms to higher energy states, but not enough to remove an electron from an atom. Now the thermal energy associated with a nuclear fusion fire is much more than enough to completely remove all electrons from the atoms of the gas. In this state the gas is fully ionized and is termed a plasma. Once kindled, the natural tendency for a fusion fire is to expand—thereby tending to extinguish itself. So the reactions must be at least partially contained if energy is to be derived. The time of containment of the fusion is called the confinement time.

The basic idea in any burning process is to derive more energy than is required to kindle the fire. For example, if a gas stove is lit by a match, hopefully, more energy is produced from the burning gas than was provided by the match. As already mentioned, it is necessary that some ignition temperature be achieved before this process can proceed. However, although this is a necessary condition, it is not sufficient because there are other criteria. (Anyone who has tried to set fire to a pile of damp leaves knows this.) The temperature, the confinement time, and the density of the ions in the plasma are the crucial characteristics which determine whether or not the fusion process in question will be a useful energy source. The British physicist J. D. Lawson has deduced that the product of ion density and confinement time must be at least 10^{14} sec/cm^3 for systems utilizing the deuterium-tritium reaction. For example, if the plasma used has an ion density of 10^{15} per cm^3, then it must be contained for least 0.1 sec. All research on fusion energy sources is centered on maximizing this product of ion density and confinement time and creating the required ignition temperature. How do you contain a substance at a temperature which far exceeds the vaporization temperature of any known material? Quite obviously the plasma cannot be contained by a pipe

*"To Imitate the Sun" is a 33-minute movie which covers two decades of controlled thermonuclear research. It is available on loan from Audio-Visual Branch, Division of Public Information, U.S.A.E.C., Washington, D.C. 20545.

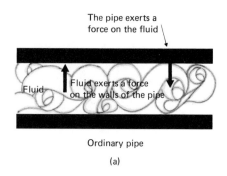

The pipe exerts a
force on the fluid

Fluid

Fluid exerts a force
on the walls of the pipe

Ordinary pipe

(a)

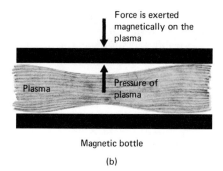

Force is exerted
magnetically on the
plasma

Plasma

Pressure of
plasma

Magnetic bottle

(b)

Fig. 3 (a) In an ordinary pipe, the fluid is contained by a force exerted by the walls of the pipe. (b) A plasma is contained by a magnetic force on the ions.

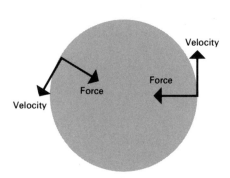

Velocity

Velocity

Force

Force

Fig. 4 An object will always travel in a circular path when the net force is perpendicular to its velocity.

as is the case for liquid sodium in a breeder reactor. Rather, the force that a pipe would exert on the plasma to balance the force exerted by the plasma on the pipe is simulated with a magnetic force. These systems are referred to as magnetic bottles (Fig. 3). All sorts of exotic configurations have been devised to try to achieve this containment and all are based on the same physical principle. The idea is to have moving charged particles interact with an imposed magnetic field (or fields) in such a way that a force is exerted on the ions so as to contain the plasma. In this way, the plasma can hopefully be kept completely away from any confining structure which would vaporize upon contact with it.

This confinement can be understood more quantitatively in the following manner. When the net force on an object is always perpendicular to its instantaneous velocity, the object will travel in a circular path (Fig. 4). Examples of this behavior include the circular orbit of a satellite about the earth and the movement of an object attached to a string when it is swung in a circle. Now if a charged particle moves in a direction perpendicular to a magnetic field, the particle will experience a force perpendicular to both the magnetic-field and its velocity (Fig. 5). Hence it moves in a circular path.

The strength of the magnetic force is given by

$$F_{magnetic} = Bqv$$

where B is the strength of the magnetic field, q is the amount of charge, and v is the velocity. Using Newton's second law, this can be equated to mass times acceleration to yield

$$Bqv = \frac{mv^2}{R} \tag{1}$$

where R is the radius of the orbit. Solving Eq. 1 for R yields

$$R = \frac{mv}{Bq}$$

The important feature of this relation is that the radius is inversely proportional to the strength of the magnetic field. If the field is increased, the radius is made smaller. It is this principle which is used to regulate both the density and temperature of ions in a plasma. The particle is thus trapped in a circle about the lines

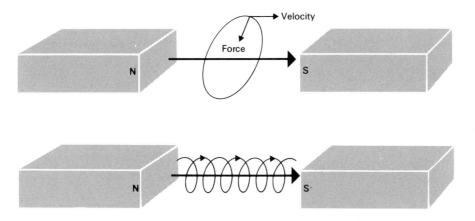

Fig. 5 The magnetic field is directed from the N to the S pole. The charge moves at right angles to the field and revolves in a circle. This is the principle employed in cyclotrons used to accelerate nuclei like protons and alpha-particles to very high speeds.

Fig. 6 A charged particle in a magnetic field moves in a helical path.

of the magnetic field. If the velocity is not perpendicular to the magnetic field, then the force is still perpendicular to the velocity and the field and the particle will now move in a helical path (Fig. 6).

This principle is simple enough. However, instabilities develop for which there is no explanation in simple terms. These instabilities lead to leakage of the plasma ions from the magnetic field and prevent achievement of the necessary confinement time.

Figures 7 and 8 show the principles of containment in two of the most promising types of fusion machines. The Tokamak is a Russian development. The Russians announced in 1969 that a Tokamak had achieved a confinement time of 0.02 seconds using an ion density of $7 \times 10^{13}/\text{cm}^3$. This is still well below the Lawson criterion for a useful energy source, but represents a major breakthrough in the state-of-the-art. A team of British physicists later confirmed these results with measurements taken on the Russian machine. It was realized in the United States that the Princeton stellarator could be converted to the Tokamak principle. This was accomplished and successful operation began in July 1970. Again the Russian results were confirmed. There is now a total of five Tokamaks either in operation or under construction in the United States (ORMAK at Oak Ridge, MIT, University of Texas, Gulf General Atomic, and Princeton). With the Tokamak results, there is hope for overcoming fundamental problems that currently prevent construction of a practical energy source.

An innovative scheme* for igniting the fusion reaction which avoids the magnetic confinement principle utilizes a device called a laser.† The lasers of interest for the fusion process produce an extremely energetic beam of electromagnetic radiation. This radiation is often visible. However, in two features it

---

*"Fusion by Laser" Moshe J. Lubin and Arthur P. Fraas, *Scientific American* **224**, No. 6, 21 (June 1971).

†Laser is a word derived from the first letter of the words—Light Amplification (by) Stimulated Emission (of) Radiation.

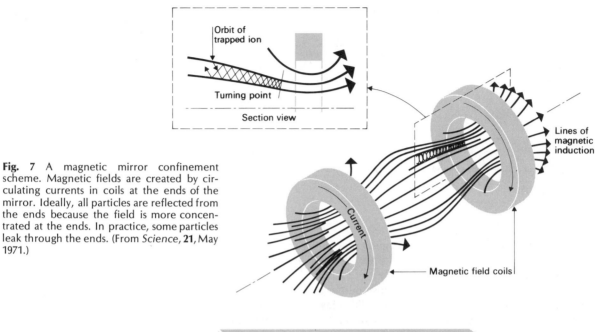

Fig. 7 A magnetic mirror confinement scheme. Magnetic fields are created by circulating currents in coils at the ends of the mirror. Ideally, all particles are reflected from the ends because the field is more concentrated at the ends. In practice, some particles leak through the ends. (From *Science*, **21**, May 1971.)

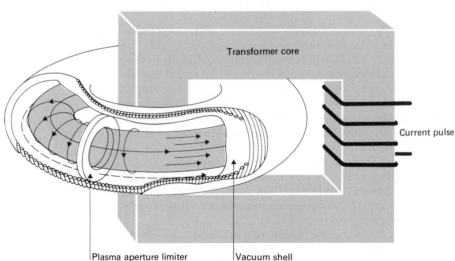

Fig. 8 The Tokamak confinement geometry. The plasma constitutes the secondary winding of a transformer. When a current is sent through the primary winding, a large current is induced in the plasma. This current heats the plasma producing the required ignition temperature. (From *Science* **21**, May, 1971.)

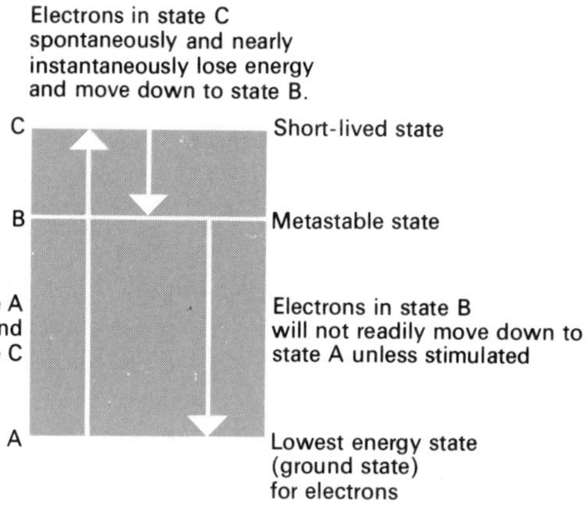

Fig. 9 Energy steps involved in a 3-level laser.

Electrons in state C spontaneously and nearly instantaneously lose energy and move down to state B.

C — Short-lived state

B — Metastable state

Electrons in state A absorb energy and move up to state C

Electrons in state B will not readily move down to state A unless stimulated

A — Lowest energy state (ground state) for electrons

is very different from a beam of light from the sun or from an incandescent light bulb, for example. First, it is composed almost entirely of waves having the same wavelength; if it were visible, it would appear as a pure color. Second, the individual waves in the beam are all in phase. This means that the peaks and valleys of the individual waves all line up. Radiation of this type is said to be coherent. Radiation from an incandescent light bulb is incoherent because there is a random relationship between the peaks and valleys of the individual waves. The mechanism for emission still evolves from a transition from a higher energy state to a lower energy state. However, the multitude of the transitions take place only after stimulation. Electrons from atoms in a gas or solid are "pumped" by absorption of electromagnetic or electric energy from their lowest energy state to a higher short-lived energy state (Fig. 9). This higher state is short-lived in the sense that electrons spontaneously and nearly instantaneously lose energy and drop to a lower long-lived or metastable state. Electrons are very stable in this metastable state and will not readily return to their lowest energy state unless further stimulated. This stimulation is achieved by a photon having just the energy difference between the metastable and lowest energy state. The emitted photon is exactly in phase with the one that stimulated the emission. Thus there is no guarantee that this emitted photon will emerge from the laser for its energy may be absorbed by an electron in the lowest state which will then move up to the metastable state. A net output occurs when there are more transitions down from the metastable state than there are up from the lowest energy state to the metastable state. This situation will result if there are more electrons in the metastable state than in the ground state, and is called a population inversion. Population inversion is one of the major "tricks" in making the laser work.

Neon/helium gas lasers with continuous power outputs of 1 or 2 milliwatts are commonly used in light experiments in an elementary physics laboratory. The lasers used in fusion research produce extremely energetic pulses of radiation. For example, the energy in a pulse may be 200 joules and the pulse may last for only 10^{-10} seconds. This corresponds to an instantaneous power of

$$\text{power} = \frac{\text{energy}}{\text{time}} = \frac{2 \times 10^2 \, \text{J}}{10^{-10} \, \text{sec}}$$

$$= 2 \times 10^{12} \, \text{W}$$

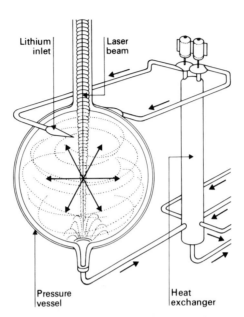

Fig. 10 Possible scheme for instigating a fusion process with a laser.

Thus the power is enormous. The lasers may be either of the gas or solid type. Carbon dioxide is often used in gas lasers.* A glass with a distribution of neodymium atoms is often used in solid lasers. It is because of the huge power outputs from a laser that it is useful for igniting a fusion reaction. The laser would simply be focused on a suitable sample of a fusible material such as a frozen deuterium-tritium pellet which would absorb the energy and, hopefully, instigate the fusion process. A proposed scheme is shown in Fig. 10. There is much work to be done on this scheme before it is practical. It is, however, being vigorously investigated.

Even if the fundamental aspects of the fusion process are solved, there are many engineering problems with no obvious solutions. Structural damage and induced radioactivity due to intense neutron bombardment will be a particularly difficult problem to solve. If both the fundamental and engineering problems can be solved, the first generation of practical fusion devices for producing electric energy will probably use the fusion reactions as a source of thermal energy for a conventional steam turbine system (Fig. 11). Later versions may use the MHD principle (see the final section of this chapter dealing with magnetohydrodynamics). Deuterium and tritium would probably be used as fuel.

$$d + t \rightarrow \alpha + n + 17.6 \text{ MeV}$$

Since the neutron possesses the bulk of the 17.6 MeV of energy, it is a matter, then, of extracting this energy, converting it to thermal energy, and transferring it to the turbine. This can be done using liquid lithium (or some lithium compound) which absorbs the neutrons via nuclear reactions.

---

*"Controlled Fusion: Plasma Heating with Lasers," *Science* **167**, 1112 (20 February 1970).

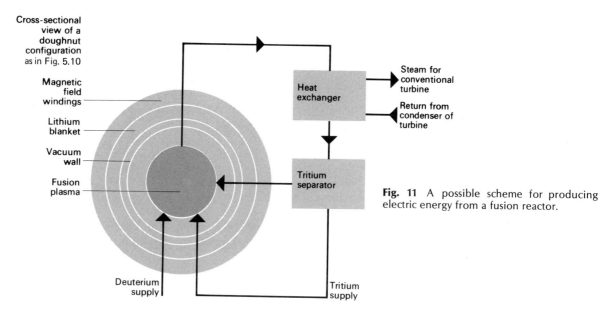

Cross-sectional
view of a
doughnut
configuration
as in Fig. 5.10

Magnetic
field
windings

Lithium
blanket

Vacuum
wall

Fusion
plasma

Heat
exchanger

Steam for
conventional
turbine

Return from
condenser of
turbine

Tritium
separator

Deuterium
supply

Tritium
supply

Fig. 11 A possible scheme for producing electric energy from a fusion reactor.

$$_3^6Li_3 + {}_0^1n_1 \rightarrow \underbrace{{}_1^3H_2}_{\text{tritium}} + \underbrace{{}_2^4He_2}_{\substack{\text{alpha} \\ \text{particle}}} + 4.8 \text{ MeV}$$

$$_3^7Li_4 + {}_0^1n_1 \rightarrow {}_1^3H_2 + {}_2^4He_2 + {}_0^1n_1 - 2.5 \text{ MeV}$$

The tritium produced could then be extracted and recirculated back to the plasma for fuel.

Environmental Considerations and Outlook

The fusion reactor system appears to solve one of the major drawbacks of the fission reactor, namely, the production of large quantities of radioactive wastes in the form of fission fragments. The alpha particles produced are stable against nuclear decay. In atomic form the helium is inert and therefore harmless. In addition it can be recovered, condensed to a liquid, and used as a low temperature coolant (4.2K) in a variety of industries. The tritium is recycled and used as fuel. However, the tritium is not recycled immediately. It must be reclaimed from the coolant and this may require a hold-up time of about a day. As a result, about 10^8 curies of tritium may be held up in the process. This is comparable to the amount of radioactivity in the fission products of a breeder reactor. Tritium emits only a low-energy beta particle so there is little problem in shielding it. But tritium, like hydrogen, is difficult to contain and will diffuse to some extent through the walls of whatever is containing it. Since tritium has the chemical characteristics of hydrogen, it can combine with oxygen to form "water," T_2O, and in this state is biologically hazardous. The tritium problems are not without solution, but nonetheless constitute the major radioactivity situation in fusion reactors which will utilize the d–t or d–d reactions.

Apart from this problem with tritium, the nuclear fusion reactor is vastly superior to the nuclear breeder reactor which is the only other presently known solution to long-term demands for energy. The fusion reactor avoids the problems of storage of radioactive fission fragments, requires no critical mass of fuel which might lead to a nuclear explosion, extends the lifetime of fuel supplies from tens of thousands of years to millions of years, makes no demands on a type of fuel which may have other national priorities, and will probably be more thermodynamically efficient. Like the breeder reactor, it emits no particulates or chemical compounds to the atmosphere. Looking well into the future, the fusion reactor offers the possibility of direct energy conversion into electric energy.* If this is achieved, the efficiency could rise to, perhaps, 90% and the thermal pollution problem using steam turbines would be eliminated. The high temperature technology gained in the development could be used in a multitude of ways.

Although the Tokamaks are an exciting prospect at this time and research is forging ahead on this type of machine, there is no guarantee that it is the machine that will ultimately be used. The fundamental breakthroughs discovered with the Tokamak have contributed to development of completely different versions. Even if practical feasibility is demonstrated by any fusion system by, say, 1975 it will be at least 2000 before a practical system is in use. The breeder reactor is well ahead of this pace and our government at this time is directing its energy budget money primarily to the nuclear breeder reactor program. The level of funding for nuclear fusion research is about $30,000,000 per year and hasn't changed significantly in the last 10 years. The Russians, on the other hand, fund their fusion efforts at about three times this rate and have, accordingly, about three times the number of scientists working. It is interesting and gratifying, though, to see the cooperation in fusion research at the international level.

SOLAR ENERGY

Motivation

The sun has been described as a giant fusion reactor continuously supplying the earth with solar (electromagnetic) energy. No mention has been made, though, about utilizing this energy other than for producing fossil fuels on a very long-term basis and for photosynthesis. Since the earth is some 93,000,000 miles from this gigantic furnace, we may think it to be an insignificant source of energy when compared to the earthbound fission and fusion reactors. Yet if this solar energy could be harnessed, it could satisfy the world's energy demands for thousands of years.† It is easy to see why this is true.

---

*"Prospects for Fusion Power," *Scientific American*, Feb. 1971.

†"Energy and Power," W. H. Freeman and Co., San Francisco (1971).

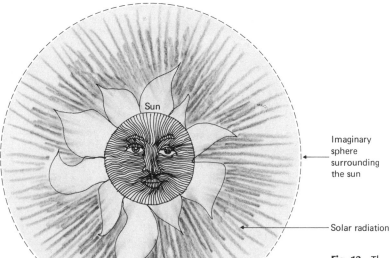

Sun

Imaginary sphere surrounding the sun

Solar radiation

Fig. 12 The rate of emission of radiation from the sun is fixed. But it becomes less concentrated as it propagates out from the sun.

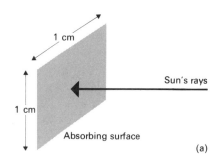

1 cm

Sun's rays

1 cm

Absorbing surface

(a)

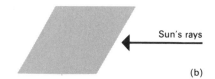

Sun's rays

(b)

Fig. 13 (a) The energy absorbed by a surface is maximum if the surface is perpendicular to the sun's rays. (b) The energy absorbed by a surface is zero if the radiation is parallel to the surface.

The power available at the sun's surface is essentially limitless (4×10^{20} megawatts: Exercise 8). But as the energy propagates out, it spreads and becomes less concentrated (Fig. 12). It is more meaningful, then, to talk about the concentration of radiation at some distance from the sun rather than the energy available at the source. Just outside the earth's atmosphere, the average solar radiation amounts to about 0.14 watts per square centimeter if the surface intercepting the radiation is perpendicular to the sun's rays (Fig. 13). The atmosphere absorbs and reflects nearly 50% of the radiation so that at the earth's surface the concentration of energy is about 0.07 W/cm^2. A power per unit area of 0.07 W/cm^2 does not seem like a lot. However, the solar power impinging on a square area only 12 inches on a side is about 60 W and this is equivalent to the electric power required to operate a typical light bulb. It is impossible, though, to convert all this solar energy into electric energy. Typical efficiencies are around 10%. To achieve 60 watts of electric power from a conversion of solar power at 10% efficiency would require a solar power input of 600 watts. A square receiving area of about 38 inches (or about 3 feet) on a side would be required and this begins to get large. Still if the solar power falling on 1% of the area of the United States were converted to electric power at an efficiency of only 1%, the power produced would exceed the present electric power output of all generators in the United States (Exercise 9). If all 50 states in the United States were of equal area, the equivalent of one-half of one state would be sacrificed as a solar energy receiver. Obviously this is a price which has to be considered carefully.

A solar power system to be competitive with conventional systems would have to produce about 1000 megawatts of electric power. Hence any system, regardless of how the conversion of electric energy is effected, necessarily

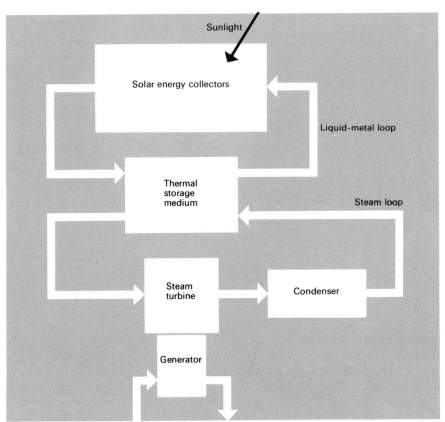

Fig. 14 The major components of a solar-electric power plant.

requires a collecting area of about 7 miles on a side.* An obvious way of achieving this conversion is to concentrate the solar energy by means of lenses or mirrors and convert it to thermal energy. The thermal energy could then be channeled into a conventional steam turbine system. This ploy, however, is faced with the fact that the solar energy is not available on a 24-hour basis whereas the demands for electric energy are. There are several possible ways to circumvent this difficulty. It is conceivable that the electric energy could be stored using batteries. Or some of the electric energy could be used to pump water to an elevated reservoir. Then, as energy was needed, the water could be used in an artificial hydroelectric system. A possibility that is being seriously considered is to use the electric energy output to produce a fuel which could then be transported and used as needed. One such scheme thermally dissociates water into hydrogen

*"Is it Time for a New Look at Solar Energy?" Aden Baker Meinel and Marjorie Pettit Meinel, *Bulletin of Atomic Scientists* **27**, No. 8, 32 (October, 1971).

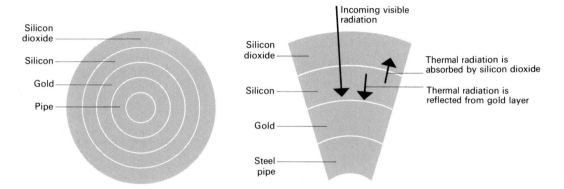

and oxygen.* The hydrogen is then used as an efficient, nonpolluting fuel in a fuel cell, for example.† A system given considerable thought and deemed practical would convert the solar energy to thermal energy and store it in a thermal reservoir for use in a conventional steam turbine electric system.‡

A Solar-thermal-electric Energy System

The proposed solar-thermal-electric energy conversion system is shown in Fig. 14. It differs from a conventional fossil fuel system only in the way the thermal energy is produced. The solar radiation falling on an area of about 14 square kilometers (5.4 square miles) would be converted to thermal energy and transferred to a heat reservoir which furnishes energy for the boiler of a steam turbine. The process presents some challenging scientific and engineering problems. For example, saying that a certain amount of solar energy falls on a surface does not mean that it absorbs and retains all the energy. The surface will reflect some radiation, some will be conducted away, and some will be radiated. The reflected radiation can be minimized by using low reflecting surfaces and the energy conducted away can be minimized by using a vacuum barrier, as is done in a thermos bottle. The radiated energy is another matter, though, because every object at a temperature above 0 K radiates electromagnetic energy. Some clever scheme must be devised to prevent this radiation from escaping. The proposed scheme makes use of the fact that some materials will readily absorb or reflect some wavelengths, but readily transmit others. Glass, for example, absorbs ultraviolet radiation but transmits visible radiation. The incoming radiation is mostly visible. About 90% of it has wavelengths less than 1.4 μm. The radiation from the collector at about 900 K is mostly infrared radiation. The collector would be coated with thin layers of selected materials (Fig. 15) and the solar radiation would be concentrated

Fig. 15 Schematic diagram of the solar radiation absorbers. The visible radiation is absorbed in the silicon layer. It is reradiated mostly as infrared radiation which is reflected from the gold surface and absorbed by the silicon dioxide layer. The metallic gold layer also provides thermal contact between the silicon and pipe.

---

*"Solar Power," Norman C. Ford and Joseph W. Kane, *Bulletin of Atomic Scientists* **27**, No. 8, 27 (October 1971).

†Fuel cells are discussed in Chapter 6.

‡"Physics Looks at Solar Energy," Aden Baker Meinel and Marjorie Pettit Meinel, *Physics Today* **25**, No. 2, 44 (February 1972).

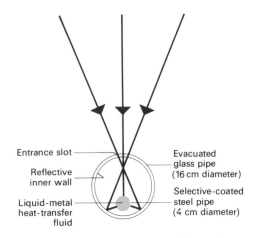

Fig. 16 Schematic depiction of the way the solar radiation would be focused on the pipe containing the heat transferrant. The evacuated glass pipe prevents heat from being conducted away from the heat transfer pipe.

Fig. 17 Map showing the possible areas for a solar energy reserve.

on it by either lenses or reflectors (Fig. 16). The top layer made from silicon dioxide (SiO_2) is transparent to visible radiation. The middle layer, made of silicon, readily absorbs all radiation less than 1.4 μm in wavelength. The bottom layer, made of gold, is highly reflecting for infrared radiation. The top layer is strongly absorbing and would entrap the infrared radiation causing a rise in the temperature. The layered coating would be placed on a steel pipe containing a heat transferrant such as liquid sodium which would convey the heat to a thermal storage area. The thermal energy is stored as latent heat. The principle is as follows. When a solid changes to a liquid, thermal energy has to be added. This is called the heat of melting. It amounts to 80 calories per gram for changing ice to water. When a liquid changes to a solid, thermal energy must be extracted. This is called the heat of fusion. Again, it amounts to 80 calories per gram for changing water to ice. The idea in the solar energy system would be to use thermal energy derived from solar energy to melt a solid. Certain salts have been proposed. Then the energy could be extracted by changing the liquid back to a solid. The thermal storage area would also function as a heat reservoir for those times when the solar radiation was absent. For this reason, the areas selected should not be costly and should be favored with a maximum amount of sunlight. There are vast areas in southwestern United States which satisfy these requirements (Fig. 17). Like any other electric generating system, these solar "farms" (Fig. 18) would not be without environmental impact. They would be subject to the thermal waste energy problem like any steam turbine system, they could upset the local desert ecology, and they could conceivably produce local climate changes. However, in an era of fossil fuel generating systems with their environmental problems, solar energy systems such as these must be considered as viable alternatives.

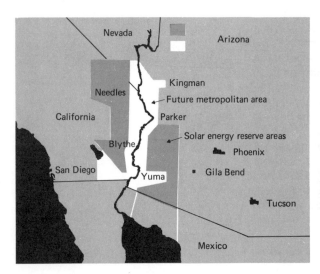

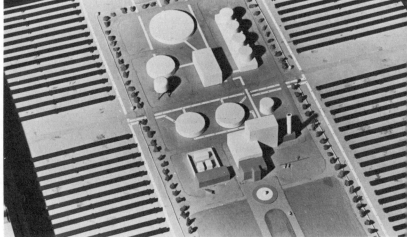

Fig. 18 Model of a solar energy "farm." The solar energy collecting panels are shown as dark, parallel lines at the sides of the photograph. (Courtesy of Marjorie P. and Aden B. Meinel.)

A Solar-electric Energy System

Solar energy can be converted to electric potential energy with a conversion efficiency of about 10% via devices known as solar cells or solar-energy converters. The physics of these devices is difficult but basically they develop an electric potential like an ordinary battery used in a transistor radio. The solar cell converts electromagnetic energy, the battery converts chemical energy. Like a battery, the electric potential developed depends on the physical and chemical characteristics of the device. The potential of a single cell like that used in a flashlight is fixed at 1.5 volts regardless of its size. Depending on the material used, a single solar cell will generate an electric potential of about 0.6 volts. To obtain a large potential and large electric capacity from either batteries or solar cells a great number of the cells must be connected together (Fig. 19). The space program uses large arrays of

(a)

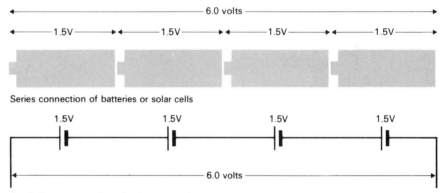

Series connection of batteries or solar cells

Symbolic representation of series connection

(b)

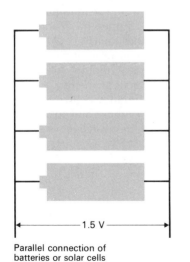

Parallel connection of
batteries or solar cells

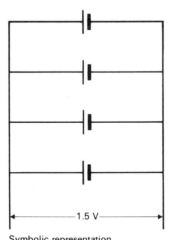

Symbolic representation
of parallel connection

Fig. 19 (a) Series connection of batteries or solar cells. The net electric potential is the sum of the potentials of the individual components. (b) Parallel connection of batteries or solar cells. The electric potential is unchanged but the parallel combination of four components yields four times as much energy as a single component.

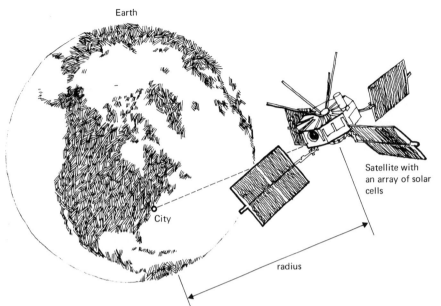

Earth

Satellite with
an array of solar
cells

City

radius

Fig. 20 The satellite must maintain its position
over the city it is serving.

solar cells manufactured in the form of very thin sheets to power equipment on
board such satellites as the Explorer series and Telstar. The technology of this
process was greatly enhanced as a result of the space efforts. Rather than use a
large area solar-energy collector from which thermal energy is derived, one could
use a large area array of solar cells which could convert solar energy directly to
electric energy. Such a system would still have the land-use problem of any other
solar-energy collector. But if an array of solar cells could be put in orbit about
the earth, this problem could be avoided. Such a system using an array 5 miles on
a side has been proposed. The electric energy generated would be transmitted to
earth by microwaves (Fig. 20). The receiver would be an open-mesh antenna
six miles in diameter which would be elevated enough that animals could graze
and feed underneath. Such a system could generate and transmit enough electric
power to supply the needs of a city like New York. At present, though, solar cells
are much too costly for large-scale electric energy production whether used in
an orbited system or on earth. Nevertheless, four U.S. corporations with expertise
in program planning, systems engineering, microwave systems, and solar cells
have combined their talents to investigate the feasibility of these satellite systems.*

GEOTHERMAL ENERGY

Geothermal energy, as the name implies, is simply the internal thermal energy of
the earth. It is not a new source and is not going to contribute significantly to

Nuclear News, May 1972.

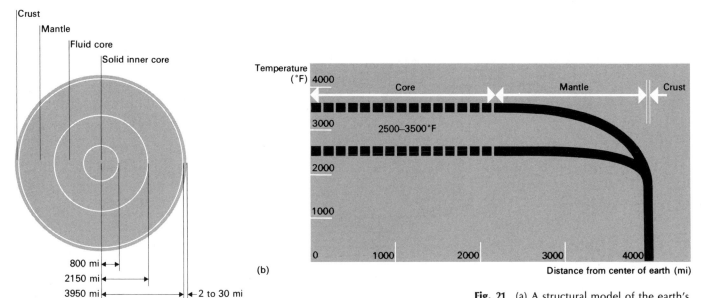

(b)

Fig. 21 (a) A structural model of the earth's interior. (b) Approximate temperature profile of the earth's interior.

satisfying rising energy demands. However, the drive to seek less polluting sources for conversion to electric energy has attracted new attention to it.

Saying that geothermal energy is not going to be competitive does not mean that the supplies are limited. Rather, it means that most of the sources are too inaccessible to be practical. Figure 21 shows a model of a cross-sectional view of the earth's interior and approximate temperature distribution. The extremely interesting feature is that the temperature of the crust 20 to 30 miles below the surface is nearly the same as the earth's interior. It is believed that it is energy released in decay of radioactive nuclei, in particular ^{238}U, ^{232}Th, and ^{40}K, which gives rise to these high temperatures in the crust. It is estimated that the thermal energy stored to a depth of six miles is equivalent to the energy obtainable from the combustion of 900 trillion tons of coal (Exercise 20). But like solar energy, this energy is so dispersed that it is economically unattractive. There are, however, areas where the hot molten rock (magma) from the core is forced up through structural defects and either surfaces in the form of volcanoes or is trapped close to the surface. These masses trapped close to the surface produce geysers and fumaroles (Fig. 22), concentrated sources of hot water or steam which can be tapped and used to heat the boiler of a conventional steam turbine. Thus there is available a natural thermal energy source which does not have the particulate and radioactivity pollution aspects of fossil and nuclear plants. Since this is also a thermodynamic process, there is still the problem of thermal pollution. It is also true that all water is not directly suited as a thermal source because of high mineral and salt content. Removal of these contaminants is a possibility but a more likely solution would be to use a heat exchanger so that the water from the geothermal source does not actually come into contact with the turbine parts. Although these minerals are

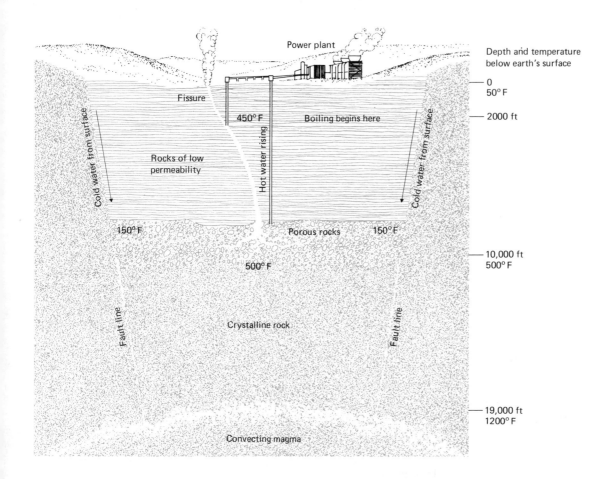

Depth and temperature
below earth's surface

0
50° F

2000 ft

10,000 ft
500° F

19,000 ft
1200° F

Fig. 22   A geological model of a geyser. Surface water flows down into the earth through faults where it is trapped in a porous layer which is bounded by crystalline rock on the lower side and a low porosity layer on the upper side. The water is heated by the magma, pressure builds up, and the water rises through a fissure. When the pressure is released, the water begins to boil and vaporize, and steam and hot water emerge at the earth's surface. If a natural vent for the water is not available, it is sometimes possible to drill a line to the reservoir. Adapted from original material from Fortune Magazine by Max Gschwind; June 1969.

contaminants as far as operation of the turbine is concerned, they could possibly be recovered and sold to appropriate markets.

These hot water and steam sources are normally found in regions of extensive geologic activity. Such regions in the U.S. are located in the furthermost western states and many are on federally owned land. In these areas there is an estimated 1,350,000 acres underlaid with thermal energy sources with a capacity for 15,000 to 30,000 megawatts of electric power. The only area commercially developed thus far in the United States is in the Big Geyser region of Northern California. Pacific Gas and Electric Company (P. G. and E.) has operated an 84,000 kilowatt plant since 1960. The cost of electricity is competitive with other more conventional means and P.G. and E. plans a total of 600 megawatts by 1975. The Imperial Valley of

California has been suggested as another possibility with an estimated capacity in the thousands of megawatts.

Seeking a geothermal source is very much like seeking an oil or natural gas well. There is an element of risk involved and incentives must be provided. The petroleum industry has long enjoyed a favorable tax structure in the form of depletion allowances. However, the federal government still considers geothermal sources in the same class as water wells, and consequently developers are unwilling to embark on any large-scale venture because of the high risk and unfavorable tax structure. There is also no clear-cut policy on the leasing of federal lands for geothermal development. Another feature of President Nixon's energy policy for 1972 is the request of the Secretary of the Interior to immediately expedite competitive lease sales of federal lands for geothermal developments. If these things are instigated, the potential forecast for geothermal energy may be realized.

Another possibility is the use of nuclear explosives to open large cavities in deep, very hot rock formations in the earth. Water could then be pumped into these cavities, vaporized, and the steam collected to drive a turbine. If this proves feasible, geothermal energy could become competitive with conventional sources.

MAGNETOHYDRODYNAMICS

"Today the Subcommittee on Minerals, Materials, and Fuels of the Senate Interior Committee is holding phase II of an exploratory and investigative hearing on MHD, which is a shorthand term for magnetohydrodynamics. *Magnetohydrodynamics is a process for production of electrical energy without going through the relatively inefficient process of boiling water to produce steam to turn generators.*"

"MHD has several *potential advantages*: One, it can use a variety of fuels, including *low-grade coals* that are found in abundance in many of the Western States, including my own state of Utah. Two, it is reported to be *relatively free of air pollution and thermal pollution*, although this point has not been as fully developed as I hope it will be. Three, MHD is said to be substantially *more efficient*, when used in the large-scale facilities, than are conventional steam plants and thus it should be able to produce electrical energy at lower costs."

From: Subcommittee on Minerals, Materials, and Fuels of the Committee on Interior and Insular Affairs, February 23, 1970. Opening remarks by Senator Frank Moss, Chairman.

Any quantity of charge in motion constitutes an electric current (see Chapter 2). A flow of gas or a stream of liquid could produce an electric current if the constituents were charged. A jet engine (Fig. 23) produces a stream of high speed gases but little electric current because there is little net charge associated with the atoms and molecules of the gas. However, if a gas were purposely "seeded" with

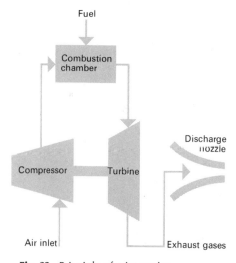

Fig. 23 Principle of a jet engine.

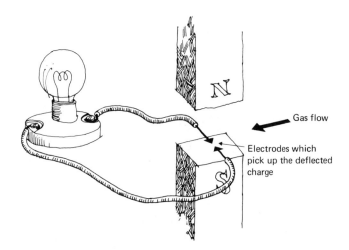

Gas flow

Electrodes which
pick up the deflected
charge

Fig. 24 Principle of the magnetohydrody-
namic electric generating process.

an appropriate material such as potassium or a potassium compound, it would be
highly conductive. To be of practical value, the large electric current associated
with a gas must somehow be transferred to the wires in a house, for example. Such
a transfer can be accomplished using a magnetic field and the principle that a
moving charge experiences a force in a magnetic field (Fig. 24). This is the principle
of magnetohydrodynamics commonly abbreviated MHD. Figure 25 shows the
complete MHD power cycle.

    MHD is attractive because it can possibly operate with low-grade fuel such as
coal at temperatures between 4000 and 5000°F. This high initial temperature increases the thermodynamic efficiency substantially over that of a conventional
steam turbine electric generating system. The efficiency can be further improved
by using the thermal energy available after passing through the MHD generator
to heat the boiler of a conventional steam turbine system. Using this technique,
we can anticipate efficiencies approaching 60%. Not only would this process
significantly reduce thermal pollution, but it would also mean direct rejection of
the thermal energy to the atmosphere without passing through a body of water.

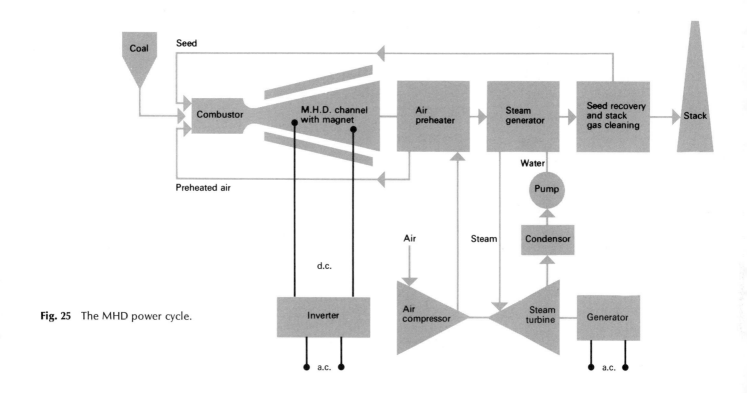

Fig. 25 The MHD power cycle.

The MHD process using coal for fuel and air for the oxygen would still produce sulfur and nitrogen oxides and particulates which are regarded as pollutants. Nitrogen oxides would be produced in large quantities because of the high temperatures involved. However, the seed used to increase the conductivity is so expensive (about the price of the coal) that it will have to be recovered. At the same time the seed is recovered it may be practical to recover the sulfur and nitrogen oxides and convert them to sulfuric and nitric acids which are salable products.

MHD power systems producing 20 megawatts of electric power have been built. These are not of the same magnitude as conventional 1000-megawatt systems but they are surely not laboratory novelties. Those in MHD research believe that 1000-megawatt MHD systems can be built which will be more than competitive, both environmentally and economically, with conventional systems. They admit that technological problems await the development. Estimated costs for MHD development are in the $50–100 million range which puts the federal government in the position of being the only possible supporter. Although the federal government recognizes some possibility for MHD and is nominally supporting its development, it has chosen to support the breeder reactor program as the most likely new technology for electric power.

REFERENCES

Breeder Reactors

"Fast Breeder Reactors," Glenn T. Seaborg and Justin L. Bloom, *Scientific American* **223**, No. 5 (November 1970).

"Third Generation of Breeder Reactors," T. R. Bump, *Scientific American* **218**, (December 1968).

"The Next Step is the Breeder Reactor," *Fortune* (March 1967).

"Energy from Breeder Reactors," Floyd L. Culler, Jr. and William O. Harms, *Physics Today*, Vol. 25, No. 5, 28 (May 1972).

Fusion Energy

"The Hot New Promise of Thermonuclear Power," Tom Alexander, *Fortune* **81**, No. 6, 94 (1970).

"The Prospects of Fusion Power," William C. Gough and Bernard J. Eastlund, *Scientific American* **224**, No. 2, 50 (February 1971).

"Controlled Nuclear Fusion: Status and Outlook," David J. Rose, *Science* **172**, No. 3985, 797 (21 May 1971).

"Fusion by Lasers," Moshe J. Lubin and Arthur P. Fraas, *Scientific American* **224**, No. 6, 21 (June 1971).

"Controlled Fusion: Plasma Heating with Lasers," *Science* **167**, 1112 (20 February 1970).

"The Tokamak Approach in Fusion Research," Bruno Coppi and Jan Rem, *Scientific American* **227**, No. 1, 65 (July 1972).

Solar Energy

"The Direct Conversion of Solar Light Energy into Electricity," C. Greaves, *Physics Education* **5**, No. 2, 100 (March 1970).

"Is It Time For a New Look at Solar Energy?" Aden Baker Meinel and Marjorie Pettit Meinel, *Bulletin of the Atomic Scientists* **27**, No. 8, 32 (October 1971).

"New Ways to More Power with Less Pollution," Lawrence Lessing, *Fortune* (November 1970).

"Solar Power," Norman C. Ford and Joseph W. Kane, *Bulletin of the Atomic Scientists* **27**, No. 8, 27 (October, 1971).

Geothermal Energy

"The Earth's Heat: A New Power Source," *Science News* **98**, 415 (November 28, 1970).

"Power from the Earth's Own Heat," Lawrence Lessing, *Fortune* (June 1969).

Geothermal Research Development, Report No. 1160, Calendar No. 1177, 91st Congress, 2nd Session, September 4, 1970.

Geothermal Energy, Donald E. White, U.S. Geological Survey Circular 519 (1965).

"Geothermal Power," Joseph Barnea, *Scientific American* **226**, No. 1, 70 (1972).

"Power from the Earth," David Fenner and Joseph Klarmann, *Environment* **13**, No. 10, 19 (December 1971).

Magnetohydrodynamics

"New Ways to More Power with Less Pollution," Lawrence Lessing, *Fortune* (November 1970).

Magnetohydrodynamics (MHD) December 18, 1969 and *Magnetohydrodynamics (MHD)* February 23, 1970, from Hearing before the Subcommittee on Minerals, Materials, and Fuels of the Committee on Interior and Insular Affairs, United States Senate, Ninety-first Congress, U.S. Government Printing Office, Washington, D.C.

QUESTIONS

1. A laser emits a very narrow beam of radiation which proceeds nearly unde-flected in the atmosphere. Can you think of some applications of this feature? Can you think of some laser applications which might utilize its large energy concentration?

2. In what ways is a solar energy farm similar to a crop farm?

3. The fusion and fission reactions both produce nuclei with greater binding energy per nucleon. These processes terminate at the peak in the binding energy per nucleon curve (Fig. 18 in Chapter 4). What is the mass number for this peak and what significance might this mass number have for the most likely component of stellar material?

4. What are some environmental questions that might be raised when solar energy is considered as a large-scale energy source for producing electricity?

5. Compare the priorities of using land for solar farms as against highways and roads.

6. We talk of a heat of fusion when water is changed to ice. Is this heat that must be added or extracted when water turns to ice? Is this fusion in any way like nuclear fusion?

7. A TV picture is produced by a beam of electrons striking the screen. The electrons can be moved back and forth across the screen by a magnetic force

similar to that used in confinement of a plasma. Try to figure out how this is done.

8. Water which is trapped deep inside the earth may be at substantially higher temperature than 212°F and still it does not boil until it gets to the surface of the earth. Why?

9. What are some environmental problems that might be associated with large-scale electric energy generation by the MHD method?

10. What might be some of the hazards associated with the mining of uranium?

11. What role does the half-life of ^{239}U and ^{239}Np play in the breeding process shown in Fig. 1?

EXERCISES

Easy to Reasonable

Exercise 1. Calculate the efficiencies of the breeder electric power plants listed in Table 1. How does the efficiency of the proposed Demo No. 1 unit compare with the efficiency of a conventional LWR power plant?

Exercise 2. Suppose that you wanted to build a device that would convert solar radiation to electric power. If the radiation amounts to 0.1 W/cm^2 and you could convert the solar energy to electric energy with an efficiency of 1%, how big a square would be required to produce 1000 W of electric power?

Exercise 3. What is the instantaneous power of a laser that produces 100 J of energy in one-billionth of a second?

Exercise 4. Early versions of the Tokamak fusion devices typically produced plasmas of 10^{13} particles per cm^3, temperatures up to 10,000,000 K and confinement times of about 0.03 sec. How close are these machines to satisfying the Lawson criterion for a useful energy source?

Exercise 5. A person traveling in a circle while on a merry-go-round is analogous to a charged particle traveling in a circle in a magnetic field. What is the force that keeps the person on the merry-go-round traveling in a circle?

Exercise 6. In what ways are experiments for obtaining information about the interiors of an atom and the earth similar?

Exercise 7. Sodium has only one stable isotope, $^{23}_{11}$Na$_{12}$. How could $^{22}_{11}$Na$_{11}$ and $^{24}_{11}$Na$_{13}$ be made by bombarding $^{23}_{11}$Na$_{12}$ with neutrons? One of these nuclei is a β^+ emitter, the other is a β^- emitter. Can you make the identification?

Exercise 8. Assuming that the solar radiation received outside the earth's atmosphere is 0.14 W/cm^2, show that the total solar power emanating from the sun is 4 × 10^{20} MW. You may assume that the distance from the sun to the earth is 9.3 × 10^7 mi.

Reasonable to Difficult

Exercise 9. The solar radiation at the earth's surface is about 0.07 W/cm^2.

a) How much radiation is received by an area equal to 1% of the area of the United States?

b) If this radiant energy is converted to electric energy with an efficiency of 1%, show that it would exceed the present electric generating capacity of the United States.

Exercise 10. The temperature of the core of the earth is not accurately known, but the general consensus is that it is less than $10,000°F$. The variation of temperature with depth into the earth's crust is fairly well known, $+1°F/100$ feet. If this variation were to continue, what would you expect for the temperature of the earth's core? This illustrates the danger of extrapolating a straight line too far.

Exercise 11.

a) In an ordinary nuclear reactor using ^{235}U for fuel, what is the minimum number of neutrons that can be released in a fission reaction necessary to maintain a self-sustaining process?

b) In a breeder reactor using ^{239}Pu for fuel and ^{238}U for fertile nuclei, what is the minimum number of neutrons released in a fission reaction necessary to maintain a self-sustaining process as well as replacement of the spent fuel?

Exercise 12. Elementary school students often demonstrate the magnitude of solar energy by constructing a solar cooker for "hot dogs." It consists of a bowl-shaped structure of aluminum foil. The bowl shape allows the radiation to be focused on the "hot dog."

a) Suppose that 10% of the radiation falling on the reflector gets absorbed by the "hot dog." How big a surface area would be required to achieve a power of 100 W on the "hot dog"?

b) Suppose that a 1000-W household appliance cooks a "hot dog" in 5 min. How long would it take the solar cooker to achieve the same result?

Exercise 13. The carbon-nitrogen-oxygen cycle is a major energy mechanism in very massive stars. This starts by bombarding $^{12}_{6}C_{6}$ with protons. One nucleus has been left out in each step of the cycle shown below. Determine this nucleus

$$^{12}_{6}C_{6} + p \quad \longrightarrow \quad \underline{\hspace{3cm}} + \gamma$$

$$\underline{\hspace{3cm}} \quad \longrightarrow \quad ^{13}_{6}C_{7} + \beta^{+} + v$$

$$^{13}_{6}C_{7} + \underline{\hspace{2cm}} \quad \longrightarrow \quad ^{14}_{7}N_{7} + \gamma$$

$$\underline{\hspace{3cm}} + p \quad \longrightarrow \quad ^{15}_{8}O_{7} + \gamma$$

$$\underline{\hspace{3cm}} \quad \longrightarrow \quad ^{15}_{7}N_{8} + \beta^{+} + v$$

$$^{15}_{7}N_{8} + p \quad \longrightarrow \quad \underline{\hspace{3cm}} + ^{4}_{2}He_{2} + \gamma$$

Exercise 14. Show that a total of 26.73 MeV of energy is released in the proton-proton cycle. You may use the following data for the masses.

	Mass (amu)
proton	1.007825
deuteron	2.014102
^{3}He	3.016030
^{4}He	4.002603
electron	0.000549

1 amu = 931.48 MeV

Exercise 15. For two protons to fuse together they must come within about 10^{-12} cm of each other. But by virtue of the associated charge, there is a mutual repulsive force which tends to prevent their coming together. The energy associated with this force is $E = e^2/R$ (ergs), where e is the charge (4.8×10^{-10} electrostatic units) and R is the separation (centimeters) of the charge. If this energy is supplied thermally, it is proportional to the temperature, $E = (3/2) kT$ where $k = 1.38 \times 10^{-16}$ ergs/K. Show that T must be of the order of hundreds of millions of degrees.

Exercise 16. Helium gas that is commonly used to inflate balloons is obtained as a by-product in the extraction of natural gas from the interior of the earth. Propose a mechanism for producing helium in the earth knowing that ^{238}U and ^{232}Th are also contained within the earth.

Exercise 17. Fill in the steps for generating fissionable ^{233}U starting from ^{232}Th.

$$^{232}_{90}\text{Th}_{142} + ^{1}_{0}\text{n}_1 \longrightarrow$$

$$\longrightarrow \quad + \beta^- + \bar{\nu}$$

$$\longrightarrow \quad ^{233}_{92}\text{U}_{141} + \beta^- + \bar{\nu}$$

Exercise 18. According to the theory of nuclear fission, $^{231}_{92}\text{U}_{139}$ would be a suitable fuel for a nuclear chain reaction. It could be produced very much like ^{233}U and ^{239}Pu.

$$^{230}_{90}\text{Th}_{140} + ^{1}_{0}\text{n}_1 \rightarrow ^{231}_{90}\text{Th}_{141}$$

$$^{231}_{90}\text{Th}_{141} \rightarrow ^{231}_{91}\text{Pa}_{140} + \beta^- + \bar{\nu}$$

$$^{231}_{91}\text{Pa}_{140} \rightarrow ^{231}_{92}\text{U}_{139} + \beta^- + \bar{\nu}$$

Even though the process is possible, it is impractical. Think of some reasons why.

Exercise 19.   The volume of water in the oceans works out to be about 10^{21} pints. Suppose that you dipped one pint from any ocean and you made the oxygen atom of each molecule in this pint radioactive so that you could later identify it. Now suppose that you dumped the pint back into the ocean and thoroughly mixed it with the total 10^{21} pints in the oceans. If you were to again dip one pint of water, how many radioactive molecules from the original pint would you expect to have?

Exercise 20.   Deep mining operations have shown that the temperature increases about 10°F for each 1000 feet of depth into the earth's crust.

a) Show that the temperature at a depth of 6 mi is about 400°F.

b) Assuming that the United States is a rectangle 1000 by 3000 mi, show that there is about 10^{18} ft^3 of crust contained down to a depth of 6 mi.

c) Assuming that the density of the crust is about 100 lb/ft^3, show that the weight of this amount of crust is about 10^{20} lb.

d) Assuming that the specific heat of the crust is about 1 Btu per pound per 1°F change in temperature, show that about 10^{22} Btu of thermal energy would be released if the temperature of this amount of crust were cooled 100°F.

e) Show that this amount of energy is equivalent to the burning of about 10^{15} tons of coal (100 trillion tons). This should convince you of the magnitude of energy contained in the earth's crust.

Difficult to Impossible

Exercise 21.   The earth is approximately a sphere having a radius of 6500 km. About two thirds of the surface is covered with oceans.

a) Show that the oceans cover an area of about 1.2×10^8 km^2.

b) The average depth of the ocean is about 1.2 km. Show that the volume of the water in the ocean is about 1.4×10^{23} cm^3 and the mass is about 1.4×10^{23} g.

c) The water molecule is H_2O having a molecular weight of about 18. Eighteen grams of water contain Avogadro's number, 6.023×10^{23}, of water molecules. Show that there are about 4.7×10^{45} water molecules and about 9.4×10^{45} hydrogen atoms in the oceans.

d) If 1 out of every 7000 of these hydrogen atoms is deuterium, show that there are about 1.3×10^{42} deuterium atoms in the oceans.

e) If all these deuterium atoms were used in d–d fusion reactions and each reaction released about 4 MeV, show that the energy available is about 4×10^8 Q where 1 Q $= 10^{18}$ Btu. How does this energy compare with that available from fossil fuels?

Exercise 22.   The fission of ^{239}Pu by a neutron can be symbolized as

^{239}Pu + n → Energy (Q) + Fission Products (F) + ηn

η denotes the number of neutrons produced. One of these neutrons is required to maintain a self-sustaining process. The remaining neutrons can be used to make new fuel by bombarding ^{238}U. This can be symbolized as

$$(\eta - 1)\, n + (\eta - 1)\, ^{238}U \rightarrow (\eta - 1)\, ^{239}Pu$$

  a) Add these two equations and show that

$$(\eta - 1)\, ^{238}U \rightarrow Q + F + (\eta - 2)\, ^{239}Pu.$$

    b) Show that there must be at least 2 neutrons (that is, $\eta = 2$) produced in each fission of ^{239}Pu to maintain the self-sustaining process as well as replacement of the original ^{239}Pu nucleus.

Exercise 23. The article "New Ways to More Power with Less Pollution," *Fortune* **82**, Volume 11, 78 (1970) reports that a satellite carrying an array of solar cells will orbit the earth 22,300 miles above the equator. The reason is that the satellite at this altitude will be synchronized with the rotation of the earth and therefore will always remain at the same position with respect to the city to which it supplies power (Fig. 20). The gravitational force between the earth and satellite keeps it in orbit. Using Newton's second law, we can write

$$\underbrace{\text{gravitational force on satellite}}_{F_{gravity}} \quad \underset{=}{\text{equals}} \quad \underbrace{\text{mass of satellite times acceleration}}_{m_s \cdot a}$$

mass of satellite cancels.

$$\frac{GM_{earth}\, \cancel{m_s}}{R^2} = \cancel{m_s}\, a$$

where

$$G = \text{gravitational constant} = 6.67 \times 10^{-11}\, \frac{\text{newton} \cdot m^2}{kg^2}$$

and

$$M_{earth} = \text{mass of earth} = 5.983 \times 10^{24}\, kg$$

The acceleration can be written in terms of the speed, V, so that

$$\frac{GM_{earth}}{R^2} = \frac{V^2}{R}$$

In one period, T, the satellite travels a distance equal to the circumference of a circle of radius R; that is, distance $= 2\pi R$. Hence the speed is

$$\text{speed} = \frac{\text{distance}}{\text{time}}$$

$$V = \frac{2\pi R}{T}$$

Using this expression for V, we can write

$$R = \left(\frac{GM_{earth}\,T^2}{4\pi^2}\right)^{1/3} \longleftarrow \text{cube root}$$

From this you can calculate the radius of an orbit for some given period T. Note that the mass of the satellite does not enter into the calculation. Show that if $T = 24$ hours, then $R = 26,000$ miles from the center of the earth or about 22,000 miles from the equator. (Be careful of the units.) This principle was employed in the SYNCOM satellite used for communications.

6

MOTOR VEHICLES: PROBLEMS AND ALTERNATIVES

Photograph courtesy of Federal Highway Commission

PROBLEMS

Carbon monoxide, nitrogen oxides, and hydrocarbons comprise the second category of air pollutants mentioned in Chapter 2. These are liberated primarily in the energy conversion process in the internal combustion (IC) engine in motor vehicles (automobiles, trucks, buses). Everyone has probably been warned of the hazards of carbon monoxide. It is colorless, odorless, and tasteless and concentrations of only 0.1% (by volume) produce unconsciousness in 1 hour and death in 4 hours. This is why it is extremely dangerous to run an IC engine in a closed garage or to tolerate a faulty exhaust system in an automobile or to burn a charcoal grill in closed quarters such as a tent. Nitrogen oxides are also poisonous but not nearly to the extent of carbon monoxide. Hydrocarbons comprise such a broad class of materials that it is impossible to make any general statement about their toxicity.

Interestingly, it is not the direct effect of nitrogen oxides and hydrocarbons on plant and animal life which causes concern. Rather it is the effect of chemical reaction products producing photochemical "smog" which is the major concern. It is important to note the difference between *photochemical smog*, which is derived from nitrogen oxides and hydrocarbons, and London smog*, which is due mainly to sulfur oxides and particulates from coal burning processes. Nearly every major city in the world is suffering from the effects of one or both of these smog types. This suffering appears as economic damage, plant and animal impairment, and possible alteration of weather. The effects generally attributed to smog are:

1. characteristic damage to vegetation
2. eye irritation
3. respiratory distress and even death
4. reduction in visibility
5. objectionable odors
6. excessive cracking of rubber products and damage to other materials

Having pointed out some of the problems with motor vehicle emissions, let us look at how these evolve from the IC engine.

MACROSCOPIC DESCRIPTION
OF THE IC ENGINE

Most people know that the engine of a motor vehicle usually resides under some cover either in the front or rear of the vehicle. They would probably know that the pollutants of concern are emitted to the environment mainly through the vehicle's exhaust system. Given the opportunity to view the total system, they would soon discover that it is a highly complex facility. The engine performs a variety of functions such as running an air conditioner, electric generator, wind-

---

*The word smog was actually derived from smoke and fog and was used to designate the type of polluted conditions that often occur in London.

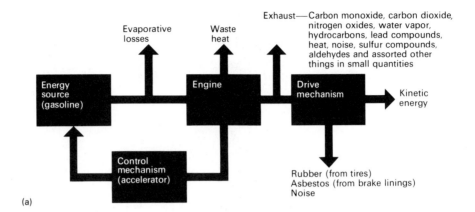

(a)

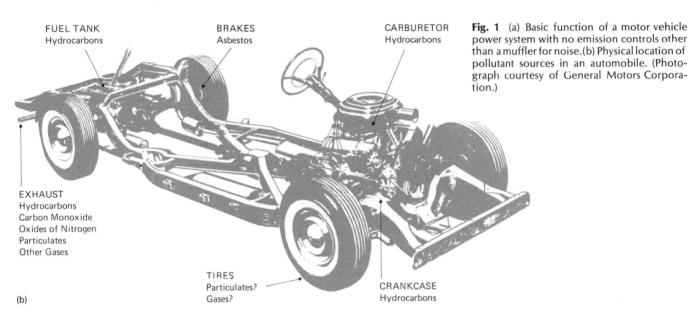

FUEL TANK
Hydrocarbons

BRAKES
Asbestos

CARBURETOR
Hydrocarbons

EXHAUST
Hydrocarbons
Carbon Monoxide
Oxides of Nitrogen
Particulates
Other Gases

TIRES
Particulates?
Gases?

CRANKCASE
Hydrocarbons

(b)

Fig. 1 (a) Basic function of a motor vehicle power system with no emission controls other than a muffler for noise.(b) Physical location of pollutant sources in an automobile. (Photograph courtesy of General Motors Corporation.)

shield washers and wipers, and hydraulic pumps; but its main function is to convert the intrinsic potential energy of gasoline into kinetic energy of the car. The model of a motor vehicle shown in Fig. 1a allows us to see the magnitude of the emissions without getting into the details of the IC engine. Figure 1b shows the point of origin of the various emissions. Let us take a typical trip in an average car with no emission controls by "dumping" 10 gallons of gasoline into the tank and seeing how it is disposed.

Gasoline tank

Amount of fuel inserted = 10 gal

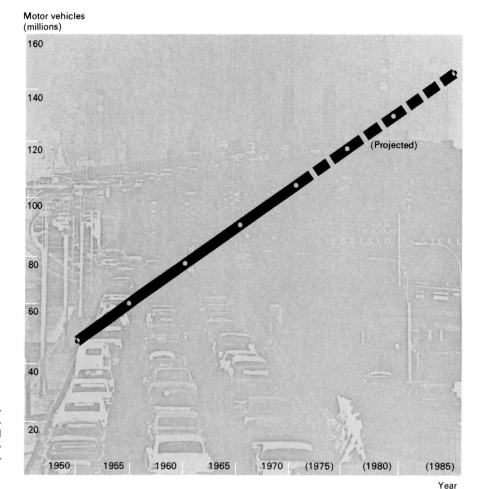

Motor vehicles (millions)

Fig. 2 United States motor vehicle registration. (From U. S. Bureau of the Census, *Statistical Abstract of the United States: 1971* (92nd ed.), Washington, D.C., 1971, p. 535. (Photograph courtesy of the Environmental Protection agency and Black Star: Dan McCoy).

Year

Amount of fuel lost through evaporation = 0.08 gal

Amount of fuel available for engine = 9.92 gal

Engine

Energy available to engine = amount of gasoline × heat of combustion
= 9.92 × 119,000 Btu
= 1,180,480 Btu

Energy available to the vehicle = energy available to engine × efficiency
= 1,180,480 × (0.20)
= 236,096 Btu

Exhaust

Material	Amount (lb)*
Carbon monoxide	32
Hydrocarbons	2–4
Nitrogen oxides	2–8
Aldehydes	0.18
Sulfur compounds	0.17
Organic acids	0.02
Ammonia	0.02
Solids (zinc, metallic oxides, carbon)	0.003

*These amounts will vary significantly for different vehicles.

These emissions do not seem exorbitant, but the average car will use about 750 gallons of gasoline per year and there are about 100 million motor vehicles in the United States. It is this volume which accounts for the magnitude of pollution attributed to motor vehicles. And there appears to be no abatement of the number of vehicles (Fig. 2). The rate of growth of motor vehicle registration and petroleum consumption both exceed the growth of population (4% per year for motor vehicle registration versus 1.3% per year for population; see Exercise 5).

PHYSICAL PRINCIPLE OF THE IC ENGINE

The conventional internal combustion and steam engines are identical, in principle, in terms of energy conversion. Both convert the random kinetic energy of a gas into linear energy of a piston and then into rotational energy of some drive shaft (Fig. 3). The high temperature of the gas which reflects the magnitude of the random kinetic energy of the molecules is achieved by igniting a gasoline vapor-air mixture in the combustion chamber of the engine. The IC cycle usually involves four strokes of the piston, two down and two up. Two valves are required for the four-stroke cycle to get the gasoline vapor-air mixture into and out of the combustion chamber and a spark plug is required for ignition* (Fig. 4a). Most of the auxiliary equipment seen in the engine compartment of an automobile engine is associated with the proper timing of these operations and with the production of the gasoline vapor-air mixture.

The thermodynamic cycle of the internal combustion engine is shown in Fig. 4b). Intake of the vapor-air mixture takes place at atmospheric pressure ($0 \rightarrow 1$). The mixture is compressed adiabatically from V_1 to V_2 ($1 \rightarrow 2$). The power stroke constitutes an adiabatic expansion from volume V_2 to V_1 ($3 \rightarrow 4$). The step from $4 \rightarrow 1 \rightarrow 0$ corresponds to loss of heat in the exhaust process. The net work, W, is just the "area" enclosed by the thermodynamic path and the efficiency, by definition, is

$$\epsilon = \frac{\text{Work done}}{\text{Heat added}} = \frac{W}{Q}$$

*A diesel engine used in some cars and trucks is also a type of internal combustion engine and it does not require a spark plug for ignition of the vapor-air mixture (Exercise 7).

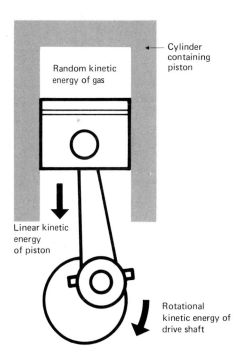

Fig. 3 The energy conversion process in an internal combustion engine.

Random kinetic energy of gas

Cylinder containing piston

Linear kinetic energy of piston

Rotational kinetic energy of drive shaft

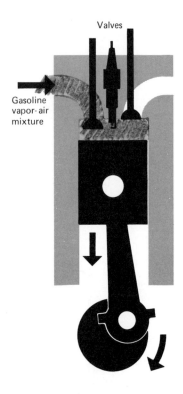
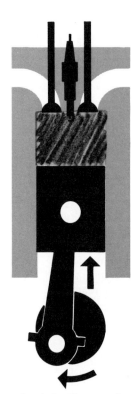

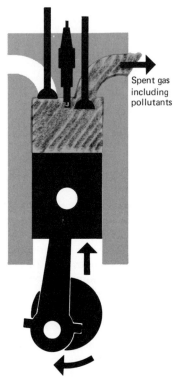

Valves

Gasoline vapor-air mixture

Spent gas including pollutants

Intake stroke. A vapor-air mixture is pulled in from the carburetor through the open intake valve.

(a)

Compression stroke. The vapor-air mixture is compressed by the upward moving piston. Both valves are closed.

Power stroke. At the top of the compression stroke, the sparkplug ignites the vapor-air mixture, forcing the piston down to produce linear kinetic energy.

Exhaust stroke. At the bottom of the power stroke, the exhaust valve opens and the spent gas is forced out through the exhaust system.

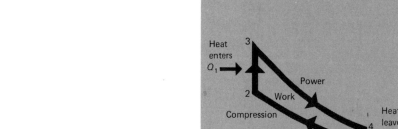

(b)

Fig. 4 (a) The physical cycle of the internal combustion engine. (b) The thermodynamic cycle of the internal combustion engine.

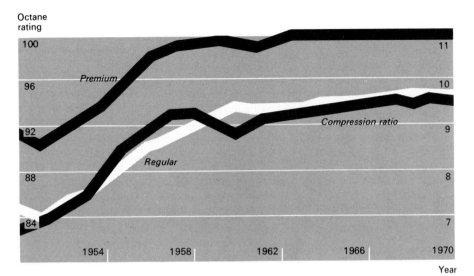

Fig. 5 Average octane number and compression ratio trends. These values will decrease in the 1970's as unleaded gasoline and lower compression engines are introduced. (From *Air Pollution—1970*, Part 3, 91st Congress, Second Session on S.3229, S.3466, and S.3546, March 24 and 25, 1970. U.S. Government Printing Office, Washington, D.C.)

This turns out to depend only on the volumes V_1 and V_2 and the properties of the gas. Formally

$$\epsilon = 1 - \frac{1}{r^{\gamma - 1}}$$

where $r = V_1/V_2$ is the compression ratio and γ is a number, typically 1.4, which depends on the type of gas. For example, if the compression ratio is 9 (Fig. 5), the *theoretical* efficiency of the engine is 0.59 or 59%.

The pollutants credited to the IC engine are left-over combustion products which are forced out of the engine in the exhaust cycle. To see the origin of these pollutants, we must now look at the combustion process at the atomic level.

MICROSCOPIC ORIGIN OF POLLUTANTS

A combination of hydrogen and carbon atoms forming a molecule is appropriately called a hydrocarbon. A class of hydrocarbons known as alkanes has the chemical form $C_n H_{2n+2}$, where n is a number from 1 to 10. For example, if $n = 5$, the molecule is $C_5 H_{12}$, called pentane. Gasoline with no special-purpose additives is a mixture of alkanes with from 5 to 10 carbon atoms. Octane, $C_8 H_{18}$ is a common molecular constituent. Although the expression $C_8 H_{18}$ tells you the composition of the molecule, it does not describe the atomic structure within the molecule, on which the physical properties, such as the burning characteristics, are dependent. There are more than a dozen compounds, all having different characteristics and names, which share the formula $C_8 H_{18}$. One of these, isooctane, is particularly useful as an automotive fuel. Isooctane has become the standard for comparison of other automotive fuels. The reason is as follows. With some gasolines, ignition occurs more as an explosion rather than a smooth burning. This not only reduces the efficiency, but also produces an audible "knock" in the engine. A fuel of pure isooctane produces little knock while one of pure heptane ($C_7 H_{16}$) knocks badly. The octane rating of a gasoline is a measure of the knock produced when used as an automotive fuel. Isooctane and heptane have been assigned octane ratings of 100 and 0, respectively. A mixture of 90% isooctane and 10% heptane would have an octane rating of 90. When any other fuel, regardless of its chemical composition, is burned in a special engine and produces the same knock as 90% isooctane and 10% heptane, it would also have an octane rating of 90. Since the inception of this procedure, fuels have been developed with anti-knock properties superior to isooctane and these will have octane ratings greater than 100.

The pollutants evolve when the fuel, whatever it may be, is burned in the engine. Like the burning of coal in an electric power plant, the pollutants would be minimal if: (1) the combustion were complete and (2) there were no side effects. To illustrate, consider the combustion of isooctane which we might consider the standard fuel. Isooctane ideally reacts with oxygen in the following way:

$$2\,C_8 H_{18} + 25\,O_2 \rightarrow 16\,CO_2 + 18\,H_2O$$

The CO_2 is harmless (except for long-term climatic effects already discussed) as is the water vapor. However, the reaction may be incomplete producing carbon monoxide. For example,

$$2\,C_8 H_{18} + 25\,O_2 \rightarrow 14\,CO_2 + 2\,CO + O_2 + 18\,H_2O$$

Oxygen for combustion is obtained from air which typically contains 78% nitrogen and 21% oxygen. The nitrogen does not enter into the burning process, but does react with the oxygen at high temperatures in the engine. The reactions which produce the nitrogen oxide pollutants are

$$N_2 + 2\,O_2 \rightarrow 2NO_2 \text{ (nitrogen dioxide)}$$

$$2N + O_2 \rightarrow 2NO \text{ (nitric oxide)}$$

Other oxides are produced and normally these, as well as NO and NO_2, are lumped together for environmental purposes and called nitrogen oxides symbolized NO_x.

The efficiency of the IC engine is directly related to the octane rating of the fuel and to the compression ratio. Throughout the years, the trend has been to increase both of these factors (Fig. 5). One way of producing a better fuel is to find molecular structures with better burning characteristics. This has, in fact, been done but is an expensive proposition. The most common method of achieving this desired result is to add a compound called tetraethyl lead, $(C_2H_5)_4Pb$, in amounts of about 2 milliliters per gallon. Although this addition increases the octane rating, provides some lubrication, and reduces valve burning it produces deposits such as metallic lead and lead oxides which remain in the engine fouling the valves and spark plugs. To combat this, ethylene dibromide ($C_2H_4Br_2$) was added to convert the lead to lead bromide (Pb Br) which escapes as a gas out the exhaust. Present day gasolines also contain antioxidants, metal deactivators, anti-rust and anti-icing compounds, detergents, and lubricants. So modern gasolines are far from the ideal isooctane suggested earlier.

THE ORIGIN OF PHOTOCHEMICAL SMOG

Although both nitric oxide (NO) and nitrogen dioxide (NO_2) are produced in the IC engine, the combustion conditions strongly favor the production of NO. Since there is considerable oxygen in air, this NO could conceivably be converted to NO_2. However, most of the oxygen is in the form of molecules containing two atoms (that is, O_2), and the reaction of NO with O_2 is weak. On the other hand, the atomic oxygen and ozone (O_3) in the air react very strongly with NO.

$$NO + O \rightarrow NO_2$$

$$NO + O_3 \rightarrow NO_2 + O_2$$

Still, unless something else is involved, these reactions would not be important because of the low concentrations of O and O_3. The initial "something else" and crucial link is the breakdown of nitrogen dioxide by sunlight

$$NO_2 + \text{solar energy} \rightarrow NO + O$$
$$\text{atomic oxygen}$$

More than 99% of the atomic oxygen combines with molecular oxygen as follows:

$$O \quad + \quad O_2 \quad \rightarrow \quad O_3$$
$$\text{atomic oxygen} \quad \text{ordinary} \quad \text{ozone}$$
$$\text{oxygen molecule}$$

However, some atomic oxygen reacts with hydrocarbons in a series of complex chemical reactions: NO_2 and O_3 are regenerated; some very undesirable molecules, namely formaldehydes, peroxyacyl nitrates (PAN), and acroleins, are produced; and the original high NO concentration is diminished. Figure 6 shows how this takes place in a large city. These processes are responsible for the characteristic haze over many cities (Fig. 7) and it is the chemicals that produce the effects mentioned in the first section.

Table 1 summarizes the characteristics of photochemical smog and gives the corresponding London smog data for comparison.

Table 1. Some distinguishing characteristics of photochemical (or Los Angeles) and London smogs.

Characteristic	Photochemical (or Los Angeles)	London
Major fuels involved	petroleum	coal and petroleum products
Principal constituents	O_3, NO, NO_2, CO, organic products	particulates, CO, sulfur compounds
Types of reactions	photochemical and thermal	thermal
Time of maximum occurrence	midday	early morning
Principal effects	eye irritation	bronchical irritation, coughing
Visibility	about 0.5 to 1 mile	less than 100 yards

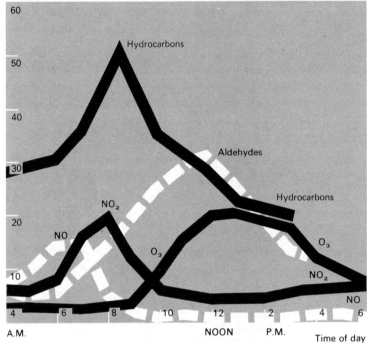

Fig. 6 Hourly variations of the constituents of an intense photochemical smog. (From "Photochemistry of Air Pollution" by Philip A. Leighton, Academic Press, Inc.).

Fig. 7 A visual illustration of clear conditions and an intense smog formation in Los Angeles. (United Press International Photograph. Reproduced by permission.)

Cities are naturally the areas where smog tends to be problematical for it is here that industry and large concentrations of motor vehicles exist. In fact, every major city in the world has smog problems. Yet the West Coast of the United States is particularly troubled. The reason is not that it is more industrialized than many other areas, but that it is subject to a characteristic weather condition (see Chapter 3) which produces a temperature inversion on about 100 days of the year. The situation in Los Angeles is complicated by mountains and hills to the north, east, and south which prevent flushing of the smog trapped by the temperature inversion.

SOME SPECIFIC EFFECTS OF AIR POLLUTANTS
OF CONCERN AND EXTENT OF POLLUTION

Some specific effects of air pollutants of concern are briefly summarized in Table 2. The data were taken from the Air Quality Criteria Handbooks published by the U.S. Department of Health, Education, and Welfare and The Environmental Protection Agency. These handbooks contain substantive information on environmental effects of air pollutants. Using these as criteria, the Environmental Protection Agency has arrived at the standards shown in Table 3. These standards represent what E.P.A. considers adequate for the protection of the general public. The standards are being questioned and, in many instances, by reliable authorities, but it is fair to say that the general consensus is that they are very stringent.

Table 2. Brief summary of air pollutant effects

Effects associated with oxidant concentrations in photochemical smog

Effect	Exposure ppm	$\mu g/m^3$	Duration	Comment
Vegetation damage	0.05	100	4 hours	leaf injury to sensitive species
Eye irritation	Exceeding		peak values	Such a peak value would be
	0.15	200		expected to be associated with a maximum hourly average concentration of 50 to 100 $\mu g/m^3$ (0.025 to 0.05 ppm).
Impaired performance of student athletes	.03–0.3	60–590	1 hour	exposure for 1 hour prior to race

Effects associated with carbon monoxide.

Effect	Exposure	Duration	Comment
Bodily damage	30	8 hours or more	impairment of visual and mental acuity
	200	2 to 4 hours	tightness across the forehead, possible slight headache
	500	2 to 4 hours	severe headache, weakness, nausea, dimness of vision, possibility of collapse
	1,000	2 to 3 hours	rapid pulse rate, coma with intermittent convulsions
	2,000	1 to 2 hours	death

238

Leaf of an ozone sensitive tobacco variety shows white spots characteristic of air-pollutant damage called weather fleck. (Photograph courtesy of the U.S. Department of Agriculture.)

Ozone damage to White Cascade petunia. (Photograph courtesy of the U.S. Department of Agriculture.)

Table 3. Proposed federal standards for carbon monoxide, photochemical oxidants, nitrogen oxides, and hydrocarbons. (From *Federal Register*, **36**, # 21, January 30, 1971.)

Pollutant	Maximum concentration per cubic meter	Approximate concentration (ppm)	Averaging period
Carbon monoxide	10 mg	9	8 hr
	15 mg	13	1 hr
Photochemical oxidants	125 μg	0.06	1 hr
Nitrogen oxides	100 μg	0.08	annual arithmetic mean
	250 μg	0.21	24 hr
Hydrocarbons	125 μg	0.2	maximum three-hour concentration during the morning rush hours

Fig. 8 Number of days per year of adverse levels of air pollution indices in Los Angeles County from 1955 to 1968. Adverse levels defined by: Ei—eye irritation recorded; Vsby—visibility less than three miles and relative humidity less than 70%; Oxidant—exceeded 0.15 ppm for 1 hour; NO_2—exceeded 0.25 ppm for 1 hour; CO—exceeded 30 ppm for 8 hours. (Data from *Profile of Air Pollution Control in Los Angeles County*, E. E. Lemke, G. Thomas, and W. E. Zwiacher (Eds.), Los Angeles County Air Pollution Control District, Los Angeles, 1969.) (Photograph courtesy of County of Los Angeles Air Pollution Control District and the Environmental Protection Agency.)

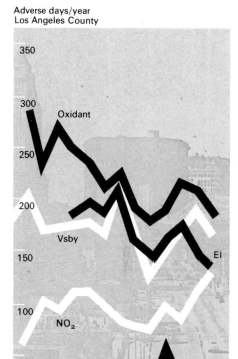

California, Los Angeles in particular, has come to be identified with air pollution in this country because of its characteristic weather patterns, style of growth, and terrain. Figure 8 reveals the seriousness of the situation in Los Angeles County. It is also true that California is the leader in air pollution research and in enactment of controls for eliminating these conditions. Los Angeles, however, is not the only city with air pollution problems. Table 4 shows that all major cities in the United States have similar problems and will have trouble meeting the standards set by E.P.A.

Table 4. Air pollution data for six representative United States cities (1968). From U.S. Bureau of the Census, *Statistical Abstract of the United States: 1971* (92nd edition.) Washington, D.C. 1971, p. 170.

	NO		NO_2		CO		Total oxidants		Hydrocarbons	
	Maximum 24 hour (ppm)	Yearly average (ppm)	Maximum 24 hour (ppm)	Yearly average (ppm)	Maximum 24 hour (ppm)	Yearly average (ppm)	Maximum 24 hour (ppm)	Yearly average (ppm)	Maximum 24 hour (ppm)	Yearly average (ppm)
Chicago	0.23	0.07	0.10	0.05	16	6.2	0.11	0.02	5.4	2.9
Cincinnati	Insufficient data		0.10	0.03	32	5.6	Insufficient data		5.6	2.6
Philadelphia	0.37	0.05	0.09	0.04	Insufficient data		0.08	0.02	4.8	2.2
Denver	0.21	0.04	0.12	0.04	21	5.4	0.08	0.03	6.4	2.9
St. Louis	0.13	0.03	0.05	0.02	9	4.6	0.05	0.02	9.8	3.4
Washington, D.C.	0.31	0.04	0.08	0.05	14	3.4	0.10	0.03	5.7	2.2

EMISSIONS CONTROL PROGRAM

The Federal Emission Standards

If air quality is to be improved in critical areas and the stringent federal standards are to be met, then motor vehicle emissions, in particular those from the automobile, must be controlled. Such a program was begun in California in 1961 and nationwide in 1968. Both programs have called for a gradual decrease in allowable automobile emissions over a period of a few years. Because of the increasing acuteness of air pollution problems, the federal timetable has been subject to change. The schedule as legislated by amendments to the 1970 Clean Air Act is as shown in Table 5. The 1975 values for CO and HC are 90% reductions from the 1970 values. The 1976 value for NO_x is a 90% reduction from the 1971 value. There is, at present, no proven technology available for meeting the 1975–76 standards. It is important to realize that even if these maximum emission requirements were met, the federal air quality criteria could be exceeded if there were sufficient automobiles. The only solution then would be to limit the number of vehicles.

Table 5. Federal timetable for maximum allowable automobile emissions expressed in grams per mile.*
(See, for example, "Pollution-free Power for the Automobile," *Environmental Science and Technology*, **6**, 6, 512 (June 1972).)

Pollutant/Year	1975	1976	Uncontrolled vehicle
Carbon monoxide	3.4		125
Hydrocarbons	0.41		16.8
Nitrogen oxides	3.0	0.4	6.

*Measuring procedure for CO and hydrocarbon described in *Federal Register*, Nov. 10, 1970. Measuring procedure for NO_x not yet established.

The Approach to Meeting the Federal Standards

There are several possible approaches to meeting the maximum emission requirements. These would include:

1. Modification of the present IC engine.

2. Development of a new fuel for the present IC engine.

3. Removal of the pollutants from IC engine effluents.

4. Development of a new power plant to replace the IC engine.

All of these are being pursued, but the major thrust, initially, is toward the third.

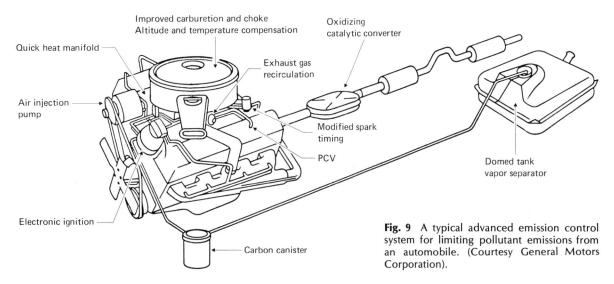

Improved carburetion and choke
Altitude and temperature compensation

Oxidizing
catalytic converter

Quick heat manifold

Exhaust gas
recirculation

Air injection
pump

Modified spark
timing

PCV

Electronic ignition

Carbon canister

Domed tank
vapor separator

Fig. 9 A typical advanced emission control system for limiting pollutant emissions from an automobile. (Courtesy General Motors Corporation).

Removal of Pollutants from the IC Engine Effluents

Figure 9 shows some of the devices used on automobiles to curb pollutant emissions. All of the carbon monoxide and nitrogen oxides and 60% of the hydrocarbons generated are combustion products. These would simply escape out the exhaust system in the absence of any controls. Thus most of the controls, the catalytic converter, for example, are strategically located along the exit route of the exhaust gases. Twenty percent of the hydrocarbons released are a result of evaporation of gasoline from the fuel tank and carburetor. Controls for these emissions, the carbon canister, for example, are placed in the fuel supply system. The remaining 20% of the hydrocarbon emissions result from a process called crankcase blowby which occurs within the engine. Ordinarily these emissions escape to the environment. Their control is effected by recirculation back into the air intake system of the engine. Let us now look a little more deeply into these control devices.

The device used to convert the linear kinetic energy of the piston to rotational energy is referred to as the crankshaft. Every piston (normally 4, 6, or 8 pistons) is connected to this shaft by a connecting rod (Fig. 10). The crankcase (Fig. 1b) serves as a reservoir of oil for lubricating the crankshaft, connecting rods, and other parts of the engine. Proper operation of the engine requires a tight, but moving, seal between the piston and the cylinder walls. There is, however, a certain amount of vapor which leaks out of the combustion chamber into the crankcase. This is the origin of crankcase blowby. It was standard practice until 1961 to simply vent the crankcase to the atmosphere allowing these emissions to escape. However, the emissions can simply be recycled back into the air intake system of the engine (Fig. 11). California required such controls in 1961 and in 1963 all American cars had them as standard equipment.

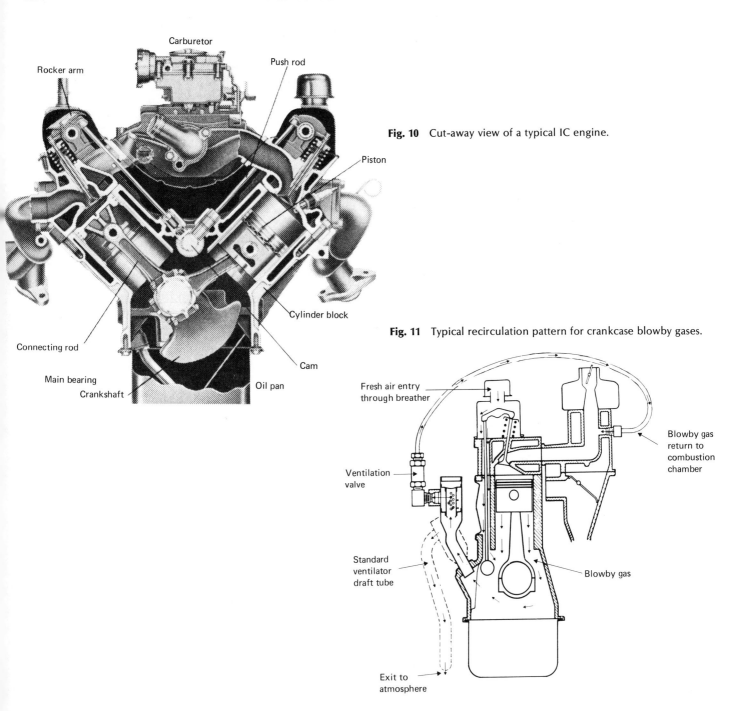

Fig. 10 Cut-away view of a typical IC engine.

Fig. 11 Typical recirculation pattern for crankcase blowby gases.

The characteristic smell associated with the filling of the fuel tank in an automobile is a graphic illustration of the evaporative ability of gasoline. And the fact the smell is more noticeable on a hot day indicates that the evaporation is temperature dependent. So long as the fuel tank is vented to the atmosphere, gasoline vapor, and therefore hydrocarbons, will escape to the environment. Significant evaporative losses also occur at the carburetor. The function of the carburetor is to convert the liquid gasoline to an air-gasoline vapor mixture. During operation of the engine, this mixture is naturally injected into the combustion chambers and ignited. However, when the engine is shut down, the temperature of the carburetor will rise from the operating temperature of about 150°F to about 200°F because the cooling system is off. As a result, the liquid fuel bowl of the carburetor becomes hot and the fuel may even boil. If the carburetor is vented to the atmosphere, these vapors escape to the environment.

There are two methods being used to control these evaporative losses. One utilizes the crankcase as a storage volume for the vapors. Then the vapor is pumped out of the storage into a condenser which returns liquid fuel to the gasoline tank. The second method utilizes a filter, namely activated carbon, that traps the vapor, and holds it until it can be fed back into the carburetion system.

Attack on reduction of the remaining emissions must take place at the engine's exhaust. Exactly how this is to be accomplished is not finalized. Two mechanisms, thermal reactors and catalytic converters, are under development. The basic idea in both is to convert the pollutants via chemical reactions to harmless carbon dioxide (CO_2), water vapor (H_2O), and molecular nitrogen (N_2). Thermal reactors will accommodate only hydrocarbons and carbon monoxide. The catalytic principle will work for all, but a specific device is required for the nitrogen oxides.

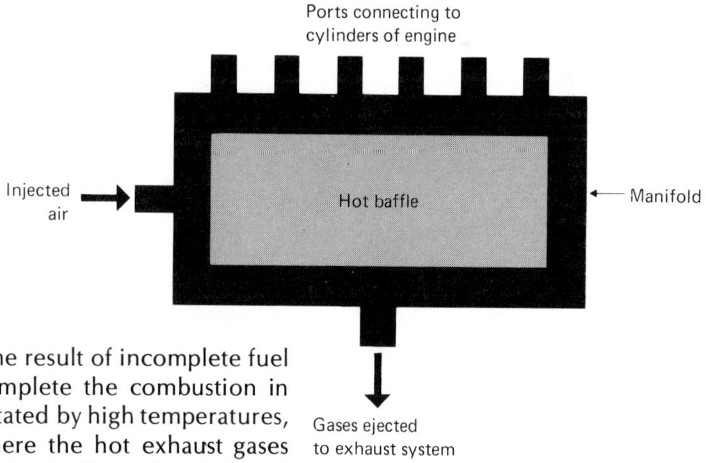

Fig. 12 Schematic diagram of a thermal converter. The thermal converter consists basically of a heated baffle placed inside the manifold of the engine. The manifold is a multiply connected chamber that channels the exhaust gases from each cylinder into the exhaust system. The exhaust gases in the presence of air injected into the manifold are oxidized in the high temperature zone of the reactor created by the hot baffle.

Since carbon monoxide and hydrocarbons are the result of incomplete fuel combustion, the idea in a thermal reactor is to complete the combustion in another area of the engine. Since combustion is facilitated by high temperatures, the hottest area, namely the exhaust manifold where the hot exhaust gases escape, is chosen. The thermal reactor is primarily a hot baffle and is designed to fit into the manifold (Fig. 12). Oxygen required for the combustion is pumped into the reactor from the atmosphere. The effectiveness of the system can be

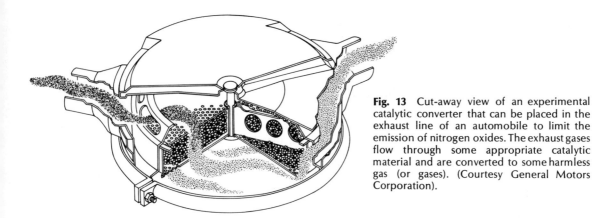

Fig. 13 Cut-away view of an experimental catalytic converter that can be placed in the exhaust line of an automobile to limit the emission of nitrogen oxides. The exhaust gases flow through some appropriate catalytic material and are converted to some harmless gas (or gases). (Courtesy General Motors Corporation).

improved dramatically by recirculating the exhaust gases. Coupled with a vapor-air mixture rich with fuel, this also inhibits NO_x formation.

A catalyst in chemistry is a substance that modifies the rate of a chemical reaction but is not consumed in the process. The modification may be either a decrease or increase in the rate. The idea in a catalytic converter is to promote the rate of conversion of the pollutants to harmless substances through the use of an appropriate catalyst (Fig. 13). What catalyst may be used and how it functions is proprietary because of the obvious possible financial gains accruing to the developer of a successful system. Metals such as platinum and chromium are often suggested. For purposes of illustration, let us consider the decomposition of NO into harmless N_2 and O_2 using a platinum (Pt) catalyst. The chemistry is only symbolic. At room temperature (300K) the reaction $NO \xrightarrow{slow} \frac{1}{2} N_2 + \frac{1}{2} O_2$ proceeds slowly. Suppose, though, that the NO molecule was first bound up by a Pt atom which then proceeded to a transition state which, in turn, readily decomposed. Symbolically,

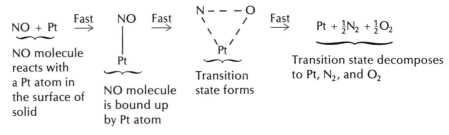

Note that the platinum atom was not consumed in the process.

Prototype emission control systems which allow the 1975-76 standards to be met have been built. But a prototype and its mass-produced counterpart are far different. And to enable such a system to function for the 50,000–100,000 mile lifetime of the automobile will require a higher degree of maintenance than the majority of people have been willing to assume responsibility for in the past. Not only will the maintenance be costly ($80–$600 on a new car) but costly inspection systems will be required. One of the most serious drawbacks to the proposed system is that conventional leaded gasoline poisons the catalytic converters and renders them ineffective. If the lead is taken out of the fuel, the octane rating drops as does the engine efficiency. Higher octane, unleaded gasolines can be made but they are more expensive (2–4 cents/gallon). In spite of the cost, it appears that unleaded gasolines are forthcoming and that automobile manufacturers will design engines to accommodate them efficiently.

SOME POSSIBLE ALTERNATIVES TO THE CONVENTIONAL IC ENGINE

Required Attributes of a Motive Power Source

No one questions the convenience and reliability of the modern automobile with its IC engine. Yet, if this engine is, in the meantime, rendering the air environment unsuitable for living, we can't help but wonder why it is so difficult to replace this engine with a nonpolluting counterpart. One of the main reasons is economics. An engine with low emissions is useless if no one can afford to buy it. There are, however, some intrinsic characteristics that we demand of an automobile and the IC engine is one of the very few that can do the job. It is easy to assess these features using Newton's laws of motion to provide the framework.

There are many forces that act on a moving car. Those which affect its forward motion fall into three types. The first is the force that actually causes the car to move forward. This is analogous to the force that causes you to walk on a floor. When you push on the floor with your shoe, the floor exerts an equal but oppositely directed force on your shoe which causes you to move. The force of the floor on the shoe is simply the reaction (Newton's third law) to the force of the shoe on the floor. An automobile engine in conjunction with a gear coupling mechanism to the wheels causes the wheels to exert a force on the road. The road reacts with an equal but oppositely directed force which moves the car. The second force is the drag force. It is similar to the drag force on a particular object in the atmosphere. It arises from a variety of rolling, sliding, and air resistance frictional forces. Experimentally, it is found that this force is often proportional to the mass and velocity of the car. The third force is the gravitational force. It is this force which causes the car to roll downhill with the engine turned off. For simplicity, we will neglect this force in our discussion and assume that the car is on flat ground moving in a straight line. Having made these assumptions, we find that the forces affecting the forward motion of the car can be depicted in Fig. 14. The net force on the car is

$$F_{car} - F_{drag}$$

This can be equated to mass times acceleration using Newton's second law

$$F_{car} - F_{drag} = ma$$

Cartoon by L.D. Warren, *Cincinnati Enquirer.* Reproduced by permission.

Fig. 14 The motion of a car on level ground is affected by the force that the road exerts on the car F_{car} and the drag force (F_{drag}) which retards its forward motion.

Assuming that the drag force is proportional to mass and speed, we observe that the force on the car becomes

$$F_{car} = \underset{\substack{\text{constant} \\ \text{of} \\ \text{proportionality}}}{k} \quad m \quad \underset{\text{speed}}{v} \quad + \quad ma$$

This expression says that the car must provide a force which both accelerates the car and overcomes dissipative forces. The work done by any force is, by definition, the product of force times distance.

$$W = F \cdot d$$

The power or rate of doing work is

$$P = \frac{W}{t} = F\,\frac{d}{t}$$

But d/t is the speed so that

$$P = F \cdot v$$

Hence the rate of work, or power, done by the car is

$$P_{car} = F_{car} \cdot v$$
$$= kmv^2 + mav$$

This result shows that the engine must provide power for acceleration and maintaining the speed. Once the car achieves constant velocity, its acceleration is zero. Hence the power developed by the car is

$$P_{car} = kmv^2$$

This means that the power depends strongly on the speed of the car. For example, it takes four times more power to maintain a speed of 60 miles per hour than to maintain one of 30 miles per hour. The power also depends on the mass of the car. More massive cars require more power to maintain a given constant speed than less massive cars. It is convenient to talk about a specific power, which is the power divided by mass (or the power divided by weight, if you prefer). Specific power is then entirely analogous to specific heat, which is an energy divided by mass times a temperature change. We express the specific power as

$$\frac{P}{m} = kv^2$$

This relation is the first of two which determine the required attributes of a car. It says that once you pick the speed for the car (that is, v), then there is a certain minimum specific power required. The higher the speed demanded, the greater the specific power required.

The second requirement relates to the amount of energy needed, which can be determined by calculating the work done by the car. We can obtain the work done by the car from $W = Pt$ and $t = d/v$. Thus

$$W_{car} = P_{car} \cdot t = P_{car}\left(\frac{d}{v}\right)$$

$$= kmv^2\left(\frac{d}{v}\right)$$

$$= kmvd$$

The work done by the car is derived from energy from the gasoline. The expression says what we might have anticipated from experience: that the fuel required depends directly on the length of the trip (that is, on d), and that the greater the speed, the more fuel is required. Again, it is convenient to talk about an energy per unit mass (or an energy per unit weight) which would be

$$\frac{W_{car}}{m} = kvd$$

The important considerations for a car are: (1) specific power and (2) specific energy. The equations developed here for these two considerations apply when the car is on level ground moving at constant velocity. Both depend on the speed and design of the car and the length of the trip. In the U.S., cars have been developed which travel great distances at high speeds and which therefore require high specific power and high specific energy. The conventional internal combustion engine turns out to be one of the very few power sources that will satisfy both the specific power and specific energy requirements for the type of driving we have demanded. This is shown in Fig. 15. The sources which are competitive with the IC engine in terms of having both high specific power and specific energy are the gas turbine and external combustion engine (piston steam engine or steam turbine, for example). These particular sources are potentially capable of producing both the long-distance and the high-speed driving that the IC engine presently accomplishes. Fuel cells, on the other hand have high specific energy and low specific power. Hence they could be used for long trips but the speed would be limited. This is also true for many batteries. Even though a source is capable

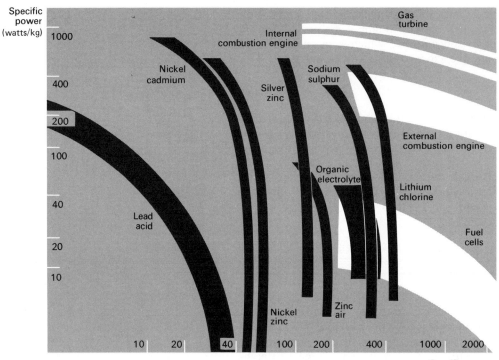

Fig. 15 The relation between the specific power and specific energy for the power sources that might be suitable for an automobile. The areas which are shaded are battery sources. The area occupied by each source on the figure represents the variation in attributes of different systems of the same type. (Figure adapted from "Steam Cars," S. W. Gouse, *Science Journal*, January 1970.)

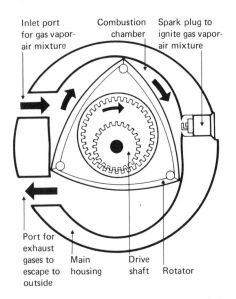

Fig. 16 The essential components of the Wankel engine.

of a certain function from the standpoint of power and energy, economics and technical difficulties often render it unacceptable. Fuel cells, for example, are very expensive at the present time and there are technical difficulties with the gas turbine.

The Wankel Engine

The Wankel engine* (Fig. 16) is one possibility for replacing the conventional IC engine. The Wankel engine does not constitute a revolutionary new principle for it is also a four-cycle, gasoline fuel, internal combustion device operating in a thermodynamic cycle much like the conventional IC engine. Its type, but not necessarily quantity, of emissions, then, would be no different. Because it is a lower compression engine than the conventional IC engine, it produces more hydrocarbons and carbon monoxide but less nitrogen oxides. Technologically it is easier to get rid of the hydrocarbons and carbon monoxide than nitrogen oxides which makes this low compression feature an asset. Furthermore it has the extremely attractive feature of being able to get an equivalent amount of

—————————

*After its inventor, Felix Wankel.

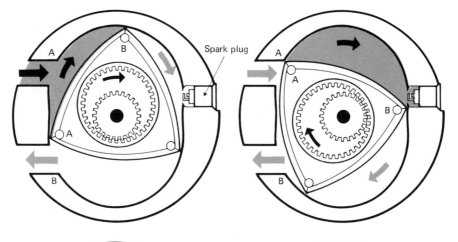

1. Intake

Gasoline vapor-air mixture is injected between the chamber walls and the triangular rotor.

Spark plug

2. Compression

When points A of the rotor and chamber align, the intake port is closed off and the gas trapped in the chamber begins to be compressed.

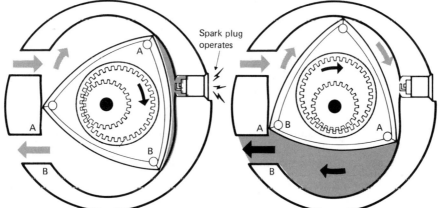

3. Power

At maximum compression, the sparkplug operates, initiating an explosion which forces the rotor on in its clockwise rotation. The gas expands and its internal energy decreases.

Spark plug operates

4. Exhaust

When points B of the rotor and chamber align, the exhaust port is open and the spent gas is ejected from the engine.

horsepower in a much smaller volume with much less weight. For example, a standard American 195-horsepower V–8 engine weighs about 600 pounds and occupies 15 cubic feet; whereas a 185-horsepower Wankel engine weighs about 225 pounds and occupies 5 cubic feet. It is these features which provide space for thermal reactors and catalytic converters; a space not available in contemporary automobiles. Furthermore, because of its simplicity it can conceivably be manufactured for half the cost of a conventional engine and the savings used for pollution abatement.

The engine sequence involves four operations: (1) intake, (2) compression, (3) power, (4) exhaust. But each operation is produced by a revolving three-sided rotor rather than by a piston as in a conventional engine. This cycle is shown in Fig. 17. The engine is obviously very simple. It has only two major moving parts—the rotor and driveshaft. In all, there are about 600 total parts, 150 of which move; compare this to about 1000 total parts, 390 of which move, for an equivalent

Fig. 17 The operating cycle of the Wankel engine.

V–8 engine. Unlike the conventional IC engine, the Wankel engine has no valves and it operates on low-octane fuel. All these features add up to low maintenance costs. The major problem has been unreliability of the seal between the rotor and combustion chamber walls. The initial versions have also been inefficient, but this is expected to improve.

The Wankel engine is definitely not a laboratory toy. It is used commercially by the Japanese and Germans. American industry is testing it in prototype vehicles. It may well be the IC engine of the future.

The Gas Turbine

Cleaning the IC engine, be it reciprocating or rotary, to meet the federal standards is a difficult, if not formidable, task. Yet with the tremendous investment the automotive industry has in this engine, this is the natural approach and, considering their resources, the probability of success is high. There must be, then, incentives for change other than low-emission characteristics, incentives such as lower cost, greater reliability, higher performance, and longer maintenance intervals. These are the features which induced the aircraft industry to change from piston to jet engines and the railroad industry to turn from steam to diesel engines. In these respects, the gas turbine is a most interesting prospect. The gas turbine has specific power and energy characteristics superior to the IC engine (Fig. 15) and its reliability is proven in jet aircraft engines where it is not uncommon to have a trouble-free life of 15,000–20,000 hours between overhauls.

The turbine principle (Fig. 18) is quite simple. Air is taken in from the environment, compressed to several times atmospheric pressure, and fed to a combustion chamber where it is mixed with a fuel such as kerosene or gasoline and ignited. The hot (about 1000°C) expanding gas impinges on the turbine blades producing rotational motion. The turbine is efficient, but develops significant power only

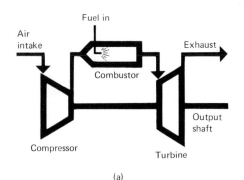

Fig. 18 (a) Schematic illustration of the gas turbine. (b) A simplified cross section of a commercial gas turbine.

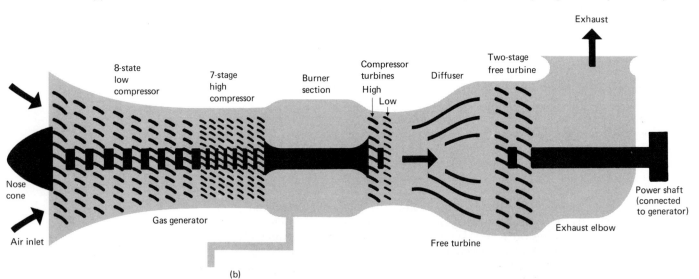

The characteristics of a gas turbine engine are best suited for large trucks and buses. The truck and bus shown here are prototypes powered by a gas turbine engine. (Photographs courtesy of the General Motors Corporation.)

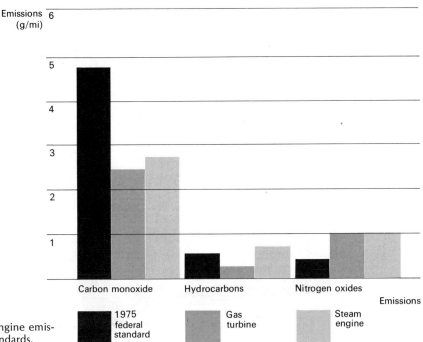

Fig. 19 Comparison of various engine emissions with the 1975–76 federal standards.

at high speeds. This is a serious drawback for it means that complex, expensive variable-speed mechanisms are required between the turbine and drive wheels to accommodate the "stop and go" type of travel demanded by automobile operators. The turbine also has poor gas mileage at low speeds. The reason that turbines are readily adapted to aircraft (and racing cars) is that they run, essentially, at constant speed. Buses and heavy trucks are much less "stop and go" vehicles which makes them the most likely initial candidates for turbine power.

The turbine's emission characteristics for carbon monoxide and hydrocarbons are excellent (Fig. 19). This is a reflection of the good combustion properties. There is some inconsistency in the reported nitrogen oxide emissions. Those shown in Fig. 19 do not meet in 1975 federal standards, but they are no worse than an uncontrolled IC engine. Other sources report the NO_x emissions as both lower and higher than reported in Fig. 19. Since combustion takes place at high temperatures, nitrogen oxides should be formed in abundance. Hence some control such as that required for the IC engine will be needed.

The contemporary automobile owes its favorable price to some 65 years of development. Initial turbine models will no doubt be expensive partly because high-temperature materials are required. Costs will probably be twice that of a standard automobile. The turbine, itself, is extremely simple and should be relatively inexpensive when mass produced. It is the control equipment which is expensive and requires research and development in order for the system to be competitive. There are reputable people who believe that the turbine will be the standard automobile power plant in the 1980's and that the IC engine will be the oddity that the piston engine is for aircraft.

The Steam Engine

From the standpoint of specific power and specific energy, external combustion engines* are an obvious alternative to the IC engine (Fig. 15). The steam engine or steam turbine discussed in Chapter 3 is the most common example and is the one external combustion engine most often considered. There is nothing new about this concept. Commercial steam powered vehicles were produced early in the automobile industry and this lasted until as late as 1930. Although they were not widely accepted, they have some desirable characteristics such as simplicity, reliability, excellent speed-torque relationship, and low emissions. It is, however, only the low emissions property which has spurred renewed interest in the 1970s (Fig. 19). Since combustion takes place at essentially atmospheric pressure and low temperature, it is fairly complete which accounts for the low emissions. There are several disadvantages which make it unattractive as an outright alternative to the IC engine. These include:

1. bulky boilers and pipes which require space not readily available in an automobile. (This problem has been alleviated somewhat with low-volume "flash boilers.")
2. some working fluids, such as water, which freeze at some required working temperatures
3. the danger associated with the high boiler pressure and temperature
4. start-up times of the order of minutes except with "flash boiler" systems which start up in about 20 seconds.
5. inadequate repair and service facilities (this would also apply to all other alternatives to the conventional IC engine)
6. the expense of the engine

However, special purpose applications, such as in buses, trucks and boats, may develop.

The Electric Vehicle

All vehicles discussed thus far utilize a fossil fuel as the primary energy source. The electric vehicle utilizes electric energy which is stored in a battery or generated on board by a fuel cell. An electric motor converts the electric energy to mechanical energy. Like the steam engine, the principle is not new. Special purpose electric vehicles have been around as long as the conventional auto-

---

*These engines are external in the sense that combustion takes place outside the working element (for example, a piston) of the engine. Energy derived from the combustion is transferred to the working fluid of the engine.

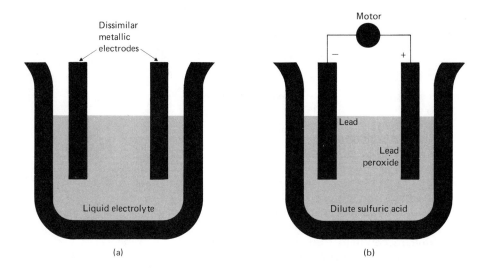

Fig. 20 (a) Essential elements of a storage battery. (b) The common lead-acid battery used in the conventional automobile.

mobile. However, it is the development of more energetic batteries and the fact that they have low, nonpolluting emissions which have attracted new interest.

An elementary storage battery consists of two dissimilar metallic plates (electrodes) immersed in a liquid solution (electrolyte) as shown in Fig. 20. Chemical energy is stored by connecting the electrodes to an electric source such as a generator. This process is referred to as charging. When the battery is charged, an electric potential develops across the electrodes. If the electrodes are then connected to a device such as a motor, a current flows as a result of transformation of chemical energy to electric energy. The amount of storable energy depends on the types and sizes of the electrodes and the electrolyte and the chemistry of their interaction.

The common battery used in the automobile employs lead (Pb) and lead peroxide (PbO_2) for the electrodes and weak sulfuric acid (H_2SO_4) for the electrolyte. The hydrogen atoms in some of the acid molecules will separate leaving SO_4. In the process, the SO_4 captures the electron from each hydrogen atom and becomes charged negatively. The dissociated charged components of the acid molecule are called ions and are depicted as H^+ and SO_4^{2-}.

$$H_2SO_4 \rightarrow 2H^+ + SO_4^{2-}$$

When a device is connected across the electrodes, the lead on the negative electrode reacts with SO_4^{2-} producing lead sulfate and releasing the two electrons that were captured by SO_4.

$$Pb + SO_4^{2-} \rightarrow \underset{\substack{\text{lead} \\ \text{sulfate}}}{PbSO_4} + \underset{\substack{\text{2 released} \\ \text{electrons}}}{2e^-}$$

At the same time, the lead peroxide on the positive electrode reacts with the H^+ and SO_4^{2-} ions and the two electrons to form lead sulfate again and water

$$PbO_2 + 4H^+ + SO_4^{2-} + 2e^- \rightarrow PbSO_4 + 2H_2O$$
$$\text{water}$$

The net result is that two electrons have been released from the negative electrode and transferred to the positive electrode; this gives rise to a current in the external device. Each transfer of two electrons involves the loss of two acid molecules and the creation of two water molecules. As the acid molecules are depleted, the battery gets weaker and is said to "go dead." Energy is restored to the battery by connecting the electrodes to an external source of electric energy and thus reversing the chemical reaction.

There are many other electrode and electrolyte variations. The basic principle, though, is the same as for the lead-acid battery even though the chemical reactions may be more complex.

No motive power source is more attractive than the battery–electric motor combination from the standpoint of emissions. Its emissions are nearly zero. However, it is the specific power and specific energy requirements, as revealed in Fig. 15, which limit their utility. For example, conventional lead-acid batteries have specific energies of about 10 watt-hours per pound. Gasoline, as utilized in the IC engine, has 1000 watt-hours per pound or about 100 times as much. This means that, for the same weight, batteries have 100 times less energy. So if a car could go 300 miles at say 60 miles per hour on a tank of gasoline it could only go 3 miles at the same speed using batteries of the same weight as the gasoline. Certainly, additional energy could be obtained by utilizing more batteries, but this also becomes more difficult. For example, a tank of gasoline is about 15 gallons which weighs about 100 pounds (Exercise 10). Thus the amount of batteries required to produce the same energy would be 10,000 pounds. Of course if the speed requirements were eased, the energy requirements would also decrease. However, the electric vehicle with proven batteries is relegated to the class of special purpose vehicles useful for short trips (Fig. 21).

Experimental batteries with specific energies of 100 watt-hours per pound are being built. Like any new product they are expensive, but if their technical worth proves out, the price will surely decrease. These batteries could increase the range to 150 miles in urban traffic on a single charge which would handle all but major interstate trips. There is reason to believe that cells with 300 watt-hours per pound are feasible but the timetable for development is around the year 2000.

The type of battery used in a flashlight also converts chemical energy to electric energy. These batteries (or cells) are not readily recharged because the chemicals are depleted in the discharge. A fuel cell operates much like these batteries except that fuel for the chemical reactions is supplied continuously. The most common fuel is hydrogen and oxygen. There are a variety of electrolytes. Sulfuric acid will work and has been used. It suffices to demonstrate the principle.

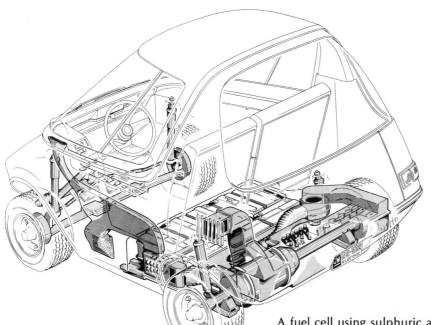

Fig. 21 An experimental, two-passenger, special-purpose vehicle intended for limited urban transportation. The vehicle is powered with an 84-volt power battery pack made up of special lightweight lead-acid batteries. It has a range of 58 miles at 25 miles per hour, and can accelerate from zero to 30 miles per hour in 12 seconds. (Courtesy of the General Motors Corporation.)

A fuel cell using sulphuric acid for the electrolyte is shown in Fig. 22. As in the lead-acid battery, some of the sulphuric acid molecules separate into $2H^+$ and SO_4^{2-}. Hydrogen fuel then combines with SO_4^{2-} to form H_2SO_4; electrons are released in the process.

$$2H_2 + 2SO_4^{2-} \rightarrow 2H_2SO_4 + 4e^-$$

Oxygen combines with the electrons and hydrogen ions to form water

$$2H_2^{2+} + O_2 + 4e^- \rightarrow 2H_2O$$

The net result is the transfer of four electrons between the electrodes.

As seen in Fig. 15, the specific energy of fuel cells is quite good which means that energy is available for trips comparable to the conventional automobile. However, the specific power is low which greatly limits the speed. Their high cost is also a limiting feature.

The resolution of any confrontation usually involves compromises (or trade-offs). Such may be the case with the automobile-environment confrontation. The electric vehicle alone will not replace the automobile in time to meet the 1975–76 federal standards, and it may not be possible to clean up the IC engine in the same time. So it may be possible to make a hybrid electric motor–IC engine combination which would combine the low emission characteristic of the electric vehicle with the desirable characteristics of the standard automobile. Two schemes have been proposed. One would use an IC engine to run a generator which would recharge the batteries of the vehicle. The other would use a low-power IC engine as the main source and an electric motor to supply power at the peak periods, such as starting and passing. This type of system could reduce emissions by 50% compared

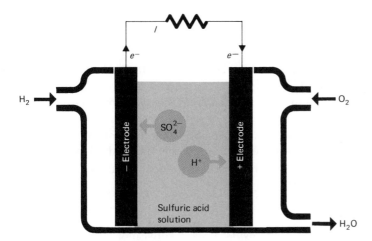

Fig. 22 A fuel cell using sulfuric acid for the electrolyte and hydrogen and oxygen for fuel.

to a standard automobile. The hybrid system would be a stop-gap measure, but it could buy time until an all electric system is feasible. A turbine-electric hybrid system would also be a viable alternative.

Even if electric vehicles were technically feasible, there are other environmental objections. The batteries would have to be recharged using electric energy from a conventional power plant. So although there were no emissions from the vehicle, there would be undesirable particulate and sulfur oxide emissions at the power plant. One can argue, though, that these emissions from a stationary source are potentially much easier to control than those from a multitude of automobiles. A more serious objection is: Why increase the demands for electric energy when shortages already exist? It is hard to argue against this, but the situation could well change if fusion energy became a practical source (Exercise 9).

REFERENCES

General

The Search for a Low-emission Vehicle, Staff Report prepared for the Committee on Commerce, United States Senate, 91st Congress, 1st Session.

"Cars and Air Pollution," Laura Fermi, *Bulletin of the Atomic Scientists* **25**, No. 8, 35 (October 1969).

"How Clean a Car?," John B. Heywood, *M.I.T. Technology Review*, p. 21 (June 1971).

"What future for the auto?," *Science News* **99**, 329 (May 15, 1971).

"Air Pollution and Public Health," Walsh McDermott, *Scientific American* **210**, No. 1, 25 (January 1964).

Photochemical smog

"Atmospheric Photochemistry," Richard D. Cadle and Eric R. Allen, *Science* **167**, No. 3916, 243 (16 February 1970).

"The Control of Air Pollution," A. J. Haagen-Smit, *Scientific American* **210**, No. 1, 25 (January 1964).

Air Quality Criteria for Photochemical Oxidants, National Air Pollution Control Administration Publication (hereafter referred to as NAPCA) AP-63, March 1970.

"Cleaning Our Environment: The Chemical Basis for Action," *American Chemical Society*, Washington, D.C., 1969.

Effects of Air Pollution

Air Quality Criteria for Carbon Monoxide, NAPCA Publication AP-62.

Air Quality Criteria for Photochemical Oxidants, NAPCA Publication AP-63, March 1970.

Air Quality Criteria for Hydrocarbons, NAPCA Publication AP-64, March 1970.

Air Quality Criteria for Nitrogen Oxides, NAPCA Publication AP-84, January 1971.

Emission Controls

Control Techniques for Carbon Monoxide, Nitrogen Oxide, and Hydrocarbon Emissions from Mobile Sources, NAPCA Publication AP-66, March 1970.

A Total Exhaust Emission Control System, E. N. Cantwell, Air Pollution Control Association, September 20–22, 1970, Montebello, Quebec.

The Wankel Engine

"Rotary Engines," Wallace Chinitz, *Scientific American* **220**, 90 (February 1969).

"The Little Engine That Could be an Answer to Pollution," George Alexander, *New York Times Magazine*, October 3, 1971.

"The Wankel Engine," David E. Cole, *Scientific American* **227**, 14 (August 1972).

The Gas Turbine

"Gas Turbines for Land Transport," *Science Journal*, April 1970, p. 54.

The Steam Engine

"Steam Cars," S. W. Gouse, *Science Journal*, January 1970, p. 50.

"Stanley's Dream Makes a Comeback," Edward Gross, *Science News* **96**, 247 (September 20, 1969).

"Minto's Unique Steamless 'Steam Car'," E. F. Lindsley, *Popular Science* **197**, No. 4, 51 (October 1970).

The Electric Vehicle

"Electric Vehicles," Mary Lee, *Science Journal*, March 1967, p. 35.

Federal Low-Emission Vehicle Procurement Act, Serial No. 91–51, U.S. Government Printing Office, Washington, D.C.

"Electric Cars: The Battery Problem," Victor Wouk, *Bulletin of the Atomic Scientists* **27**, No. 4, 19 (April 1971).

Interesting short articles from Science News

"Practical Problems Remain" **100**, 8 (1971).

"Clean Air: An R & D Gap" **99**, 177 (1971).

"Restricting Autos" **98**, 163 (1970).

"Industry Focuses on Unleaded Gases" **98**, 71 (1970).

"Getting the Lead Out" **97**, 167 (1970).

"Worse, not Better" **99**, 80 (1971).

Movies

"Air Pollution and Plant Life, TF–102," 16 mm Sound, Color, 20 minutes. Free loan from National Audiovisual Center, N.A.R.S., GSA, Washington, D.C. 20409.

The following films are available on free loan from National Medical Audio-Visual Center, Station K, Atlanta, Georgia.

"Sources of Air Pollution," 16 mm Sound, 5 minutes; Order No. MIS–677.

"Effects of Air Pollution," 16 mm Sound, 5 minutes; Order No. MIS–678.

"Control of Air Pollution," 16 mm Sound, 5 minutes; Order No. MIS–676

"Breathe at Your Own Risk," 16 mm Sound, black and white, 58 minutes; Order No. MIS–679.

"Pall Over Our Cities," 16 mm Sound, black and white, 15 minutes; Order No. MIS–985.

"Combustion in Action," Sound, Color, 19 minutes—Available from Film Library, General Motors Building, Detroit, Michigan 48202.

QUESTIONS

1. Discuss the automobile as a social force in the United States.
2. Some toy cars are driven by comparatively large, massive wheels that can be set into motion by first rapidly pushing the car along some surface. These cars

travel a surprisingly long distance. What energy transformations are involved and why do the cars go surprisingly far? Do you think that this principle has any merit for full-size cars? (See "A Lift for the Auto," *Environment* **13**, No. 10, 35 (December 1971).)

3. The idea of a relative-biological-effectiveness was introduced for nuclear radiations. Discuss how a similar scheme could be used for air pollutants. Is this actually taken into account in the federal air pollutant standards?

4. Any burning of hydrocarbons that uses air for the oxygen supply will produce the same pollutants attributed to the internal combustion engine. What are some burning processes in our environment where this is problematical?

5. The efficiency of the internal combustion engine rose a few percent between 1920 and 1970. However, the average number of miles a car goes on a gallon of gasoline went down during this period. Why?

6. More than half the gasoline consumed by automobiles was in trips of less than 3 miles with generally less than 2 passengers per trip. Discuss some effective ways of eliminating many of these trips.

7. To what extent do you think the automobile is an indicator of manhood to a male driver?

8. The motor vehicle standards for 1975–1976 were invoked in 1971. These standards were originally intended for 1980. One of the big hopes for meeting the 1980 standards was the possibility of a completely new power plant to replace the IC engine. What effect do you think this new schedule has on these plans for a new power plant?

EXERCISES

Easy to Reasonable

Exercise 1.   What is the octane rating of a gasoline composed of equal parts of isooctane and heptane?

Exercise 2.   Is the breakup of NO_2 into NO and O in the presence of sunlight an exoergic or endoergic reaction? Explain.

Exercise 3.   The complete oxidation of heptane (C_7H_{16}) produces only carbon dioxide (CO_2) and water (H_2O). Knowing this, complete the reaction below

$$C_7H_{16} + 11\,O_2 \rightarrow$$

Exercise 4.   The two reactions below can be used to convert SO_2 to SO_3. The end product of the first reaction is used as part of the input for the second reaction.

$$2NO + O_2 \rightarrow 2NO_2$$

$$2NO_2 + 2SO_2 \rightarrow 2NO + 2SO_3$$

How many molecules of NO and NO_2 are consumed in the execution of these two reactions? What, then, is the role of NO and NO_2 in the reactions?

Exercise 5. Using Fig. 2, show that motor vehicle registration in the United States grows at a rate of about 4% per year.

Exercise 6. When viewed with ordinary light, nitrogen dioxide gas has a characteristic reddish-brown color. What colors (or wavelengths) would you expect, then to contribute mainly to the breakdown of NO_2 into NO and O by light? Knowing that NO does not readily absorb sunlight, what color would you expect it to have when viewed by ordinary light?

Exercise 7. A diesel engine operates, in principle, like an IC engine. It utilizes a kerosene-air vapor mixture for fuel, but does not use a spark plug for ignition. Rather it relies on bringing the mixture to a temperature high enough that ignition occurs spontaneously. What physical principle is involved in achieving this ignition temperature?

Exercise 8. Why would a thermal reactor not be effective for removing nitrogen oxides?

Exercise 9. The transformation of energy from fossil fuels to mechanical energy in an automobile via electric generation and battery-powered motors involves several steps with characteristic efficiencies.

1. Power plant-efficiency: 40%

2. 10% energy loss through transformers and transmission to the city

3. 10% energy loss through transformers and distribution within the city

4. Battery-charging efficiency: 80%

5. Motor-efficiency: 90%

Show that the overall efficiency is 23%. How does this compare with the efficiency of a typical automobile?

Exercise 10. "A pint's a pound the world around" is a very old but useful mnemonic. It is, of course, not true for every substance, but it is nearly true for water and for a liquid like gasoline whose density is close to that of water.

a) Show that 15 gallons of water weigh about 120 pounds.

b) Give an argument in support of why you would expect 15 gallons of gasoline to weigh less than 15 gallons of water.

Exercise 11. When a lead-acid storage battery discharges, a sulphuric acid molecule (H_2SO_4) is replaced by a water molecule (H_2O).

a) What would you expect to happen to the density of the electrolyte as the battery discharges?

b) What use can be made of this result?

Reasonable to Difficult

Exercise 12. It is important to realize that automobile emissions depend on how a vehicle is driven. And this depends on where the vehicle is, whether it is in city or rural areas, for example. So an emission standard of so many grams per mile must imply some sort of a "typical" mile. Discuss what you would consider to be the "typical" driving trip which would be appropriate for evaluating the emissions of an automobile. What is the average speed for your "typical" trip? Number of stops? Time?

Exercise 13. Compare the wasted motions in a conventional IC engine with those in a Wankel engine.

Exercise 14. Carbon monoxide is emitted from an uncontrolled IC engine at a rate of about 125 grams per mile (see Table 5). Estimate how many miles an average car travels for each ten gallons of gasoline consumed and calculate the number of pounds of carbon monoxide emitted in the process. Show that your answer is consistent with the stated on p. 231. (For conversion of grams to pounds, use 1000g $=$ 2.2 lb).

Exercise 15. If $\gamma = 1.5$, the theoretical efficiency of an internal combustion engine is

$$\epsilon = \left(1 - \frac{1}{\sqrt{r}}\right) \times 100\%$$

where $\sqrt{r}$ is the square root of the compression ratio.

a) What is the theoretical efficiency if $r = 1$? Does this agree with your physical intuition knowing basically how an internal combustion engine operates?

b) Determine the theoretical efficiency for $r = 1, 4, 9,$ and 16 and make a sketch of efficiency versus compression ratio.

c) Compare the gain in efficiency for going from $r = 4$ to $r = 9$ with that going from $r = 9$ to $r = 16$.

d) In what ways is this situation with the internal combustion engine similar to that for the steam turbine discussion in Chapter 3?

Difficult to Impossible

Exercise 16. A possible way of rendering NO and NO_2 harmless is to convert them to molecular nitrogen (N_2) and water vapor (H_2O) through thermal reactions with ammonia (NH_3). The reactions would be of the form

$$a \cdot NO + b \cdot NH_3 \rightarrow c \cdot N_2 + d \cdot H_2O$$
$$e \cdot NO_2 + f \cdot NH_3 \rightarrow g \cdot N_2 + h \cdot H_2O$$

where $a, b, c, d, e, f, g,$ and h are whole numbers. Figure out the lowest set of whole numbers which would satisfy these equations.

Exercise 17. Potassium hydroxide (KOH) is sometimes used as an electrolyte in a modern fuel cell. The dissociation of KOH is in the form of K^+ and OH^- ions. Deduce reactions similar to those which occur from the use of sulphuric acid for the electrolyte to show how the fuel cell operates.

SOUND FROM MOTOR VEHICLES: AN EXERCISE IN NOISE POLLUTION

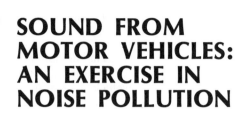

Photograph courtesy of the Environmental Protection agency and Black Star: Dan McCoy.

BACKGROUND INFORMATION

Thus far the pollution aspects of motor vehicles have centered on molecular particle emissions. Noise,* though not particulate in nature, is a by-product of motor vehicle operation and is increasingly regarded as a pollutant because of its magnitude and its potential effect on human health and well-being. Motor vehicles produce the greatest noise effect on the population (Table 1). In this connection it is especially important to consider motorcycles, trucks, buses, and sports cars even though their numbers are small compared to automobiles. Noises from these vehicles emanate primarily from the exhaust, engine, transmission, tires, and body vibrations. Noise pollution from automobiles would be intolerable had not manufacturers, almost from inception, installed noise mufflers to stifle the explosive-like sounds resulting from the combustion process.

The effects on human health of sulfur dioxide and carbon monoxide emissions have focused on physical damage. Yet these have profound psychological and social implications as well. Facing a smog-ridden city has to be a demoralizing experience, and provides an example of the kind of effects of which are the most far-reaching. That is not to say that sound cannot produce physical damage. There is documented evidence of loss or impairment of hearing from exposure to excessive noise levels.† And there are reports of noise effects on such things as blood pressure and heartbeat which could lead to a variety of physiological effects.‡ It is, however, the interference with work performance, interference with efficiency, interference with sound communication, and the outright annoyance produced by noise which lead to the psychological and mental health effects of prime concern. Legislative bodies and courts of justice are becoming more aware of these problems and action is being taken. For example, an Elizabeth, New Jersey Board of Education was awarded $164,119 for damages caused by noise coming from new highways.§ The Secretary of Labor in 1969 used, for the first time, existing power to set standards for industrial noise for firms holding government contracts in excess of $10,000. ‖ Industrial noise affects nearly 60% of the nation's work force.

Table 1. Sources of bothersome noises from a survey of Los Angeles, Boston, and New York in 1966. (From "Noise in Urban and Suburban Areas: Results of Field Studies," U.S. Department of Housing and Urban Development, Washington, D.C., 20410.)

Source of bother	Percentage of population affected
Traffic	15.4%
Children/neighbors	13.9%
Other	6.2%
Planes	5.4%
Animals	5.0%
Sirens/horns	3.9%
Industry	2.7%
Sonic boom	2.3%
Motorcycles	2.3%
Passersby	2.3%
Trains	0.8%
Not bothered	39.8%

---

*In physics, noise is any disturbance that obscures the clarity of a signal. Extraneous light entering an astronomer's telescope would be noise in this sense. Or interference of a TV signal by electromagnetic radiation from a thunderstorm, automobile, or another transmitter constitutes noise. And noises of these types are becoming categorized as pollutants. Usually, though, noise is associated with interference of sound waves. It is this type of noise that is of interest in this discussion.

†See for example, *Noise and Man*, by William Burns, J. B. Lippincott Co., Philadelphia, 1968.

‡"Air Pollution — 1970," Part 3, 91st Congress, Second Session on S. 3229, S. 3466, and S. 3546, March 24 and 25, 1970. U.S. Government Printing Office, Washington, D.C. 20402.

§*The New York Times*, October 3, 1971.

‖ "Guidelines to the Department of Labor's Occupational Noise Standards," U.S. Department of labor Bulletin 334 (1971).

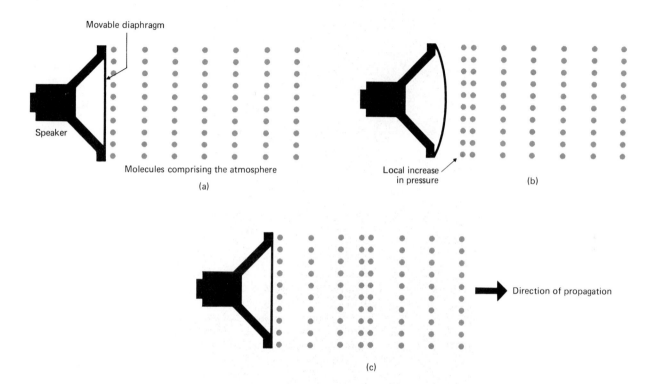

Movable diaphragm

Speaker

Molecules comprising the atmosphere

(a)

Local increase
in pressure

(b)

Direction of propagation

(c)

PHYSICAL PROPERTIES OF SOUND

Sound is a wave phenomenon that requires an elastic medium for propagation. The property of elasticity ensures that if any portion of the medium is displaced or compressed by the action of some force, the portion will return to its initial position when the force is removed. Air is an elastic medium. Solids, such as springs, are not elastic for all sizes of forces, but they are elastic for the sizes of forces involved in sound propagation. The disturbance associated with a sound wave is a change in the local pressure or density of the medium. The disturbance is propagated through collisions of atoms and molecules that comprise the medium. For example, the speaker in a radio or TV generates by the piston-like action of its diaphragm local compressions (pressure increases) and rarefactions (pressure decreases) which propagate through the air. To illustrate, consider such a speaker in an idealized "air" in which the molecules are uniformly distributed (Fig. 1a). Suppose that the speaker diaphragm is suddenly pushed out. This will cause the first layer of molecules to be compressed into the second (Fig. 1b). When the diaphragm moves back, the first layer returns and the second moves into the third

Fig. 1 (a) Schematic illustration of a speaker like that on a TV set operating in an idealized air medium. (b) When the diaphragm of the speaker moves out, it exerts a force on the first layer of molecules and moves them away from the diaphragm. This produces a local increase in the pressure of the air. (c) The energy acquired by the first layer of molecules is transmitted by collisions to other layers and the local increase in pressure propagates through the air.

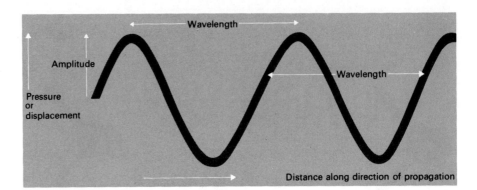

Fig. 2 A plot of the air pressure along the direction of propagation of a sound wave often takes the form shown here.

and so on. Note that the displacement of the molecules is in the direction of propagation of the disturbance. For this reason, sound is a type of *longitudinal* wave* and differs from a *transverse* water wave in which the displacement is perpendicular to the direction of propagation. If the diaphragm is oscillated back and forth, then an oscillatory pressure disturbance is produced much like the disturbance produced in a string which is oscillated up and down at one end. If the pressure variations along the direction of propagation or the displacement variations along a string are recorded, then identical types of wave patterns are produced (Fig. 2). There is a wavelength, λ, which is just the distance between two crests of the pressure wave. It is related to the speed (v) and frequency (f) of the wave by the ordinary relation.

$v = f\lambda$

---

*This type of wave can be easily demonstrated with a coiled spring such as a child's Slinky toy.

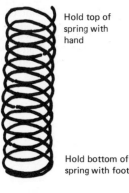

Hold top of spring with hand

Hold bottom of spring with foot

Compress a few turns and then release

Compression propagates down the spring

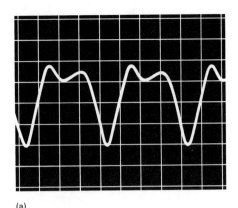

(a)

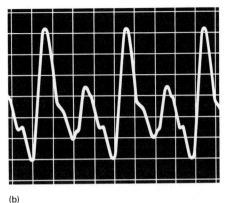

(b)

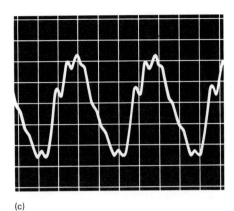

(c)

The speed depends on the physical properties of the medium and the temperature, but does not depend on frequency or wavelength. It is 332 meters per second at 0°C and increases as the temperature increases. Like any other type of wave, sound waves will reflect, refract, diffract, and interfere (see Appendix 2).

A note sounded with a musical instrument would produce a sound wave which would approximate this well-defined oscillatory type. But most sound waves would be formed from the superposition of waves of several frequencies and would be peculiarly shaped (Fig. 3). What would be considered noise would normally be even more complex. However, the speed, being independent of frequency, would be unaffected by the wave complexity.

It is important to realize the magnitude of the pressure disturbance involved in a sound wave. A good reference is atmospheric pressure. The pressure on the earth's surface due to the air atmosphere amounts to about 10^6 dynes/cm² (or about 15 lb/in²).* Pressure fluctuations near the absolute threshold for hearing are of the order of 10^{-4} dynes/cm². Ordinary conversational speech is about 10^{-1} dynes/cm² and the threshold of pain in the ear is about 200 dynes/cm². So we are talking about pressures which are extremely small when compared to atmospheric pressure. The actual atomic displacements are about 10^{-7} cm which is in the same realm as the diameter of an atom.

ENERGY AND POWER OF A SOUND WAVE

Energy and power are attributes of all wave phenomena. This is, perhaps, obvious in the case of sound since it is purely a mechanical effect. A sound wave is an energy transmitting mechanism. The rate at which energy is transmitted would simply be the amount of energy passing some point in the path of the wave per

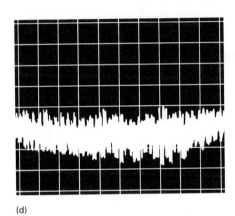

(d)

Fig. 3 An earphone like that used by hi-fi enthusiasts converts a sound pressure wave into an electrical signal which is proportional to the pressure in the wave. These signals can be displayed on an instrument called an oscilloscope. The shape of the displayed waveform is just like the actual sound wave. The displays are (a) a vowel sound sung by a human voice, (b) an alto saxophone, (c) a clarinet, and (d) random noise.

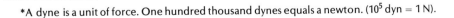

*A dyne is a unit of force. One hundred thousand dynes equals a newton. (10^5 dyn = 1 N).

Fig. 4 Measuring the rate at which energy passes some position in the path of a wave is much like measuring the rate at which energy passes by some point along the path of moving vehicles on a highway.

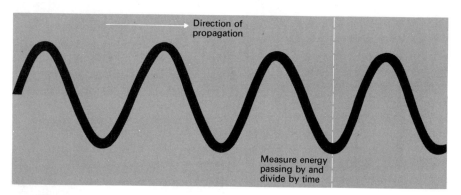

unit time (Fig. 4). This rate would just describe the power associated with the wave. The power alone, though, does not tell how the energy is spread out in space; that is, how it is concentrated. For this reason, it is appropriate to divide the power by the area over which the power is distributed to obtain what we call the intensity.

$$I = \frac{p\,(W)}{A\,(cm^2)}$$

The human ear is incredibly sensitive, not only to the lowest intensity to which it can respond but to the complete range of intensities. The human ear is capable of sensing intensities as low as 10^{-16} W/cm² and as high as 10^{-2} W/cm². The sensation of loudness is related to the intensity of the sound. This sensation is nearly proportional to the logarithm of the intensity. Because of this and the fact that the ear records such an enormous range of intensities, it is convenient to construct an intensity scale in terms of the logarithm of the intensity. The common unit is the decibel, abbreviated dB.

$$dB = 10 \log \frac{I}{I_0}$$

I_0 is a reference intensity which is taken as 10^{-16} W/cm². On this scale, conversational speech and the threshold of pain would be

$$dB = 10 \log \frac{10^{-10}}{10^{-16}} = 60 \text{ (conversational speech)}$$

$$dB = 10 \log \frac{10^{-2}}{10^{-16}} = 140 \text{ (threshold of pain)}$$

Photograph by Bob van Lindt/Editorial Photocolor Archives.

Table 2 Some common sound power levels found in a variety of environments

Source	Distance from Source	dB	Environment
		140	Threshold of pain
Hydraulic press	(3')	130	
Large pneumatic riveter	(4')		Boiler shop (maximum level)
Pneumatic chipper	(5')		
		120	
Overhead jet aircraft-4 engine	(500')		Jet engine test control room
Unmuffled motorcycle		110	Construction noise (compressors and hammers) (10')
Chipping hammer	(3')		Woodworking shop
Annealing furnace	(4')	100	Loud power mower
Average rock and roll band (often much higher)			Inside subway car
Subway train	(20')	90	Food blender
Heavy trucks	(20')		
Train whistles	(500')		Inside sedan in city traffic
Small trucks accelerating	(30')	80	Heavy traffic (25' to 50')
			Office with tabulating machines
Light trucks in city	(20')		
		70	
Autos	(20')		Average traffic (100')
Dishwashers			Accounting office
		60	
		50	Private business office
Conversational Speech	(3')		Light traffic (100')
			Average residence
		40	
		30	
			Broadcasting studio (music)
		20	
		10	
		0	

Table 2 lists some common intensity levels found in a variety of environments.

It is important to realize that small changes in intensity levels measured in decibels correspond to large changes in intensity measured in W/cm^2. For example, an increase of 10 dB at any level corresponds to a factor of 10 increase in intensity. An increase of 30 dB corresponds to a factor of 1000 increase in intensity.

One often sees intensities expressed in terms of the sound pressure level (SPL) which is defined as

$$\text{SPL (dB)} = 20 \log \frac{P}{P_0}$$

where the reference pressure, P_0, is 0.0002 dyne/cm². This appears to be a different unit than the one already discussed but it is not. It follows from the fact that sound energy is proportional to the square of the pressure amplitude. The logarithm of the square of a number is 2 times the logarithm of the number

$$\log P^2 = 2 \log P.$$

Hence the 2 that comes from the logarithm multiplies the 10 in the definition of dB to give the 20 in the definition of SPL.

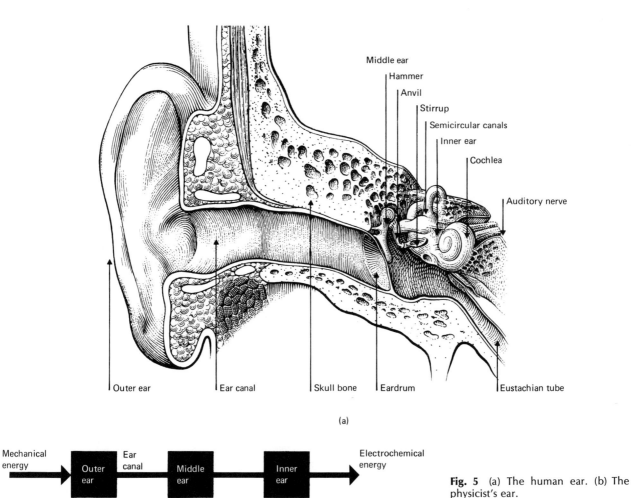

(a)

(b)

Fig. 5 (a) The human ear. (b) The physicist's ear.

THE HUMAN EAR AS A SOUND RECEIVER

We have stressed the fact that sound is a mechanical phenomenon and that it is produced by mechanical devices—engines, tires, speakers, etc. Receivers that extract energy from the sound wave would also be mechanical devices. Acoustical tile used to absorb sound converts the mechanical energy of the sound waves into thermal energy which is also mechanical in nature. The entire human body is a receptor of sound energy and because of this it is susceptible to physiological damage. It is, however, the human ear which is the receptor of prime interest. The human ear is a physiological device which has many beautiful design features. But basically it is an energy converter which converts the mechanical energy of sound into electro-chemical energy which is transmitted to the brain (Fig. 5). Sound waves enter the outer ear, travel down the ear canal, and strike a thin membrane called the eardrum. The eardrum forms a boundary between the outer ear and an *airspace* called the middle ear. Three small bones, the hammer, anvil, and stirrup, transmit vibrations from the eardrum to the oval window. The oval window is a membrane that transmits vibrations to the inner ear. Sound vibrations are received by nerve terminals in the cochlea in the inner ear and are transmitted electro-chemically to the brain. The actual energy conversion takes place in the inner ear which operates in a fluid. It would seem that the middle ear is superfluous and that if the ear were properly "designed," the outer and inner ear would be connected directly. However, were the middle ear not present, very little of the sound energy would actually get into the inner ear. The reason is associated with an extremely important concept called impedance matching.

A cheerleader uses a megaphone or cups her hands to get sound energy from the throat into the air where it travels to and is perceived by waiting ears. Both the megaphone and middle ear function as impedance matchers to facilitate this energy transfer. Impedance derives from impede, meaning to block or hinder. Obstructions in a pipe impede the flow of water and defects in a wire impede the flow of electric current. Every channel through which energy is transmitted impedes the transmission. The concept is quantified by defining impedance as

$$\text{impedance} = \frac{\text{effort (force or pressure)}}{\text{flow}}$$

For example, the effort in electricity is provided by a voltage and the flow is a current. The ratio of voltage and current is an electrical impedance. In sound, the effort would be a pressure and the flow would be whatever mass is moving, for example, the eardrum. Every source and receiver of energy has a characteristic impedance. Energy is most efficiently transferred from a source to a receiver when these impedances are equal; that is, when they are matched. If they are not matched, the transfer is less efficient because some energy is reflected. This mismatch is often overcome by an intermediary (megaphone or middle ear) which is referred to as an impedance matcher.

The ear is an incredible device in terms of its sensitivity and range of detectable intensities. Yet the ear, like all wave receptors, does not respond to all sound frequencies. The human ear responds to frequencies from about 20 Hz to 20,000 Hz. This is referred to as the audiofrequency range. Frequencies greater than 20,000 Hz are called ultrasound; frequencies lower than 20 Hz are called infrasound. The ear does not respond uniformly to all audiofrequencies. One reason is that the impedance matcher (middle ear) does not function the same for all frequencies. If the energy at the outer ear is compared with that actually received at the inner ear, one finds that the response (ratio of the two energies) decreases at both low and high frequencies. This response also depends somewhat on

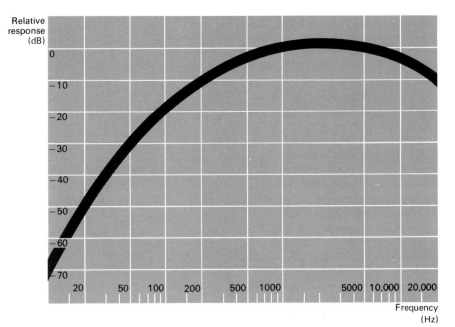

Photograph by Daniel Brody/Editorial Photocolor Archives.

Fig. 6 The relative response curve which approximates that of the human ear. Note that the horizontal scale is logarithmic.

intensity. One such scale, called the A scale, which approximates the response of the human ear is shown in Fig. 6. In terms of actual effects on hearing, it is the amount of energy actually in the workings of the ear that is important. Many meters will correct the actual intensity levels for this effect and will indicate the intensity levels as dBA rather than dB.

NOISE: MEASUREMENT AND STANDARDS

It is very satisfying whenever a concept can be quantified and a number assigned for measurement and comparison. For example, when we talk about infrasound, audiofrequencies, and ultrasound, the meaning is clear because a quantitative delineation is made in terms of frequency expressed as cycles per second. However, such a quantification of noise, is difficult, because of the subjective considerations involved. This subjectiveness is reflected in such definitions as "sound undesired by the recipient," "stench in the ear," and "sound without value." What may be of value to one person may be worthless to another. The sound of an approaching car is priceless to a person who might otherwise have been caught in its path. But it is of negative value to a sleeping person in a nearby house who was awakened by it. The main effect of noise is annoyance which obviously is difficult to quantify. One method of doing this is through a perceived noise level (PNL). This is a level (in dB) assigned to a noise judged by *normal* observers to be equally as noisy as a reference noise consisting of random frequencies centered

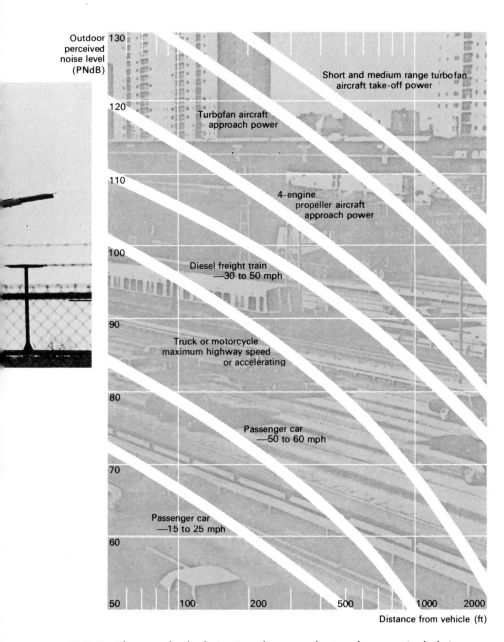

Outdoor perceived noise level (PNdB)

Short and medium range turbofan aircraft take-off power

Turbofan aircraft approach power

4-engine propeller aircraft approach power

Diesel freight train —30 to 50 mph

Truck or motorcycle maximum highway speed or accelerating

Passenger car —50 to 60 mph

Passenger car —15 to 25 mph

Distance from vehicle (ft)

Fig. 7 Perceived noise levels for various types of vehicles. (From Science, January 24, 1969.)

on 1000 Hz. The actual calculation is rather complex* and not particularly important here. It is common, though, to see noises of environmental interest expressed in PNL (dB) (Fig. 7).

Even after it is decided what noise is and how it is measured, the situation is complicated by the fact that scientists studying the effects of noise on man disagree as to the damage produced. For example, some will say that the psycho-

————————

*Noise and Man, W. Burns, op. cit.

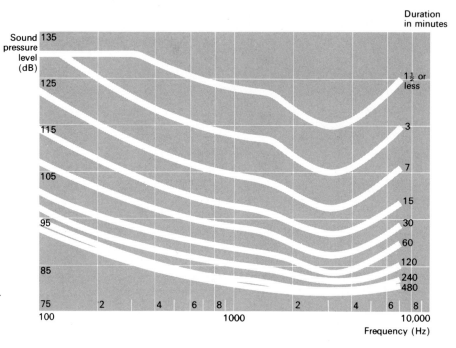

Fig. 8 Damage risk criteria for producing hearing impairment.

Table 3. Permissible noise exposures as established by the Walsh-Healey Public Contracts Act. The "slow" indication means that the recording instrument averages out high-level brief, noises such as hammering. Exposure to impulsive or impact noise should not exceed the 140 dBA peak sound pressure level.

Duration per day, hours	Sound level dBA slow
8	90
6	92
4	95
3	97
2	100
1½	102
1	105
½	110
¼ or less	115

logical and physiological responses to noise other than changes in hearing are short-lived and that they vanish with continued exposure.* Others will say that the emotional disturbance from exposure to noise threatens our good health.† There is no question, though, that excessive noise levels can produce irreversible damage to hearing and that motor vehicle noise, especially from trucks, motorcycles, sports cars, and aircraft can produce a high degree of annoyance. Therefore, maximum noise levels are being established for many segments of the population.

Actual damage to hearing has occurred primarily in industrial occupations. To protect those working in firms holding federal contracts of more than $10,000, the Department of Labor under the Walsh-Healey Public Contracts Act has set maximum permissible exposure levels. These provide a good base for comparison and are shown in Table 3.

An office with tabulating machines has a background noise level of about 80 dB (Table 2). Many factories have substantially higher background levels and therefore have trouble meeting the federal standards. Figure 8 shows another good reference for permissible noise levels, the damage risk criteria proposed by the Committee on Hearing, Bioacoustics, and Biometrics of the U.S. National Academy of Sciences–National Research Council. These data indicate noise levels at which one exposure per day is *likely* to produce damage.

---

*The Effect of Noise on Man, Karl D. Kryter, Academic Press, N.Y. (1970).

†C. D. Leake in Physiological Effects of Noise, Plenum Press, N.Y. (1970).

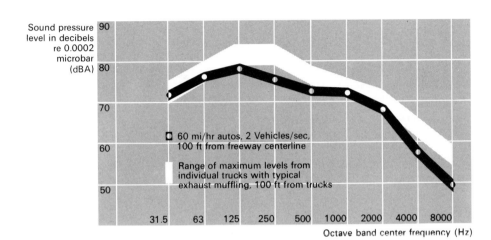

Sound pressure level in decibels re 0.0002 microbar (dBA)

□ 60 mi/hr autos, 2 Vehicles/sec, 100 ft from freeway centerline

Range of maximum levels from individual trucks with typical exhaust muffling, 100 ft from trucks

Octave band center frequency (Hz)

Fig. 9 Typical free-flowing automotive traffic noise spectra. (From *Noise in Urban and Suburban Areas: Results of Field Studies*, U.S. Department of Housing and Urban Development, Washington, D.C. (January, 1967).)

MOTOR VEHICLE NOISE: MAGNITUDE AND ABATEMENT

Although traffic noise constitutes the greatest annoyance to the general public, there are no federal standards for maximum permissible levels. Some states and many cities have enacted ordinances for permissible levels. The state of New York in 1965 enacted a law limiting the noise that a motor vehicle produces at a distance of 50 feet to 88 dB while traveling 35 miles per hour. The state of California in 1967 set a limit of 86 dB for automobiles moving 35 miles per hour and 92 dB for motorcycles and trucks weighing more than three tons. Figure 7 shows data for noise levels actually produced by different types of vehicles. It is clear that automobiles marginally satisfy these standards and that trucks and motorcycles are particularly bad. The data in Fig. 9 illustrates that this is also the case for free-flowing traffic.

Unless abated, noise pollution can only worsen because of the annual increase in motor vehicles on the highways. The possible approaches to alleviation of noise parallel those for reducing atmospheric particle-like pollutants from motor vehicles. Starting at the source, noise could be reduced by sound absorbers on the vehicle and improved design of the vehicle and tire treads. Some improvement could be made on exhaust noise mufflers in automobiles and considerable improvement could be made on those in trucks and motorcycles. The advent of steam and electric propulsion would reduce noise significantly. For example, it is reported that a Rankine steam car cuts perceived vehicular noise by a factor of 4 at peak loads.* Going away from the source, noise could be reduced through

---

*"The Search for a Low Emission Vehicle," Staff Report prepared for the Committee on Commerce, U.S. Senate, U.S. Government Printing Office, Washington, D.C., 1969.

construction of sound barriers between the highways and the populace. But probably the most efficient and logical approach is to construct underground mass transit systems and reduce the number of vehicles on the highways and streets. This would also alleviate many of the other problems associated with our transportation system based on the automobile.

REFERENCES

"Fundamentals of Noise: Measurement, Rating Schemes, and Standards," U.S. Environmental Protection Agency Publication NTID300.15 (December 31, 1971).

"Transportation Noise and Noise from Equipment Powered by Internal Combustion Engines," U.S. Environmental Protection Agency Publication NTID300.13 (December 31, 1971).

"Effects of Noise on People," U.S. Environmental Protection Agency Publication NTID300.7 (December 31, 1971).

"The Social Impact of Noise," U.S. Environmental Protection Agency Publication NTID300.11 (December 31, 1971).

"Noise," Leo L. Beranek, *Scientific American* **215**, 66 (December 1966).

"Environmental Noise Pollution: A New Threat to Sanity," Donald F. Anthrop, *Bulletin of Atomic Scientists* **25** No. 5, 11 (May 1969).

The Effects of Noise on Man, Karl D. Kryter, Academic Press, New York (1970). *Noise and Man*, William Burns, J. B. Lippincott Co., Philadelphia (1968).

Air Pollution – 1970, Part 3, U.S. Government Printing Office, Washington, D.C., 20402.

Guidelines to the Department of Labor's Occupational Noise Standards for Federal Supply Contracts, Bulletin 334, U.S. Department of Labor, Washington, D.C.

Noise: Sound Without Value, Federal Council for Science and Technology (Sept. 1968), U.S. Government Printing Office, Washington, D.C.

Noise in Urban and Suburban Areas: Results of Field Studies, U.S. Department of Housing and Urban Development, Washington, D.C., 20410.

QUESTIONS

1. Give some examples of sound which is acceptable to one person, but distastefully noisy to another.

2. When a moving, hard, steel ball makes a head-on collision with an identical stationary ball, all its energy is transferred in the collision. However, if a steel ball makes a head-on collision with either a more massive or a less massive stationary steel ball, only a portion of its energy is transferred. Suppose now that you place another stationary ball between the moving ball and

the original stationary ball, and suppose further that the moving ball is less massive than the ball furthest from it. If you want the m_3 ball to get the most of the energy from the moving ball, the middle ball must have a mass between that of m_1 and m_3. Explain. Can you see any relation between this example and the way a megaphone works?

m_1 strikes m_2. Then m_2 strikes m_3

3. Energy is propagated in the cochlea of the ear by traveling waves much in the same way that electromagnetic energy is transmitted in a cable coming from the antenna to a TV set. High sound frequencies produce waves which are dissipated near the entrance of the cochlea while low sound frequencies produce waves which travel to the end of the cochlea. The peak of the traveling wave is much sharper for high-frequency than for low-frequency sounds. Based on this, which sound intensities would you expect to be capable of greater physiological damage to the cochlea? Explain.

4. An impedance matcher such as the inner ear allows energy to be effectively transferred from a source to a receiver. Discuss some ways that an impedance mismatcher could be used effectively in solving environmental noise problems.

5. Why is automobile noise more noticeable to a driver when passing through a tunnel?

6. Knowing how the sound of your voice can be muffled with your hand, speculate on the operation of an automobile muffler placed on the exhaust of an internal combustion engine which muffles the noise accompanying the combustion of the fuel-air mixture.

7. What are some of the reasons that motorcycles are particularly noisy?

8. Identify the major sources of noise in an automobile and discuss how the noises might depend on the speed of the vehicle.

9. Discuss three features of a highway system that enter into its noise output.

10. A Helmholtz resonator is essentially a hollow, metallic container with a slender, necked opening. An intense sound can be produced by blowing into it. A Helmholtz condition sometimes occurs with a moving automobile producing an annoying sound in the interior of the vehicle. How might this condition be achieved?

EXERCISES

Easy to Reasonable

Exercise 1. At atmospheric pressure and room temperature the speed of sound in air is about 1100 ft/sec. Calculate the wavelength of sound waves of frequency 20 Hz and 20,000 Hz.

Exercise 2. A certain sound source has a power rating of 100 dB. What is the intensity of this source in W/cm^2?

Exercise 3. What is the intensity in decibels of a sound source with an intensity of $10^{-10}W/cm^2$?

Exercise 4. What is the disposition of the energy associated with a sound wave absorbed by a piece of material such as a drapery?

Exercise 5. Suppose that a single jet plane on take-off produces a sound level of 100 dB at some given distance from the runway. What is the sound level at the same position if 10 of these jets were to take-off simultaneously?

Exercise 6. The decibel unit is sometimes thought of as a measure of loudness. On this basis, will ten sources sounding simultaneously be ten times as loud as a single source? Explain.

Reasonable to Difficult

Exercise 7. The starter and timer of a 100-yd dash stand at the beginning and end of the 100-yd path. The timer starts his watch on the flash seen from a gun fired by the starter. If he were to start his watch on the sound heard from the gun, how much time would elapse between the start of the race and the start of the watch? Would this amount of time be significant in a 100-yd dash?

Exercise 8. The wall of a room typically reduces a sound level by 40 dB. What is the ratio of the intensity outside a room to that inside a room containing a sound source if the walls are "typical"?

Exercise 9. Explain the caption in the figure below.

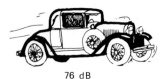

76 dB

79 dB

Exercise 10. The major individual sources of noise from a diesel truck are engine, exhaust, intake, and cooling fan. The typical noise from each of these sources is:

Engine: 83 dB   Intake: 80 dB
Exhaust: 86 dB   Cooling fan: 81 dB

a) Explain why the net noise from these four sources is not the sum of the individual noise levels, that is, 330 dB.

b) From the definition of dB sound level, show that the net noise level is 89 dB.

8

MASS TRANSPORTATION: NECESSITY AND NEW TECHNOLOGY

NECESSITY

Background Information

On a straight tonnage basis, motor vehicles account for about 60% of all air pollutants. This, coupled with their contribution to noise pollution, gives every indication that motor vehicles are threatening our physical, psychological, and, perhaps, sociological well-being. Unquestionably, something has to be done, but the entire industry has such impact on our society and is growing at such an ever-increasing (nearly exponential) rate that it will take time to effect change. Let us look at the positive and negative aspects of the motor vehicle industry and then try to make a case for mass transit* as a necessary alternative to our present major forms of urban transportation.

Transportation accounts for about 20% of the gross national product. As the GNP grows, transportation follows (Fig. 1). Motor vehicle registration is growing at a rate of about 4% per year as opposed to about 1.3% per year for the population. Note that private automobiles account for about 90% of the passenger miles. Table 1 illustrates the purposes of this travel. It is quite clear that the vast amount of travel is motivated by personal considerations, and this implies that the citizens will be extremely reluctant to relinquish their private mode of travel. It is interesting that the total highway mileage, including the interstate system, which has to accommodate these increasing numbers has leveled off between 1965 and 1968 (Fig. 2). However, about 1.5% of the U.S. land is occupied by roads. The role of the automotive industry in the consumption of natural resources (some of which are also becoming scarce) is indicated in Table 2. The growing number of fatalities resulting from motor vehicle accidents is an increasing concern. Deaths on the highways have increased from 38,137 in 1960 to 52,924 in 1968. However, on an hourly exposure basis, the motor vehicle death rate is low by comparison with some other activities. The economic liability from motor vehicle accidents is staggering. In 1968, insurance premiums amounted to $11.6 billion. Total losses paid added up to $6.6 billion.

Transportation Energy Efficiency

Conventional motor vehicles are depleting the energy resources at an incredible rate (Fig. 3). This is happening in the face of repeated warnings of limited petroleum reserves. In 1965 about 35 billion gallons of fuel were used in urban service. Sixty percent of this was consumed in trips of less than 2½ miles.

The personalness and convenience of the automobile are captivating features. But is the automobile an efficient form of transportation? Just as for an engine or a refrigerator, one has to decide what the function of an automobile is before a number can be assigned for its efficiency. The basic idea of any form of passenger transportation is to move people between two points. For freight transportation, it is weight or tonnage that is moved. Now we ask "What is the most important commodity sacrificed in achieving this desired result?" In the past we thought

---

*Mass transit is intended to mean "transportation of large *numbers* of *people* along well-defined *corridors* away from the streets in *urban* areas."

Fig. 1 Growth of gross-national product and passenger traffic in the United States.

Fig. 2 Growth of the highway system in the United States. (From the U.S. Bureau of the Census, *Statistical Abstract of the United States: 1971* (92nd edition), Washington, D. C., 1971, p. 527) Photograph Courtesy of the Federal Highway Commission.

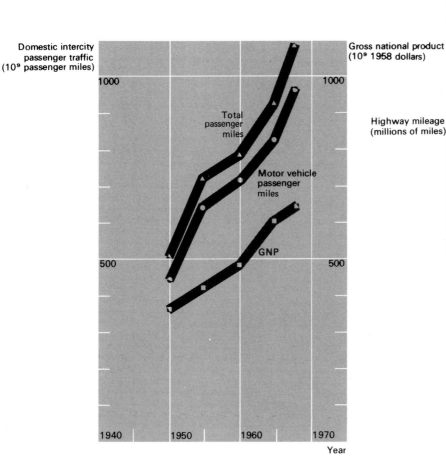

Domestic intercity passenger traffic (10⁹ passenger miles)

Gross national product (10⁹ 1958 dollars)

Highway mileage (millions of miles)

Table 1. Breakdown of passenger transport in the United States. Percentages are based on number of person-miles which accounts for number of passengers as well as length of trip. (From: Dept. of Commerce, Bureau of the Census: 1967.)

Mode		Purpose	
Auto	77%	Visits	44%
Bus	2%	Other pleasure	25%
Train	2%	Business	18%
Commercial air	17%	Other personal	13%
Other	2%		

Table 2. Consumption of selected materials by the automotive industry (1965).*

	Total consumption	Percentage consumed by automotive industry
Rubber (long tons)	2,325,117	63.1
Lead (tons)	1,241,500	51.4
Zinc (tons)	1,530,000	35.5
Steel (tons)	92,666,182	21.7

*U.S. Department of Commerce, Subpanel Reports to the Panel on Electrically Powered Vehicles, The Automobile and Air Pollution: A Program for Progress, Part II, p. 90. U.S. Government Printing Office, Washington, D.C. 1967.

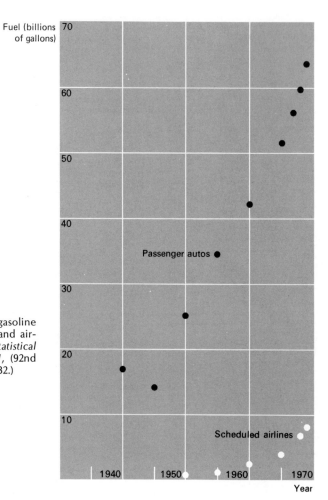

Fig. 3 Annual consumption of gasoline (including jet fuel) by automobiles and aircraft. (U.S. Bureau of the Census, *Statistical Abstract of the United States: 1971*, (92nd edition), Washington, D.C., 1971, p. 532.)

mostly in terms of time; that is, we try to get a mass (people or freight) between two points in the least amount of time. But if fuel becomes more important than time, then we try to move the most goods with the least amount of fuel (or energy). So on an energy basis, an efficiency can be defined as

$$\text{Passenger efficiency (P.E.)} = \frac{(\text{number of people}) \times (\text{number of miles})}{\text{amount of fuel used}}$$

$$= \frac{\text{passenger-miles}}{\text{gallon}}$$

In 1968 an automobile averaged 13.9 miles for each gallon of fuel consumed (this was a decrease from 15.3 miles/gallon in 1940, which also says something about efficiency). So if an automobile carries two passengers, its P.E. would be

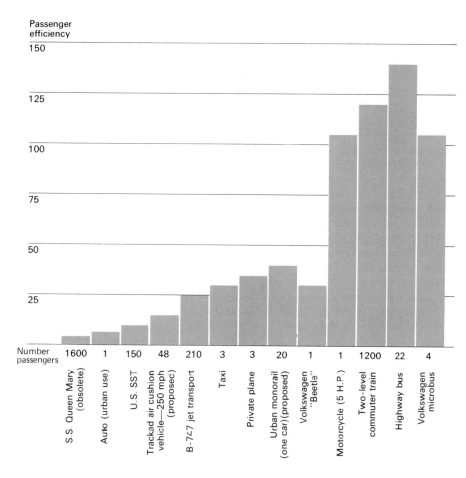

Passenger efficiency

| Number passengers | 1600 | 1 | 150 | 48 | 210 | 3 | 3 | 20 | 1 | 1 | 1200 | 22 | 4 |

S.S Queen Mary (obsolete) · Auto (urban use) · U.S. SST · Tracked air cushion vehicle—250 mph (proposec) · B-747 jet transport · Taxi · Private plane · Urban monorail (one car) (proposed) · Volkswagen "Beetle" · Motorcycle (5 H.P.) · Two-level commuter train · Highway bus · Volkswagen microbus

Fig. 4 Compilation of passenger efficiencies for various modes of transportation. (Most of these were taken from "System Energy and Future Transportation," Richard A. Rice, *Technology Review*, p. 31, January 1972. The VW Beetle and Microbus passengers were reduced from 2 to 1 and 7 to 4, respectively, from the *Technology Review* article.)

$$P.E. = 2 \text{ passengers} \cdot \frac{13.9 \text{ miles}}{\text{gallon}}$$

$$= \frac{28 \text{ passenger-miles}}{\text{gallon}}$$

An automobile in urban use typically has only one occupant. Furthermore, it consumes more fuel per mile in urban driving conditions. As a result, its passenger efficiency may be as low as 10. Figure 4 is a compilation of passenger efficiencies for a variety of systems. While the automobile is unquestionably convenient and while it has become a way of life, it clearly ranks among the lowest modes of transportation in passenger efficiency.

Prerequisites for a Mass Transit System

Mass transit is not a novel or new idea. Most major cities have a mass transit system of one form or another which is essential to the functioning of the city. New York City, for example, is absolutely dependent on its subway system in spite of the system's many failings on a human scale. However, the last urban transport system built in this country was in Philadelphia in 1907.* Cities have since relied on automobiles and complicated, expensive highway systems with their ever-increasing environmental problems as a means of mass transit. No mass transit system is clearly going to replace all functions of the motor vehicle. What it must do is allow people to get from their homes into the cities with only a short car or bus ride to an entrance station and a short walk or bus ride from an exit station. To do this, the system must be more attractive than taking the automobile from the home to the city via an expressway. This means that the transit system must be convenient, fast, reliable, clean, comfortable, safe, economical, and available around the clock. Shortcomings in just some of these necessary conditions have led to the boycott of many of the present outdated systems. However, given a transit system with the designated attractive attributes, people will use it in favor of automobiles. In Sweden, where the per capita ownership of private automobiles is second only to the United States, 90% of the people travel to downtown Stockholm by the subway system.

An effective mass transit system would be one of the very first considerations of a designer setting out to build an urban community from first principles. Properly done, it could mold the social and economic structure of the community. There certainly is opportunity for ventures of this sort, but the main problem at hand is to effect change in transit systems as they presently exist in the cities. Inherent in achieving such a goal are the enormous problems of inertia and finance. But there is evidence of motion in both sectors. The Urban Mass Transit Bill enacted in 1970 authorizes $3.1 billion in federal matching grants for urban transit systems financed over five succeeding years. An average annual expenditure of $1 billion is envisioned for the next 12 years. In contrast, the federal government spent a *total* of about $1 billion on urban transit in the past decade. There are two obvious avenues open for developing new urban mass transit systems. One is simply to update and modernize the technology of existing types. The other is to go to more revolutionary and, in the long run, more efficient systems. The San Francisco Bay Area Rapid Transit (BART) system was developed along the lines of the first approach. In connection with the second approach there are a variety of new technologies including aerial suspension of vehicles, linear induction motors, monorails, gravity-vacuum propulsion with an underground tube, and air-cushion suspension. Let us examine the features of the BART system as well as a few of the new more interesting technologies.

---

*Science **171**, 1125 (19 March 1971).

Fig. 5 The San Francisco BART system. Solid lines indicate tracks above ground. The upper fork in San Francisco is still in the planning stages.

Fig. 6 A BART car.

NEW TECHNOLOGY

The Bay Area Rapid Transit (BART) system*

The BART system became operational in 1972. Planning actually began in 1951 which gives an indication of the magnitude of the problems associated with the effort. It is 75 miles long (Fig. 5) and was financed with essentially no federal funds. In spite of many environmental arguments such as reduced air pollution and auto congestion, reduction of sprawling suburbs, and provision of jobs for inner-city ghetto residents, it was the attraction of potential profits from a rejuvenated city which mainly motivated the venture. The system is an updated version of existing rail-type technology. The cars (Fig. 6) are carpeted, well-lit, air conditioned, and free from noise. The cars average 50 miles per hour (including stops), run every 90 seconds during peak periods, and are completely controlled by a central computer. Fares are charged on a mileage basis. A passenger buys a card good for so many miles. The card is coded magnetically and computerized machines do the accounting.

It is reported that the system made an impact on the bay area through a building boom in the billions of dollars even before a train ran. Because BART

---

*An excellent film titled "BART: Vision to Reality" is available on loan from the Bay Area Rapid Transit District, 800 Madison Street, Oakland, California 94607.

is the first newly-constructed urban transit system in the United States since 1907, it will be extremely interesting to follow its environmental, social, and economic impact on the urban community that it serves.

The Gravity Vacuum Transport (GVT) System

Although there are several potential energy sources around us that contain huge reservoirs of energy, the sources are essentially of no use because they are difficult to tap. These would include the rotational kinetic energy of the earth, thunderstorms, the tides, and gravitational energy. There is, however, an excellent possibility that a gravity vacuum transport (GVT) system may exploit the gravitational potential energy source. An evacuated tube would extend down into the earth and then emerge at another point on the surface. The gravitational force of the earth on the vehicle would pull the car down to the lowest point in the tube and then it would decelerate and coast up to an exit port on the surface (Fig. 7). Some additional force such as air pressure on the back of the vehicle would be used to increase the acceleration and to overcome frictional, dissipative forces. Such a system may seem far-fetched but it has developed well beyond the science fiction stage and has the extremely attractive feature of using a free, nonpolluting energy source. The principle is quite interesting and instructive. Proponents of this type of system contend that an 8-mile trip extending 3000 feet down into the ground could be made in 3.2 minutes. This amounts to an average speed of 150 miles per hour.* Recent advances in the technology of tunnel boring using a boring tool heated to 1200°C by electric or nuclear energy or a high-energy electron beam may provide incentive for development of these gravity transport systems.†

Fig. 7 The vehicle in a GVT system is accelerated down into the earth by the gravitational force between the vehicle and the earth. Likewise, it is decelerated on its upward path by the gravitational force.

The Tracked Air Cushion Vehicle (TACV) system

The Tracked Air Cushion Vehicle (TACV) will use compressed air to provide a highly cushioned support between the vehicle and the guideway over which it runs (Fig. 8). The TACV utilizes an air propulsion mechanism such as a jet engine or air propeller. However, there are efforts to use electric energy through a linear

---

*"High Speed Tube Transportation," by L. K. Edwards, *Scientific American*, Vol. 213, No. 2, 30 (August 1965).

†"Why the U.S. Lags in Technology," by Lawrence Lessing, *Fortune*, Vol. 75, No. 4, 69 (April 1972).

Fig. 8 A Tracked Air Cushion Vehicle (TACV). (Photograph courtesy of the Garrett Corporation.)

induction motor.* The motor works on the following principle. Whenever a wire is placed in a *changing* magnetic field, an electric current is induced in the wire by virtue of electric forces on the electrons. An ordinary induction motor like that used to run an appliance, a refrigerator, for example, includes a circular armature containing electrical conductors which is free to rotate and a series of concentric windings of wire, called the field, which surround the armature (Fig. 9). The armature and the field are not connected either electrically or mechanically. When an alternating current† flows in the field windings, a changing magnetic field is produced. Consequently, a current is induced in the wires of the armature. The current in the armature produces a magnetic field. The interaction of this magnetic field with that from the field windings produces a force which causes the armature to rotate.

A linear induction motor works on the same principle, but the armature and field are linear rather than cylindrical as in an ordinary motor. In a tracked vehicle, the rails would carry an alternating current which would cause a current to be induced in wires attached to the vehicle. Interaction of the two magnetic fields would produce a linear force which would pull the vehicle along. Voltages on the rails would be of the order of 2000 volts. Speeds of up to 250 miles per hour are anticipated for vehicles of this type. A TACV system using linear induction motors is proposed to be built from Washington, D.C.'s Dulles Airport to McLean, Virginia.‡ The system would cover a distance of 13.5 miles and would be engineered for speeds up to 150 miles per hour.

Fig. 9 Schematic cross-sectional view of an induction motor.

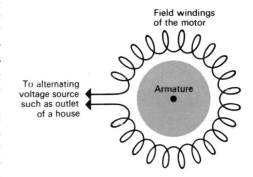

---

*Induction comes from the word induce, meaning to force or influence.

†An alternating current is one in which the charges change directions at regular intervals of time (see Chapter 2).

‡*Science News* **99**, 144 (Feb. 27, 1971).

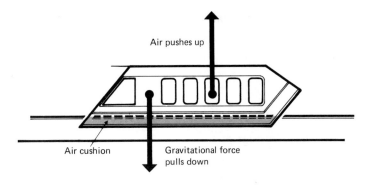

Air pushes up

Air cushion

Gravitational force
pulls down

Fig. 10 A TACV is supported by a cushion of compressed air underneath the vehicle. This principle is widely used in air tracks which produce a nearly frictionless surface for physics demonstrations.

Another means of suspension being explored uses a magnetic principle.* In the air suspension system, the weight of the vehicle is balanced by the force of the air pushing up on the vehicle (Fig. 10). In a magnetic suspension system, the weight of the vehicle is balanced by a magnetic force pushing up on the vehicle. It is similar to a repulsive force that exists between like magnetic poles. A magnetic field is created by currents in wires in both the track and the vehicle. The fields interact and a repulsive force is exerted on the vehicle. Such a system would be useful in a gravity transit system where the vehicle must run in an evacuated tube.

The Supersonic Transport

Background Information

For one reason or another, man has an incessant urge to travel the greatest possible distance in the least amount of time. The simple relation

distance = rate × time
$$d = r \cdot t$$

suggests that he seek that vehicle which maximizes his speed. The supersonic transport (SST) is just another attempt for a faster vehicle.† Until shortly after World War II speeds were limited to less than the speed of sound (about 750 miles per hour). Tremendous advances in aviation technology were made both during and after World War II and speeds greater than that for sound were attained. The pressure for extending this new technology to commercial aircraft was ever present and plans for an American SST began under the Eisenhower administration. The first funding ($11 million for a feasibility study) came in 1959. Funding continued but opposition continually mounted. In March 1971, the U.S. Senate voted (49–47) to terminate the SST program.‡ Arguments against the program centered

---

*"Why the U.S. Lags in Technology," by Lawrence Lessing, *Fortune* Vol. 75, No. 4, 69 (April 1972). "Magnetic Levitation; Tomorrow's Transportation," *Bulletin of The Atomic Scientists* **28**, No. 3, 39 (March 1972).

†The word supersonic means "greater than the speed of sound."

‡A good, but brief, account of the political aspects of the American SST program can be found in *The Congressional Digest*, December 1970.

around its being an unrealistic ordering of priorities, a multi-billion dollar gamble, and an environmental danger. In any event it was a first for terminating a program for a highly sophisticated machine which would have been a symbol of technological progress. It was a program of international significance because the British and French with their Concorde and the Russians with their TU-144 were in competition for the SST passenger. Let us look at some of the environmental reasons that entered into the arguments against the American SST program.

Prerequisites for a Supersonic Transport

The speed of sound depends on the temperature of the air and the temperature, in turn, varies with position in the atmosphere. So the speed of sound depends on where the plane is flying. It is convenient to compare the speed of the plane to the local speed of sound through the Mach number M.*

$$M = \frac{\text{speed of the plane}}{\text{speed of sound where the plane is flying}}$$

A Mach number of 2 would mean that the plane travels twice the speed of sound at the position of the plane.

If environmental, economic, and social impacts are neglected, it is easy to make plausible arguments for the basic prerequisites for the vehicle. The success of the trip depends on whether there is enough fuel, F, on board and the rate, R, at which it is consumed. So the time for the trip could be written as

$$t = \frac{F \text{ (gal)}}{R \text{ (gal/hr)}}$$

Now the speed depends on several things. First, it depends on the plane's weight, W. Secondly, it depends on the force that "lifts" the plane into the air and the force that retards the plane's forward motion. These are called the lift, L, and drag, D. The lift force is an aerodynamic effect. When air travels over the curved surface of an aircraft wing, the pressure on top of the wind becomes less than that on the bottom. Thus a pressure difference between the upper and lower wing surfaces is created which lifts the plane (Fig. 11). This is called Bernouilli's principle. The flexible top on a moving convertible-type automobile often is pushed out as a result of this effect.† The drag is simply a frictional force between the plane

---

*After Ernst Mach, Austrian physicist and philosopher 1836–1916.

†This effect can be demonstrated by hanging a strip of paper (such as newspaper) over a holder and then blowing between the strips. The two pieces will move together as a result of excess pressure on the outside.

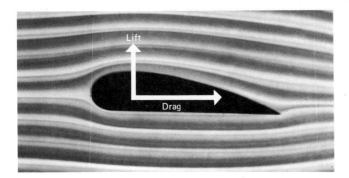

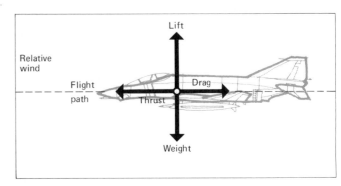

Fig. 11 Illustration of the forces on a plane in flight.

and the air. It is just like the drag force on a particulate in the atmosphere. Hence we could reasonably expect that

1. if the weight increases, the speed should decrease;
2. if the drag increases, the speed should decrease;
3. if the lift increases, the speed should increase.

Incorporating these ideas into an equation, we write

$$V \propto \frac{L}{D \cdot W}$$

Let us write the velocity in terms of the Mach number and put in a factor which makes the expression dimensionally correct and includes things that we may have neglected.

$$V = C \frac{M}{D} \cdot \frac{L}{W}$$

Using $t = F/R$ for time, we find that the expression $d = rt$ becomes

$$d = C \frac{M}{D} \cdot \frac{L}{W} \cdot \frac{F}{R}$$

Rearranging terms, we find that

$$d = C \left(\frac{M}{R} \right) \quad \cdot \quad \underset{\substack{\uparrow \\ \text{lift-to-drag} \\ \text{ratio}}}{\left(\frac{L}{D} \right)} \quad \cdot \quad \underset{\substack{\uparrow \\ \text{fuel-to-weight} \\ \text{ratio}}}{\left(\frac{F}{W} \right)}$$

A designer picks a given Mach number, say $M = 3$, and distance, say 3000 miles, and then he gets an engine which minimizes the fuel consumption rate and a plane which maximizes the lift-to-drag ratio and fuel-to-weight ratio. And this is no easy

Fig. 12 Artist's sketch of the proposed American SST. (Photograph courtesy of the Boeing Company.)

task! To get the necessary lift-to-drag, new designs are needed. That is why SST vehicles have such radical looking features (Fig. 12). To minimize the weight, a light metal is required. But this metal must have extremely good structural properties and be able to withstand the heating that results from the frictional drag force. Aluminum, which is used on subsonic aircraft, is unsuitable for Mach numbers of about 3. Expensive and rare titanium metal has been proposed.

There are many arguments, pro and con, about the need for a fleet of SST's. Certainly, political and national pride is a consideration. And one cannot argue against the economic impact that a program would have on an aircraft industry, especially one suffering from the termination of a massive space program. The economics of SST operation is a debatable question, but no doubt people would be attracted by the glamour and would probably use it in favor of conventional subsonic jet transportation. There are, however, other factors which argue against the SST, at least until more information is obtained. These would include

1. rate at which fuel is used in the face of diminishing supplies,
2. community noise and effect of the sonic boom,
3. effect of engine exhaust emissions on the atmosphere,
4. passenger and crew radiation exposure.

Let us review each of these factors.

SST Energy Efficiency

The proposed United States SST would consume fuel at the staggering rate of about 15 gallons per mile. If the plane carried 150 passengers, its passenger efficiency would be

$$\text{P.E.} = \frac{150 \text{ passengers}}{15 \text{ gallons/mile}}$$

$$= 10 \frac{\text{passenger-miles}}{\text{gallon}}$$

This is comparable to an automobile in urban use and is about one-half that of a commercial 747 jumbo jet. A single SST would consume about 100 million gallons of fuel annually. Thus a fleet of 100 SSTs would consume 10 billion gallons of fuel annually which is comparable to the entire fuel use by all aviation sources in 1970. Obviously, from the standpoint of energy efficiency, the SST ranks very low.

The Sonic Boom

Aircraft noise, especially in the vicinity of congested airports, is currently a problem. An SST at subsonic speeds creates noise comparable to a contemporary jet and would therefore contribute in the same way. Abatement of this type of noise is technologically feasible, but, like everything else, requires time, effort, and money. The sonic boom, however, is the noise problem which is associated with a plane flying faster than the local speed of sound and for which there is no obvious abatement technology. The sonic boom is an abrupt sound usually lasting for a fraction of a second following the passage of a supersonic plane. The sound wave transmitting this disturbance is called a shock wave. It is a shock in the sense that it produces significant pressure changes in the local atmosphere. For example, air pressure variations in the wave may be as large as 4 pounds per square foot. Such a variation is enormous when compared with pressures of the order of 10^{-4} pounds per square foot associated with conversational sound.

It is a popular belief that the sonic boom occurs only at the instant at which the plane achieves supersonic speed (breaking the sound barrier). However, the sonic boom persists as long as the plane flies faster than the speed of sound. Also the shock wave phenomenon is not something associated only with an object moving faster than the speed of sound. Other types of shock waves result when an object moves faster in a medium than some wave velocity characteristic of the medium. A ship moving faster than the local speed of a water wave produces a shock wave which is commonly called a bow wave.* An atomic particle—such as an electron which moves faster than light would in the medium of the electron—produces a shock wave which gives rise to Cerenkov radiation. This radiation is often seen as a characteristic blue light in the water surrounding certain types of nuclear reactors.

The physical origin of the sonic shock wave is as follows. An object moving in an air medium exerts forces on the air which produce local pressure disturbances which, in turn, propagate as sound waves with a speed characteristic of the local properties of the air. If the object is travelling slower than the waves, the waves simply move away from the object in all directions. However, if the object moves faster than the local speed of sound, the pressure disturbance can move laterally from the object, but cannot move forward. As a result, the air becomes

---

*"Formation of Shock Waves" is an excellent, Super-8, 3-minute film cartridge that shows the formation of shock waves in water. It is available from several teaching supply companies such as The Ealing Company, 2225 Massachusetts Ave., Cambridge, Massachusetts, 02140.

Fig. 13 A sonic shock wave produces density variations in the medium in which it propagates. These density variations can be detected by a process called Schlieren photography and the form of the shock wave can be delineated. This is a Schlieren photograph taken in a wind tunnel at a wind speed of Mach 2.0. The black part of the picture corresponds to the object placed in the tunnel and the lighter portions next to the black outline correspond to the shock wave. (Photograph courtesy of NASA.)

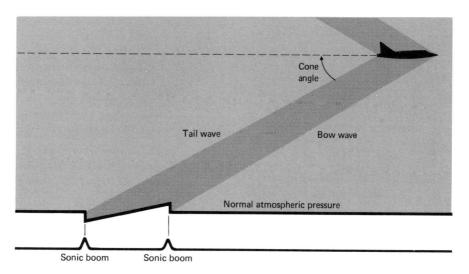

highly compressed at the front of the object and highly rarefied at the rear. These large compressions and rarefactions propagate laterally from the object (Fig. 13). The wave front of the disturbance defines the surface of a cone (Fig. 14). The faster the plane moves, the narrower the cone angle. If a shock wave is generated by a supersonic plane, the sonic boom is heard when the wave front of the disturbance reaches the ground. Typically, two booms corresponding to the compression and rarefaction parts of the disturbance are heard. Because of the shape of the pres-

Fig. 14 Outline of the wave front produced by a supersonic plane.

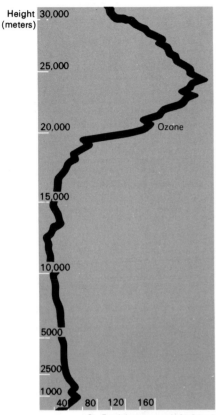

Height (meters)

O_3 Partial pressure $(10^{-9}$ bar)
1 bar = 10^6 dynes/cm^2

Fig. 15   Distribution of ozone in the atmosphere.

sure disturbance, it is often called an "N" wave. The shock wave, like any sound wave, loses energy as it propagates. So the farther the observer is from the plane, the smaller the perceived intensity. The greatest intensity would be experienced directly under the plane and it would decrease laterally along the ground. How-ever, a significant intensity may be felt for a lateral distance of 50 miles along the flight path. Hence a single plane could produce a sonic boom effect on a large segment of the population. It is estimated that even on a circuitous overland route some 35,000,000 people would experience the sonic boom.

The intensity of the sonic boom is sufficient to actually produce physical damage. Most of this is in the form of the likes of cracked windows and plaster. The effect is again one of annoyance in such things as a shaking house, being startled, and interruption of sleep, rest, conversation, and radio and TV. There is no boom level below which these annoyances will not occur. Annual paid damages due to noise effects such as these on overland flights by an SST fleet are estimated to be around $37,000,000.

The Effect of Engine Exhaust Emissions on the Atmosphere

Several questions have been raised about the effects of the SST on the atmosphere and the weather. Subjects for concern include the prospect of the formation of a permanent "cloud cover" from SST contrails and the air pollution resulting from the exhausts of the engines. The SST's may be a threat in these areas, but they are not the lone offenders. However, an SST will cruise at an altitude of 60–65,000 feet as opposed to about 35,000 feet for a conventional jet transport. Exactly how the SST could influence atmospheric conditions at these altitudes is uncertain, but worthy of consideration are the effects of possible enhancement of photo-chemical reactions, possible changes in atmospheric circulation patterns, and possible alterations in the natural stratospheric properties. Let us examine one such effect to illustrate the importance of trying to evaluate the magnitude of their combined impact.

Ozone is a molecule of oxygen containing three atoms of oxygen (O_3). There is a natural concentration of ozone in the atmosphere. Its peak concentration is in the stratosphere between 60,000 and 100,000 feet (Fig. 15). Ozone is a highly

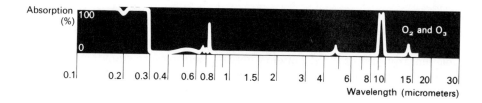

Fig. 16 Absorption characteristics of ozone for electromagnetic radiation.

efficient absorber of ultraviolet radiation (Fig. 16). Nearly all radiation less than 0.3 micrometers in wavelength is removed from solar radiation before it reaches the earth. This is extremely important because ultraviolet radiation is known to produce detrimental effects on man, skin cancer, for example. Hence life as we know it on earth would be drastically different without this ozone layer.

Ozone can be broken down into normal oxygen (O_2) and atomic oxygen (O) by absorption of ultraviolet radiation.

$$O_3 + sunlight \rightarrow O_2 + O$$

This process and its inverse go on all the time. Suppose that nitric oxide (NO) were introduced into the stratosphere from the exhaust of an SST. Then we could have

$$NO + O_3 \rightarrow NO_2 + O_2$$

followed by

$$NO_2 + O \rightarrow NO + O_2$$

Note that NO is regenerated in the process and that O_3 is depleted. That is to say, NO serves as a catalyst. So a small amount of NO could deplete a large amount of ozone. If enough were depleted, serious consequences could result.

Radiation Exposure in an SST Flight

In Chapter 4, we discussed the background of nuclear radiations to which everyone is exposed. Radiation from space is one contribution to this background. This contribution, however, varies significantly with altitude and somewhat with latitude. For example, a person at sea level in the mid-Atlantic states would receive about 28 mrem/year, but one in the Denver-Colorado Springs area would receive about 45 mrem/year.* The intensity of this radiation increases as one goes farther out into space because the greater absorption of radiation by the atmosphere occurs closer to the earth. There is no question that a plane passenger, SST or

---

*"Natural Radiation in the Urban Environment," D. B. Yeates, A. S. Goldin, and D. W. Moeller, *Radiation Safety*, Vol. 13, No. 4, 275 (July-August 1972).

otherwise, gets exposed to a higher radiation level than a person on earth* because he is at a higher altitude. However, because it flies faster, the SST is exposed for a shorter time. A person on a 2-hour SST flight at 70,000 feet would receive a radiation exposure of about 2 mrem whereas a person on an equivalent distance flight in a subsonic jet at 35,000 feet would receive 3 mrem (Exercise 11). An international jet flight crew averages about 120 flights per year. So in an SST and subsonic plane they would receive 0.24 rem/year and 0.36 rem/year, respectively. Both of these exposures are less than 10% of the 5 rem/year allowable for an individual. If one accepts this 5 rem/year as a safe level, then neither type of air travel constitutes an exposure problem.

REFERENCES
Mass Transportation

"System Energy and Future Transportation," by Richard A. Rice, *M.I.T. Technology Review* (January 1972).

"Rapid Transit: A Real Alternative to the Auto for the Bay Area?" *Science* **171**, 1125 (19 March 1971).

"How to Move People in the Cities," *Science News* **98**, 464 (December 19, 1970).

"Rapid Transit—A Prescription for Urban Growth," *Westinghouse Engineer* (January 1970).

"Systems Analyses of Urban Transportation," *Scientific American* **221**, No. 1, 19 (July 1969).

"High Speed Tube Transportation," *Scientific American* **213**, No. 2, 30 (August 1965).

Supersonic Transport

"Sonic Boom", by Herbert A. Wilson, Jr., *Scientific American* **206**, 1, 34 (January 1962).

"The Supersonic Transport," by R. L. Bisplinghoff, *Scientific American*, **210**, 6, 25 (June 1964).

"Shock Waves," by Otto Laporte, Scientific American, *181*, No. 5, 14 (May 1949).

"The SST and the National Interest," by William M. Magruder, U.S. Department of Transportation, August 12, 1970.

Economic Analysis and the Efficiency of Government, Hearings before the Subcommittee on Economy in Government of the Joint Economic Committee, Congress of the United States, 91st Congress, Second Session. Part 4—Supersonic Transport Development, May 7, 11, and 12, 1970. Superintendent of Documents, U.S. Government Printing Office, Washington, D.C.

---

*"Radiation Exposure in Air Travel," Hermann J. Schaefer, *Science* **173**, 780 (27 August, 1971).

QUESTIONS

1. Define an efficiency for freight transportation in the same manner that a passenger efficiency was defined in the text. Estimate and compare the freight efficiencies of trucks, airplanes, and trains.

2. When a taxi driver proceeds to a pick-up point, does he necessarily take the route of shortest distance? Are there other situations of environmental interest where similar options are possible?

3. The picture in Fig. 17 shows how the median strip of an expressway has been converted to a bus lane. Discuss some of the features of this system. Imagine what this scene would look like if people chose to ride the bus rather than drive. What are some ways that they could be persuaded to ride the bus?

Fig. 17 Express bus lane in the center of I–95 provides faster public transportation into down town Washington, D.C. The unimpeded movement of the buses has lured many to public transit. (Photograph courtesy of the U.S. Department of Transportation.)

4. The railroad system was developed primarily for commercial reasons. Contrast this with the motivation for the development of the highway system.

5. Make a proposal for the layout of a mass transit subway system serving a typical city. Have cities with subway systems tended to follow your scheme? What sort of a pattern for mass transportation evolves if a city relies primarily on the automobile?

6. The popularity of the automobile arises from its tremendous convenience. Some of this convenience could be retained and the passenger efficiency increased if more passengers could be handled. One scheme called Dial-A-Bus would allow a person to phone for a bus which would pick him up wherever he desired. Discuss the possibilities of this system. What other schemes can you think of which might make more efficient use of automobile-type transportation?

7. Devise a program for persuading the public to use a transit system.

8. Even though the technology exists for building an American SST, the decision was made in 1971 not to do so. What is your own personal feeling about this decision? Are there other similar technological projects that you would like to see discussed in the public domain?

9. The gravitational force between two masses is always attractive. Suppose that a material were found which experiences a repulsive force in the vicinity of an ordinary mass like earth. What use could be made of this material in a transportation system?

10. Ozone (O_3) is a minor constituent of an unpolluted atmosphere in the troposphere. However, ozone is fairly concentrated at altitudes of 12 to 18 miles (Fig. 15). Ozone is an efficient absorber of ultraviolet radiation. Discuss some of the effects that would occur on earth were this ozone not present at high altitudes.

11. It is reported (*Technology Review*, July/August 1972, p. 51) that there is a perceptible buildup of heat in the ground surrounding the subway system. One of the factors is 600 new air-conditioned cars. Why should this tend to heat up the tunnels and the surrounding ground?

EXERCISES

Easy to Reasonable

Exercise 1.   The speed of sound in the atmosphere is about 1100 ft/sec. What would the speed be in mi/hr?

Exercise 2.   What is the Mach number of a plane traveling 1750 mi/hr in air where the speed of sound is 1100 ft/sec?

Exercise 3.   What is the passenger efficiency of a car carrying 3 passengers and obtaining 20 mi/gal of gasoline?

Exercise 4. Suppose that an SST and a subsonic jet average 1000 and 500 mi/hr respectively, on a 3000-mile trip between two cities. How long does it take each plane to travel between the cities? Suppose that a person takes 1 hour to get from his house to the originating airport and 1 hour to get from the terminating airport to his place of business. Estimate his average speed from home to business.

Exercise 5. A person on a bicycle going 12 mi/hr uses about 100 Btu of energy per mile of travel. The energy equivalent of 1 gal of gasoline is about 1.3×10^5 Btu.

a) Calculate the gasoline equivalent of 100 Btu.

b) Determine the passenger efficiency for bike riding and compare it with those shown in Fig. 4.

Exercise 6. Based on your everyday experience of the relative amounts of energy expended in walking and bicycling, estimate the passenger efficiency for walking using the result of Exercise 5. What considerations other than passenger efficiency does a person make when considering a form of transportation?

Exercise 7. The sonic boom from a supersonic transport flying over land would be felt by all people within about 10 miles on each side of the flight path. Assuming that the population of the United States is uniformly distributed, about how many people would be affected in a coast-to-coast flight of an SST?

Exercise 8. A train on the BART system has the ability to accelerate to 80 mi/hr in one-half minute. Calculate the acceleration in mi/hr per second assuming that the acceleration is constant.

Exercise 9. The miles traveled by passenger trains in the United States decreased from 481,000,000 in 1945 to 90,000,000 in 1970. If this trend were to continue, estimate the year in which the number of miles traveled is zero.

Reasonable to Difficult

Exercise 10. During the peak periods, trains on the BART system will run every 90 sec.

a) How many trains will pass through a terminal each hour in a peak period?

b) A BART car holds 72 passengers and there are 6 cars in each train. What is the maximum number of passengers per hour that could be moved from a station?

Exercise 11. An SST and a B-747 subsonic jet flying at 65,000 and 30,000 ft, respectively, at a latitude of 45° are exposed to radiation levels of about 1000 and 500 microrems per hour (see *Science* **173**, 780 (27 August 1971)).

a) Calculate the time for a 3000-mile trip in each plane if the SST travels at Mach 2.7 and the B-747 at Mach 0.8. Assume that the speed of sound equals 1100 ft/sec.

b) Calculate the radiation exposures for a person in each plane.

Exercise 12. Assume that the distance between stops on a subway is one mile.

a) Calculate the time (in sec) between starting and stopping for average speeds of 20, 30, 40, 50, and 60 mi/hr.

b) Suppose that the subway train stops for 20 sec at a station. What is the average speed between *stations* if the average speed between starting and stopping is 20, 30, 40, 50, and 60 mi/hr?

c) From your results, can you see why there is a fallacy in using high-speed trains if inordinate amounts of time are lost at station stops?

Exercise 13

a) Calculate the times required for a trip of 240 mi at average speeds of 20, 30, 40, 50, and 60 mi/hr. You might want to plot these results, that is, time versus speed, just to get a picture of what is going on.

b) Calculate the time saved by going from 20 to 30, 30 to 40, 40 to 50, and 50 to 60 mi/hr. Note that in each case the change is 10 mi/hr. Again you might want to plot your results in the form of change in time versus speed. Comment on the gains made in time at the higher speeds taking into account the associated risks.

Difficult to Impossible

Exercise 14. Assume that the distance between stations on a subway is one mile. Suppose that the average train speed between actual starting and stopping is 30 mi/hr.

a) What is the time (in sec) between actual starting and stopping of the train?

b) Suppose that the train spends 60 sec at each station. What is the average speed (in mi/hr) between the time the train arrives in one station and then arrives in the next station?

c) Make a sketch of your personal idea of the speed of the train as a function of time between station arrivals.

d) Make a judgment of the importance of the average speed between starting and stopping and the time spent idle in a station.

e) Trains on the BART system are fairly small—6 cars, 72 passengers per car. Can you see the rationale for this?

Exercise 15. If the acceleration of a vehicle is constant, the speed and distance achieved starting from rest are given by $v = at$ and $d = \frac{1}{2}at^2$, where d, v, a, and t are the distance, speed, acceleration, and time, respectively (see Chapter 1).

a) Assume that the acceleration of a subway train is 4 mi/hr per second. Calculate the distance (in mi) and speed (in mi/hr) achieved in 5, 10, 15, 20, 25, and 30 sec and make a sketch of distance and speed versus time.

b) Assume that the distance between stations is one mile and the acceleration and deceleration rates are +4 mi/hr per second and −4 mi/hr per second. Make a sketch of speed versus time for the trip between stations.

c) What is the average speed for the trip?

d) Suppose that the top speed of the train is 60 mi/hr. Make a sketch of speed versus time for the trip again assuming an acceleration of 4 mi/hr per second.

Exercise 16. The overall average speed of a subway train is determined by the maximum tolerable acceleration of the train, the time stopped at the station, and the distance between stops. The top speed attained by the train is not a determining factor.

a) Give a qualitative argument why this is true.

b) Using graphs or numerical calculations, substantiate your argument as given in part (a).

9

ENERGY AND RESOURCES: REMOTE SENSING AND MATERIALS RECYCLING

Photograph courtesy of the Institute of Scrap Iron and Steel.

REMOTE SENSING

Introduction

After centuries of assuming that natural resources are limitless, giving little thought to the consequences of uncontrolled utilization rates, expanding populations, and pollution, we have come to realize the need for better management. And the same technology that must accept much of the blame for creating the problems must now play an important role in providing solutions to them. In so doing, no technology offers greater promise than that of remote (or environmental) sensing. Remote sensing is the art of acquiring information about objects or systems using remote measurements or observations. The concept is clearly not new. The eyes, ears, and olfactory nerves function as remote sensors of light, sound, and odors. A child quickly learns to sense the presence of a hot stove remotely just as a rattlesnake learns to sense the location of his supper (a mouse, for example) by detecting heat radiation. A "water witch" would argue, perhaps even show you, that he can remotely sense the presence of an underground store of water using the fork of a peach tree branch or divining rod as a sensor. The ordinary camera is the most widely used permanently recording remote sensor. The technique of remote sensing has come into prominence because of advance in the technology of sensing devices (data processors) and the vehicles (especially satellites) that carry the devices. The capabilities of remote sensing are broad and impressive. From the standpoint of the environment, the most significant aspect is the ability to obtain information on a worldwide basis. Important environmental information is obtained through the monitoring for climatic effects and pollution levels; the exploration for new sources of fuels, geothermal sources, water and food, inventories of food supplies, and natural resources; and the early warning detection of crop diseases.

Electromagnetic radiation is the single most important carrier of information for remote sensing. This would include radiation emanating from naturally emitting sources such as a heated object, as well as solar, radar, and laser radiation reflected from objects on the earth. The role of energy as the driving mechanism for all environmental systems was pointed out at the beginning of our discussion. Remote sensing enables us to see exactly how vital an element energy is in a scheme for obtaining information about the systems, mainly those associated with the earth, that it drives.

The Technique and an Illustrative Example

The principle of remote sensing emerges quite naturally from the fact that all objects emit and reflect electromagnetic radiation to varying degrees. It is the emission or reflection characteristic of an object which provides the signature that identifies it (Fig. 1). Figure 2 summarizes the types of information obtainable by sensing various portions of the electromagnetic spectrum. The part of the spectrum covered by remote sensing is still smaller as shown in Fig. 3. Nonetheless it is no trivial task to sense even this small part because no single detector responds over this range of wavelengths completely. Ordinary camera film is sensitive to

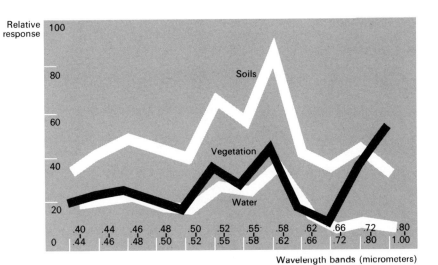

Fig. 1 Radiant energy curves for three types of materials on the earth's surface. The material can be identified on the basis of the amount and type of radiation coming from it. (Courtesy of the Laboratory for Applications of Remote Sensing, Purdue University.)

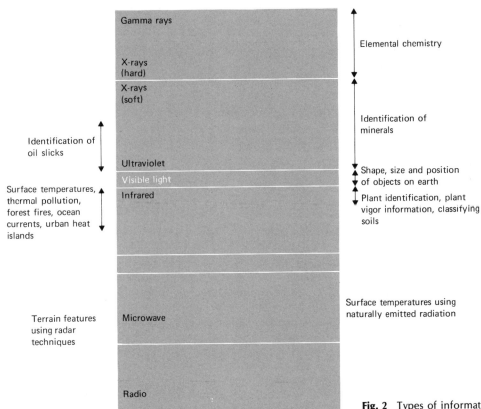

Fig. 2 Types of information obtainable from observations of various parts of the electro-magnetic spectrum.

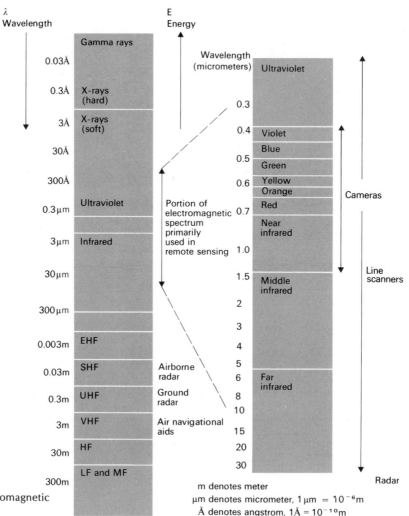

Fig. 3 The portion of the electromagnetic spectrum used in remote sensing.

m denotes meter
μm denotes micrometer, $1\,\mu m = 10^{-6}m$
Å denotes angstrom, $1\text{Å} = 10^{-10}m$

radiation from 0.4 to 0.7 micrometers, which is mostly visible. Special film will record in the infrared up to about 1 micrometer and is extremely useful because this is the range which encompasses much of the reflected solar radiation. However, even if a film were ideal in the sense that it would work over the entire wavelength region of interest, the interpretation of photographs is tedious and not readily accomplished by machine methods. Rather, the choice is often made to use a detector which records the radiation as an electrical stimulus (voltage or current) whose strength depends on the magnitude of the radiation. Two types are commonly used—the photomultiplier tube based on the photoelectric effect (see Appendix 2) and solid state detectors (Fig. 4). Insofar as it is possible, all three

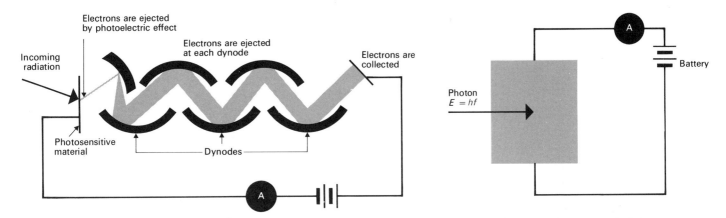

Fig. 4 Principles of the photomultiplier tube and solid state detectors used in remote sensing.

Fig. 5 The four "color" bands typically viewed in a remote sensing experiment.

	Wavelength (micrometers)	Terminology
Violet	0.4	
Indigo		Blue
Blue	0.5	
Green		Green
Yellow	0.6	
Orange		Red
Red	0.7	
		Near Infrared
	0.9	

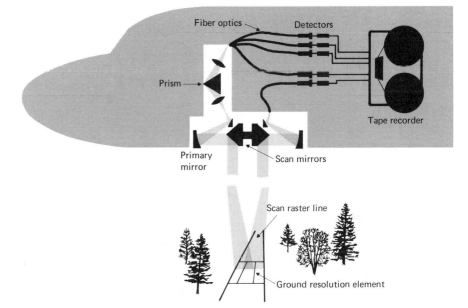

Fig. 6 Schematic layout of an airborne remote sensing system.

detectors, that is, film, solid state, and photomultipliers, are used simultaneously because some effects show up better in particular bands of the spectrum. To eliminate all wavelengths except those of interest, filters can be placed in front of a camera lens. Typically, four cameras are used in sensing the bands shown in Fig. 5. The entire group of detectors is mounted in an appropriate carrier, for example, an airplane or satellite, or on an elevated platform, and the region of interest is then scanned much as a TV camera scans for a TV picture. Data are recorded by the film and electronic data processors on board the carrier (Fig. 6). The sensors actually see a complex distribution of radiation coming from a variety of sources (Fig. 7). So the data processors must be "trained" to identify the sources

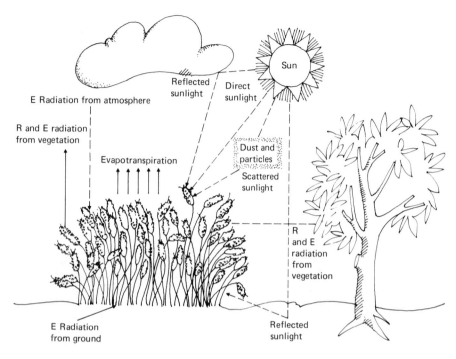

Reflected (R) and Emitted (E) radiation energy exchange in a natural environment

Fig. 7 Illustration of sources of electromagnetic radiation from objects on the earth's surface.

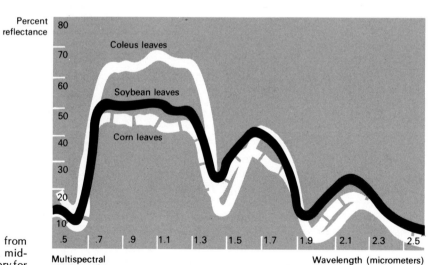

Multispectral response of different kinds of green vegetation

Fig. 8 Example of radiation patterns from three types of vegetation found on a midwestern farm. (Courtesy of the Laboratory for Applications of Remote Sensing, Purdue University.)

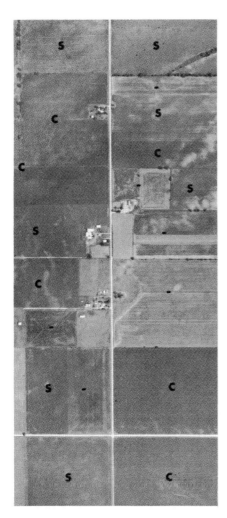 

Fig. 9 Photograph and a computer processed "picture" of a farm area in Indiana. (Courtesy of the Laboratory for Applications of Remote Sensing, Purdue University.)

by comparing their radiation patterns with radiation patterns of known types. Figure 8 shows the type of calibration that is obtained for three common crops in the midwestern United States. By comparing the relative responses, the types of crops can be distinguished. If done electronically with a computer, an area identified as growing corn will be labeled a C. Figure 9 shows an area of a midwestern farm photographed with an ordinary camera and an electronic data processor. The electronic identification is clear. Such a system can be used to inventory crops and food sources rapidly on a very large scale, perhaps even worldwide, if the technique can be extended to satellite systems. An even more impressive application is the detection and evaluation of diseased areas of an expanse of vegetation.

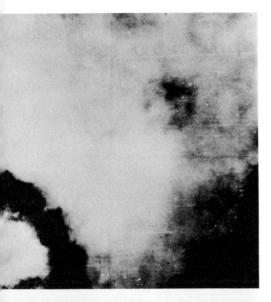

Fig. 10 Conventional and infrared photographs of a forest fire. The fire is clearly delineated when photographed with film sensitive to infrared radiation. (Courtesy of the U.S. Forestry Service.)

Other Applications

The sensing of the infrared spectrum for thermal effects is one of the most important environmental applications. A forest fire viewed from an aircraft is often completely obscured by smoke. Yet if the fire is viewed with a camera sensitive only to infrared radiation, its extent is clearly delineated (Fig. 10) which greatly facilitates the extinguishing procedures. Such a technique is used routinely by the U.S. Forest Service and has saved literally millions of dollars in timber. Figure 11 shows the pattern of the raw sewage flow into Lake Erie that resulted from a breakdown in a sewage treatment plant in Cleveland. Other environmental areas that can be sensed thermally include cities for so-called heat island effects, oceans for delineation of warm stream motion such as the Gulf Stream, geyser areas for geothermal energy sources, and the human body using the technique of thermograms.

Meteorology has been revolutionized by visible and infrared photography of the earth's atmosphere, using satellites. The Nimbus 4 satellite, for example, circles the earth in a polar orbit every 107 minutes at an altitude of about 1100 kilometers and is able to scan the entire globe in a 12-hour period. Such things as the distribution of ozone, the amount of water vapor, and the temperature are determined in the process. Other satellites scan the atmosphere for cloud formations with TV cameras. These weather maps are seen routinely on daily TV news programs throughout the country.

Tornadoes which plague the Plains and the midwestern states in the summer months are known to emit a characteristic electromagnetic radiation in the VHF* region of the spectrum. Remote monitoring of the radiation in these states could provide an early warning detector of this deadly killer.

Air and water pollution are worldwide problems and the only conceivable way of management on this scale is through satellite remote sensing. The techniques, when developed, will provide information on the nature, origin, and concentration of pollutants, as well as the development of trends and patterns with time. The techniques will be invaluable in the efforts to enforce standards invoked by the various authorities. The National Aeronautics and Space Administration's Earth Resources Technology Satellites (ERTS) put into operation in 1972 (Fig. 12) are the first attempt to do monitoring of this type on a worldwide scale. The program is intended to yield both pollution data, including oil spills in the ocean, and maximum information on types and condition of natural vegetation and crops. The ERTS satellites will be in circular, polar orbits 566 miles above the earth's surface and will pass over any given point on the earth every 18 days. The sensing will be done by TV cameras and each picture will cover an area of 100 square miles.

*VHF means Very High Frequency

Fig. 11 An infrared photograph showing the pattern of sewage flowing into Lake Erie as a result of a breakdown in a sewage treatment plant in Cleveland. (Photograph courtesy of NASA Lewis Research Center, Cleveland, Ohio.)

Fig. 12 The ERTS satellite. (Courtesy of the National Aeronautics and Space Administration, Greenbelt, Maryland.)

Remote Sensing of Air Pollutants*

The enforcement of pollution laws and the assessment of air and water quality will be increasingly dependent on the ability to detect and assess the quantities of pollutants in the environment. The conventional technique is to extract a sample (which may be a gas, liquid, or solid) and analyze it in a laboratory using a variety of techniques. Most of these involve chemical methods and go by such names as gas chromatography, spectrophotometry, flame photometry, colorimetry, atomic absorption, ultraviolet absorption, etc. The procedure is tedious and often expensive. A more satisfactory scheme would be to make the assessment without bringing the sample to the laboratory. One might think of placing the analyzer in the area to be assessed and then having the results recorded by electronic means or even transmitted to some central station. Or, in the case of the atmosphere, the instruments might be mounted in an airplane or satellite and flown through the medium to be assessed. All these techniques are, in fact, employed. For example, the Nimbus satellites have measured the ozone concentration in the stratosphere. The National Air Pollution Control Administration (now E.P.A.) has operated the Continuous Air Monitoring Program (CAMP) in six major cities for over ten years. The national Air Surveillance Network tends some 200 manned stations which collect and mail samples to a central laboratory for processing. Although all these techniques are extremely useful, there is a pressing need for more remote-type assessing procedures. For example, how do you determine the amount of particulates emerging from a smokestack without actually taking a sample of the smoke? One of the most promising technologies for this utilizes a laser.

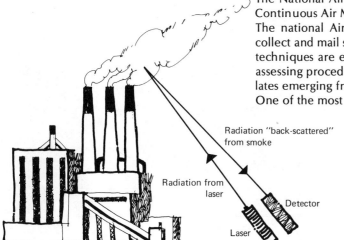

Radiation "back-scattered" from smoke

Radiation from laser

Detector

Laser

Fig. 13 A beam of radiation from a laser is directed toward particulates rising from a smokestack. The amount of radiation scattered back toward the laser is related to the particulate concentration.

As discussed in the section on fusion energy, the laser emits intense electromagnetic radiation having a nearly constant wavelength. This radiation can then be scattered from or selectively absorbed by pollutants. For example, as suggested by Fig. 13, the radiation could be scattered back toward the laser source by particulates emerging from a smokestack.† Since the amount of radiation returning is

---

*For a comprehensive treatment, see *Remote Measurement of Pollution*, NASA Publication SP-285. For sale by National Technical Information Service, Springfield, Virginia, 22151. Price: $3.

"Lidar Application in Air Pollution Research and Control", W. B. Johnson, *Journal Air Pollution Control Association* **19**, 176 (1969).

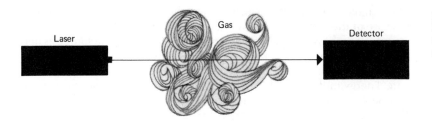

Fig. 14 Selective absorption of laser radiation by atoms or molecules in a gas can be used to determine the amount and identity of the absorbers.

related to the particulate concentration, a measure of this back-scattered radiation is a measure of the particulate concentration. The laser beam could be directed through a gas (Fig. 14). Atoms or molecules could selectively absorb the radiation. Knowing what type of molecule will absorb the frequency of the laser radiation makes identification possible. This technique is somewhat limited because of the precision of the frequency of the laser radiation. Only molecules which absorb at precisely this frequency will be affected and the number of such molecules is limited. If the frequency of the laser could be varied, and this is a possibility, the technique could be expanded.

A more promising technique utilizes the Raman effect. Some molecules will produce a shift in the frequency of the radiation that scatters from them. The magnitude of the frequency shift is characteristic of the molecule which initiates the scattering and thus provides a signature which can be used to identify the molecule. The feasibility of quantitatively assessing the amounts of SO_2 and CO_2 in the air environment has been demonstrated* and commercial instruments are becoming available.

ANOTHER CATEGORY OF REMOTE SENSORS

Introduction

Most of the remote sensors discussed thus far yield information about areas on the earth with dimensions of the order of miles and, in many cases, hundreds of miles. A camera in a satellite is certainly remote on this distance scale. Yet a detector a few inches from a source of electromagnetic radiation is remote on the scale of atomic or nuclear dimensions. There are a number of remote sensors in this second category that provide a wealth of information about the types and amounts of various pollutants and contaminants in the environment. These sensors are based on the principle that an atom, molecule, or nucleus in a state of excess energy will release electromagnetic radiation with well-defined wavelengths. Since no two

---

†"Spectroscopic Detection of SO_2 and CO_2 Molecules in Polluted Atmosphere by Laser-Raman Radar Technique," Tadao Kobayasi and Humio Inaba, *Applied Physics Letters* **17**, No. 4, 139 (Aug. 15, 1970).

atoms, molecules, or nuclei will radiate the same spectrum of wavelengths, a measure of the emitted radiation identifies the source. For example, a measure of the wavelengths of the light from a neon sign identifies the source as neon atoms. However, if a container of cold neon gas is viewed, this radiation does not appear and the reason is that all the neon atoms are in their lowest (or ground) energy state. Energy must be fed into the gas to get the atoms into states of excess energy. In a neon sign, this is done electrically by creating a discharge in the gas with an electric potential applied between two electrodes. Atomic or nuclear sensors, therefore, require that the radiation be induced by some appropriate energy source. Energetic electrons, neutrons, and X-rays are extremely useful energy sources for this purpose.

Neutron Activation Analysis*

Potassium-40 (^{40}K) is a radioactive nucleus with a mean lifetime of 1.9×10^9 years which emits a 1.460 MeV gamma photon. It is a nuclide commonly found in building materials such as bricks, concrete, mortar, and natural stone. Its radiation can be detected in the background of nearly any room and it does not take a chemical analysis of the building material to identify it. Rather, a precise measurement of the energy of its gamma photon is sufficient. And this is true for any radioactive nucleus. Such remote identification of a constituent of a substance has the obvious advantage of being nondestructive. But is is clearly limited to elements that are radioactive. Neutron activation analysis utilizes this principle by activating (that is, making radioactive) elements through bombardment with neutrons.

A neutron impinging on a nucleus can produce a variety of nuclear events such as fission and nuclear transformation. It may also just scatter like a billiard ball bouncing off another billiard ball. It may also be captured by the target nucleus (Fig. 15). In this process, called neutron capture, the new nucleus formed is an isotope of the target because both have the same number of protons. For example, the most abundant isotope of oxygen is $^{16}_{8}O_8$ containing 8 protons and 8 neutrons. If $^{16}_{8}O_8$ were to capture a neutron, it would then be $^{17}_{8}O_9$. Both $^{16}_{8}O_8$ and $^{17}_{8}O_9$ are isotopes of oxygen, and the electrically neutral atoms having these as nuclei would have the same chemical properties. Neutron activation analysis uses a source of neutrons, normally from a nuclear reactor, to impinge on nuclei of interest. The nuclei that capture neutrons are usually left in a state of excess energy. The excess energy is often released as gamma photons, a measure of which identifies the isotope.

The spectrum of gamma ray energies from different radionuclides are not the same, but the differences are often slight. It is necessary, then, to have a detector capable of distinguishing these subtle differences in energy. The solid state germanium detector is a major advancement in this technology. This detector is capa-

Neutron   Target   New nucleus formed
                 nucleus   by neutron capture

Fig. 15 Schematic illustration of the process of neutron capture. The nucleus formed from the capture of a neutron is usually unstable and releases energy in the form of gamma radiation.

---

*"Isotopes in Environmental Control," 16 mm, color, 14 min. Shows some of the ways radioactive isotopes are being used to help preserve and restore the environment. Free loan. Audio-Visual Branch, Dept. of Public Information, U.S. Atomic Energy Commission, Washington, D.C., 20545.

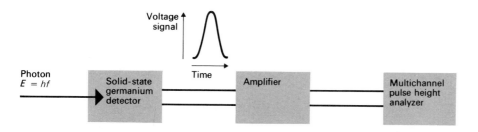

Fig. 16 Schematic illustration of a high resolution electronic system for counting and measuring the energy of gamma ray photons. The detector produces a voltage signal whose amplitude is proportional to the photon energy. This is a little like producing a pulse in a stretched string by giving it a sharp displacement at one end.

ble of distinguishing between photon energies of 1.111 and 1.114 MeV, for example. The physics of its operation is difficult, but its function is to produce a voltage signal whose strength is proportional to the energy of the photon. By counting these voltage signals, one is effectively counting the photons and determining their energies. Physicists use a so-called multichannel pulse height analyzer which automatically sorts and counts these signals. The scheme is shown in Fig. 16. The results from the multichannel pulse height analyzer can be presented as a visual display on a screen much like a TV picture, plotted as a graph, or fed to a digital computer to be analyzed for identification of the emitting source. Figure 17 shows the visual form of the output from a system like the one depicted in Fig. 16.

Although neutron activation analysis is capable of determining concentrations as low as a trillionth of a gram (10^{-12} g), it is not a trivial task. The reason is that when a sample is bombarded with neutrons, many isotopes are activated. So when the sample is placed in the gamma detecting system, photons from all these activated elements are detected, and the gamma ray spectrum which results may be highly complex as shown in Fig. 18. Individual contributions to this total spectrum must be determined by comparing known gamma ray spectra with the unknown. Sometimes this can be done visually, but it often requires complex and elegant computer programs to do the "unfolding." Figure 18 shows how the constituents of an oil slick are identified by comparing known gamma ray spectra with the spectrum obtained by activating the oil slick with neutrons.

There is much concern about the health hazard of trace amounts of metals in the environment, especially in water. This would include such things as mercury, vanadium, selenium, lead, chromium, etc. Neutron activation analysis plays an important role in determining the concentrations of these elements.

Neutron activation analysis has the distinct disadvantage of not being able to determine how the element is tied up chemically in a sample. For example, if mercury were detected in a fish by neutron activation analysis, there would be no way of determining whether the mercury was in elemental form or in the form of a compound like methyl mercury. And this knowledge is extremely important because the potential health hazard to humans is strongly dependent on the type of compound.

The intense neutron source required for activation has in the past come from nuclear reactors, a practice which has severely limited the technique as an environ-

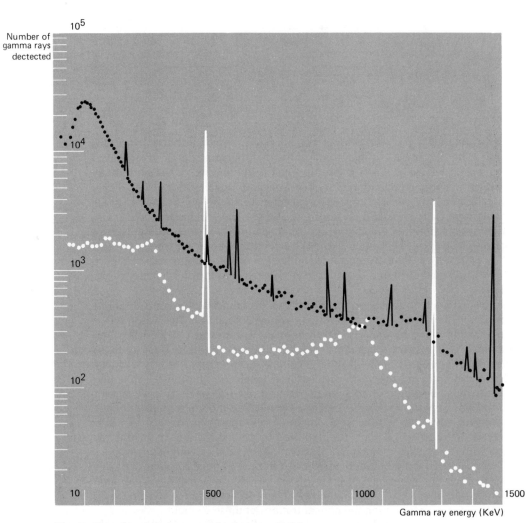

Fig. 17   The white line represents the visual form of the output from a system like that depicted in Fig. 16. A ^{22}Na (sodium) source which yields two gamma rays having energies of 511 and 1275 keV was used. These gamma rays produce two sharp peaks positioned at these energies along the horizontal axis. The black line represents a gamma ray spectrum due to background radiation in a room. The many peaks are due to a variety of sources. The pronounced peak at about 1460 keV is due to ^{40}K (potassium) which is found in nearly all building materials.

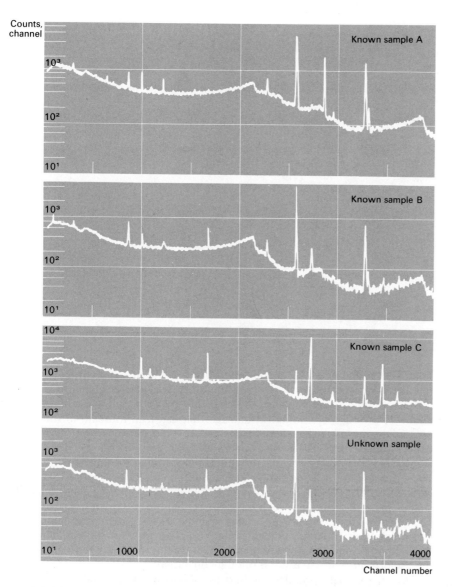

Fig. 18 Identification of constituents of an oil slick using neutron activation analysis. The trace element fingerprint of the unknown sample closely matches, and is therefore from the same source as, known sample B. (From "Trace Analysis—the Nuclear Way," *Industrial Research*, Sept. 1971, p. 48. Drawing courtesy of Joseph John, H. R. Lukens, and H. L. Schlesinger, Gulf Radiation Technology.)

mental field test service. However, intense neutron sources are now being made from artificially produced californium-252 (^{252}Ca). These portable sources greatly enhance the utility of the neutron activation technique as an environmental tool.

X-ray Fluorescence

The molecules that comprise living things such as the human body are constituted primarily from light atoms such as carbon, oxygen, nitrogen, etc. In fact, more

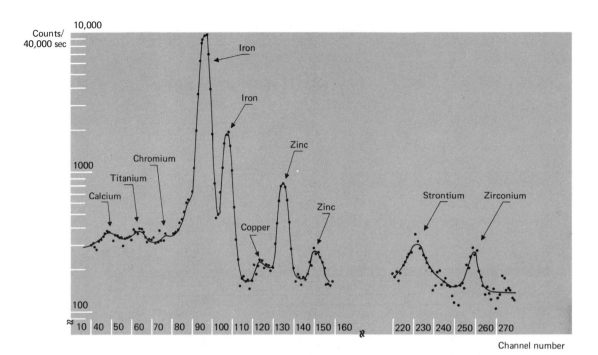

Fig. 19 Analysis of a sample of water from a river which receives effluents from a variety of industries. (Courtesy of Professor C. R. Cothern, University of Dayton.)

than 99% of the structure is composed of 12 of the first 20 light elements. The small fraction remaining is reserved for the more massive elements which are just as essential to human life. In either too small or too large amounts, these elements produce physiological damage that can lead to death in some cases. Most of the more massive elements are metals, which, because they exist in small concentrations, are called trace metals. A trace amount can be defined as an amount less than 0.01% of the structure being studied. Among the metals considered essential for living tissues are aluminum, antimony, mercury, cadmium, lead, silver, and gold.* Many of these elements are entering the human food chain from water supplies which have been contaminated by industrial processes. The effect of trace metals on human systems makes it extremely important to determine their concentrations in water and food sources. This can be done by the method of neutron activation analysis. Since this technique normally requires the use of a nuclear reactor, it is often inconvenient.

An increasingly popular method for assessing trace metal concentrations utilizes the principle of X-ray fluoresence.† The principle including the photon detection and counting facility is very similar to neutron activation analysis

---

*"Trace Elements: No Longer Good vs. Bad," *Science News* **100**, 112 (August 14, 1971).

†Most contemporary watch dials that shine in the dark work on the principle of fluorescence. It is instructive to take such a watch and illuminate it with an ultraviolet light source which stimulates the emission of visible light from the watch dial.

(Fig. 16). An energy source is used to excite *atoms* to higher energy states. The states excited involve the most tightly bound electrons near the nucleus. The energies of these excitations are in the X-ray energy range. Thus the atoms lose their excess energy by emission of X-ray photons whose energies are characteristic of the species of atoms involved. The energy is supplied by an X-ray source having energy of the order of, but somewhat larger than, the energy of the X-rays emitted. Historically, the source was a conventional one which produced X-rays by "slamming" energetic electrons into a block of metal like tungsten. Such sources tended to be bulky and cumbersome. The advent of nuclear reactors has brought the development of intense nuclear X-ray sources which are the size of a pin head. Although every spectrum of energies of X-rays from an atom is unique to the atom, the differences are slight if the atomic numbers of the atoms are close. There has been a technological breakthrough in the detection of X-ray photons using a solid state detector which has permitted greater resolution of these X-rays. These detectors are very much like the ones used to detect gamma rays in neutron activation analysis, although the X-ray detectors are usually made of silicon and the gamma ray detectors of germanium. Even with developments in sources and detectors, the technique is not sufficiently precise to distinguish all atoms. However, it does work well for atoms with atomic numbers between about 22 (titanium) and 82 (lead) which include the trace metals of interest. Current technology can detect concentrations of these trace metal contaminants as low as 20 to 30 parts per billion. Figure 19 shows a sample of water in which trace metals of iron, zinc, calcium, titanium, chromium, strontium, and zirconium were detected.

MATERIALS RECYCLING AND SOLID WASTES*

Entropy and Waste

When a person steps out of his warm slippers onto a cold bathroom floor, heat flows spontaneously from his feet to the floor and his feet feel cold. He could warm them by rubbing them briskly with his hands in which case mechanical energy would be converted to heat. These are two everyday examples of the *natural* processes involving thermal and mechanical energy. The natural flow of heat is from a hot object to a cold object and the natural transformation of energy is from mechanical to thermal. Any *unnatural* process which violates these conditions must be accompanied by a natural process which at least compensates for the effects of the unnatural process. For example, heat is removed from the inside of a refrigerator to warmer surroundings, but this is more than compensated for by the natural transformation of mechanical energy to thermal energy by the refrigerator's motor. The thermal energy of steam in a boiler is transformed into mechanical energy only because it is accompanied by the natural flow of heat

---

*"Things Worth Saving," 16 mm, color, 14 minutes. Free loan to schools and civic groups. National Center for Resource Recovery, 1211 Connecticut Ave., N.W., Washington D.C. 20036.

from the boiler to a colder reservoir. The ultimate fate of a neat, *orderly* stack of cans in a supermarket is an untidy, often repulsive, *disorderly* junkyard. The *natural* process is from a state of order to one of disorder. This and many other environmental processes and the thermodynamic examples discussed at the beginning of this section are related to an extremely important concept called entropy. Entropy can be thought of as a measure of the degree to which transformations are natural or, approaching the concept from a different stance, as a measure of the disorder of a system. Entropy increases as the disorder increases. In a natural thermodynamic process, the entropy always increases. In an unnatural thermodynamic process, the entropy always decreases. When one considers the functioning of a system in its entirety, a refrigerator, for example, the total entropy of all the processes involved cannot decrease. It may be constant, but it cannot decrease. This turns out to be another way of formulating the second law of thermodynamics.

All around us there are many things which are in a state of order. The stack of cans in the supermarket or a neat table setting at dinner or a shiny new automobile are examples. These states were achieved, however, only at the expense of using *energy*. The natural process for these systems is toward disorder and they will not be reordered without the expenditure of energy. And that is why it is difficult to get children to tidy up after dinner and why the automobile ends up in a highly disordered state in someone's backyard.

The can in the supermarket and the automobile on the road evolved from a source of nearly constant mass, the earth. Man merely redistributes and disperses this mass through use of available energy. The can and automobile are temporary intermediaries. Much of the discarded wastes are salvaged for glass, paper, steel, copper, brass, aluminum, lead, zinc, nickel, etc., and are recycled back into the manufacturing industry. Still, there is an enormous accumulation of waste in localized areas of high commodity consumption and a depletion of natural resources in other localized areas. The entropy increases in both. Such problems demand that our resources-to-junk philosophy be changed somehow by recycling more of the junk back into the commodity cycles. This requires the exploitation of several interesting physical principles. To illustrate, let us consider two major waste problems—municipal waste and junked motor vehicles.

Municipal Solid Wastes

The solid waste collected in the United States in 1968 averaged 5.32 pounds per person per day.* In 1971, New York City discarded trash at the rate of 24,000 tons each day and the rate grows at 4% per year. Throughout the country, about 90% of this trash ends up in dumps or landfills where it is covered with dirt. Ninety-four percent of the land disposal sites are considered inadequate.* There is an acute shortage of land space for waste disposal in many areas. This is another fundamental reason why alternate means of waste disposal and new recycling methods must be developed.

Cleaning Our Environment, The Chemical Basis for Action, American Chemical Society, (1969).

Table 1. Typical composition of municipal solid waste. (Source: Kaiser, E. R., "Refuse Reduction Processes," in "Proceedings, The Surgeon General's Conference on Solid Waste Management for Metropolitan Washington," U.S. Public Health Service Publication No. 1729, Government Printing Office, Washington, D.C., July 1967, p. 93.)

Physical	Weight percentage	Rough chemical	Weight percentage
Cardboard	7%	Moisture	28.0%
Newspaper	14	Carbon	25.0
Miscellaneous paper	25	Hydrogen	3.3
Plastic film	2	Oxygen	21.1
Leather, molded plastics, rubber	2	Nitrogen	0.5
Garbage	12	Sulfur	0.1
Grass and dirt	10	Glass, ceramics, etc.	9.3
Textiles	3	Metals	7.2
Wood	7	Ash, other inserts	5.5
Glass, ceramics, stones	10		100.0
Metallics	8		
Total	100		

Table 1 shows a typical composition of municipal refuse. The refuse is enough of a conglomeration that one can only wonder if anything other than burying it is possible. And even if there are constituents of interest, it is clear that some subtle physical differences will have to be exploited to effect a separation. If one looks at the chemical content, there are indications of possible gains. First of all, a large fraction of the refuse is water and this could at least be cleaned up and recycled back to the environment. Secondly, carbon and oxygen comprise nearly 50% of the chemical composition and this means energy potentialities. These solid wastes constitute a fuel with a heating value of about 5000 Btu/lb which is nearly half that of commercial coal. Furthermore it has very low sulfur content. Over 50 years ago, the city of Paris began burning garbage to generate electric energy and today enough is generated in the process to supply the needs of a city of 50,000. In addition, the rejected thermal energy provides steam heat for hundreds of Paris buildings. A similar system is being developed in this country at Palo Alto, California. It would produce 15 megawatts of electric power by burning 400 tons of solid wastes each day. Another such system is proposed to be built in northern New Jersey. It would produce 150 megawatts of power by burning 6000 tons of refuse per day. The main motivation for this proposal comes from a serious daily loss of land of about one acre per day in the New Jersey Meadowlands. Burning solid wastes just for the energy content does not fully exploit its potentialities because ashes which remain contain considerable amounts of glass and valuable metals. Furthermore the paper content is potentially more useful as a recycled paper product than as a source of fuel. To achieve separation of these components requires some very sophisticated techniques which take advantage of subtle differences in physical properties. For example, metal and glass can be separated from the main body of material because they do not dissolve in water. The system

Fig. 20 Photograph of samples of material separated from unsorted municipal solid waste by the Hydrasposal/Fibreclaim system. (Courtesy of Black Clawson Co., Middletown, Ohio.)

METALS

GLASS CULLET

FOURTH FOREST PULP

employed to separate them out would work like the cyclone separator used to separate particulates from the gaseous effluents in a smokestack. Much of the solid material in the waste is rich in iron. An ordinary "tin" can, for example, is mainly iron. Iron can be easily magnetized whereas aluminum and glass are not so easily magnetized.* The materials rich in iron can then be separated magnetically. It is not so easy to separate the nonmagnetic materials from the glass but it can be done on the basis that metals are good conductors of electricity whereas glass is not. Glass can even be separated into different color categories on the basis of different optical properties. Systems which effect separation of the wastes to the extent indicated here are very recent and small-scale activities. One such facility is Hydrasposal/Fibreclaim™ developed by the Black Clawson Company for the city of Franklin, Ohio. It is a system which not only separates the metals and glass from the wastes but produces a salable paper product (Fig. 20). It is capable of processing 150 tons of refuse each day which, on the basis of 5 pounds of solid wastes per person per day, satisfies the needs of a city of 60,000. The main motiva-

*The top of many "easy-opener" soft drink cans is aluminum, but the wall is rich in iron. It is instructive to demonstrate the different magnetic properties of the top and wall.

(a) Natural resources, manmade. Giant shredding machines, employing a system of steel "hammers," rip automobiles into fist-size pieces of scrap. Costing between $400,000 and $4 million, shredders are capable of processing from 25,000 to 250,000 tons of scrap per year. (Photograph courtesy of the Institute of Scrap Iron and Steel.) (b) This mountain of steel fragments is the remains of 30,000 old automobiles which have been processed for remelting by steel mills and foundries. (Photograph courtesy of the Institute of Scrap Iron and Steel.)

tion for the plant was that the local community that it served had run out of land-fill space for a conventional disposal system.

The Recycling of Junked Automobiles

The disposition of junked automobiles is a particularly vexing problem. About 2500 cars a day are abandoned throughout the United States. Over 100 are abandoned each day in New York City. The reason for this is economically clear. The recoverable metal in an automobile amounts to about $56* (Table 2). It takes about $51 of processing to retrieve that $56. Besides, it costs about $50 to have a disabled car towed to a processing center. In a 1965 study of an area containing 510,000 junked cars only 6% were shown to be in the hands of scrap processors. Seventy-three percent were in the hands of auto wreckers and the remaining 21% were

Table 2. Value of recoverable metals in an automobile

Metal	Lb	Price	Value
No. 2 bundle iron	2,614	$18.70 per ton	$24.44
Cast iron	429.3	42.20 per ton	9.06
Copper:			
Radiator stock	15.4	0.3275 per lb	5.04
No. 2 heavy and wire	13.8	0.396 per lb	5.46
Yellow brass solids	2.7	0.31 per lb	0.84
Zinc, die castings	54.2	0.0625 per lb	3.39
Aluminum, cast, etc.	50.6	0.124 per lb	6.27
Lead:			
Battery	20.0	1.40 per battery	1.40
Battery cable clamps	0.4	0.11 per lb	0.04
Totals	3,200.4		$55.94

Fig. 21 Location of automobile scrap shredders that are either operating, under construction, or planned as of April 1969.

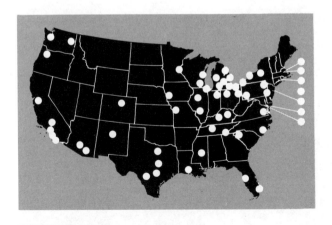

The Automobile Cycle: An Environmental and Resource Reclamation Problem, U.S. Dept. of Health, Education and Welfare (1970).

in auto graveyards.* However, if large-scale operations can be effected, the automobile scrap industry can be profitable. Such operations wherein the automobile is either shredded into pieces, incinerated, or compressed and baled are becoming commonplace (Fig. 21). The metals are then separated by a variety of sophisticated techniques that exploit differences in the physical and chemical reaction properties of the materials.

REFERENCES

Remote Sensing

Aviation Week and Space Technology 44–45 (Mar. 30, 1970).
124+ (June 22, 1970).
60–62 (Oct. 5, 1970).
44–45 (Dec. 21, 1970).
42 (Mar. 15, 1971).
130 (Apr. 29, 1968).

Bio Science **17** No. 7, 444–449 (July 1967).

Bio Science **17** No. 6, 384–390 (June 1967).

Bio Science **17** No. 1, 450–451 (July 1967).

Encyclopedia Britannica **10**, 199–200.

Fortune June 1, 1968.

National Geographic **135**, 45 (January 1969).

National Wildlife (Feb., Mar., 1971).

Remote Sensing in Ecology, Philip L. Johnson (Editor), University of Georgia Press, Athens, Georgia, 1969.

Saturday Review **54**, 53–57 (April 3, 1971).

Scientific American **218**, 54–69 (January 1968).

Science Digest **46**, 66–69, (August 1959).
63, 39–45 (January 1968).
66, 63–66 (July 1969).
69, 67–72 (February, 1971).

Materials Recycling and Solid Wastes References

Cleaning Our Environment: The Chemical Basis for Action, American Chemical Society, Washington, D.C. (1969).

---

*Cleaning Our Environment, The Chemical Basis for Action, American Chemical Society (1969).

"Physics Looks at Waste Management," David J. Rose, John H. Gibbons, and William Fulkerson, *Physics Today* **25**, No. 2, 32 (February 1972).

"Bottles, Cans, Energy," Bruce M. Hannon, *Environment* **14**, No. 2, 11 (March 1972).

"Disposal of Junked and Abandoned Motor Vehicles," printed for the use of the Committee on Public Works, 91st Congress, First Session, Superintendent of Documents, Washington, D.C.

"The Automobile Cycle: An Environmental and Resource Reclamation Problem," U.S. Department of Health, Education, and Welfare, Bureau of Solid Waste Management (1970).

"Entropy," George Gamow, in *The World of Physics*, edited by Arthur Beiser, McGraw-Hill Book Company, Inc., New York (1960).

QUESTIONS

1. What are some possible ways of remotely sensing the presence of weapons on commercial airlines passengers?

2. What are some things that people in homes could do to facilitate the recycling of household trash?

3. How could neutron activation analysis be used to determine the presence of mercury in fish?

4. What are some possible ways of preventing or, at least, discouraging the abandonment of automobiles?

5. The size and structure of a physical system is often determined by the scattering of a particle or some electromagnetic radiation from the system. For example, the size of a nucleus can be determined by the alpha particles scattered by the nucleus. One requirement is that the wavelength of the scattering radiation be about the same size as a characteristic dimension of the object investigated. What sort of electromagnetic radiation would be useful for determining the size of nuclei? atoms? molecules? airplanes? mountains?

6. Most detectors of wave phenomena are not equally sensitive to all frequencies or wavelenghts. The ear, for example, is not equally sensitive to sound frequencies between 20 and 20,000 Hz. The human eye and most camera film is not equally sensitive to all wavelengths of visible light. What color or wavelength of light would you suspect the eye to be most sensitive to? Can you think of any everyday observations to back up your answer?

7. Think of some ways that neutron activation analysis and X-ray fluorescence might be used in the solution of some of our environmental problems.

8. How could you determine the roaming habits of an animal by remote sensing?

9. Some systems are easy to reverse. A motor, for example, is an electric generator running backwards. Discuss the energy aspects of the reversibility of a heat

engine and of the freezing of water. What are some other systems that are reversible? Many systems are difficult to reverse even if ample energy is available. Reversing the aging process of a human is an example. The conversion of trash back to useful products is another example. Discuss the difficulties of reversing this system.

10. Discuss some environmental situations where the concept of entropy is at work. You might consider strip mining of coal, the aging process in humans, the deterioration of a building, or the breaking of an egg, for example.

11. Name some animals that have remote sensing capabilities for navigation.

12. Why is a setting sun particularly red on a hazy, perhaps polluted, day?

13. What are some physical characteristics of glass and metal that could be used to separate them in a trash recycling system?

EXERCISES

Easy to Reasonable

Exercise 1. $^{75}_{33}As_{42}$ is the only stable isotope of arsenic. What nucleus would be if $^{75}_{33}As_{42}$ were to capture a neutron?

Exercise 2. Why does an object made of iron, for example, change color as it is heated?

Exercise 3. How hot would an object have to be in order for the radiation to be predominantly ultraviolet?

Exercise 4. The half-life of a sample of radioactive nuclei could be defined as the time required for 50% of the sample to decay. Using this concept, determine the half-life for automobiles from the graph below.

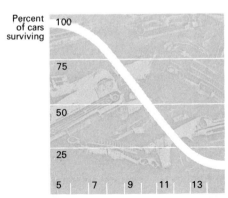

Percent of cars surviving

100

75

50

25

5   7   9   11   13

Years after original registration

Photograph courtesy of the Institute of Scrap Iron and Steel.

Exercise 5. Estimate the volume of trash accumulated and disposed of by a university student each week. Multiply this by the total number of students and figure out how many typical rooms full of trash are accumulated each week.

Exercise 6. In 1971, the city of New York disposed of 24,000 tons of trash each day. This disposal is increasing at a rate of 4% per year. If the disposal continues at this rate, how much trash will be disposed each day in the year 2000?

Exercise 7. Discarded trash in most urban communities amounts to about 5 pounds per person per day. A typical trash can holds about 35 pounds of trash.

  a) How many cans would a city of one million people fill each day?

  b) A trash can is about 3 ft high. If the cans for this city of one million people were stacked end-to-end, how many mileshigh would the stack be?

Exercise 8. On a sunny, summer day a black asphalt road often gets so warm that it is unbearable to walk on it with bare feet. Yet there is no problem walking on the white centerline of the road. Explain.

Exercise 9. Thermal radiation varies directly as the fourth power of temperature in Kelvins, that is, $R \propto T^4$. What is the ratio of radiant energies from two identical objects at 300K and 400K?

Exercise 10. It is common to go near a campfire to get warm. Would it be possible to go *near*, but not in contact with, a large block of ice to get cool? Explain.

Difficult to Impossible

Exercise 11. Affluence as measured by the gross national product per person correlates with per capita energy consumption in all countries of the world (Fig. 2 in Chapter 1). To what extent might you expect affluence to be related to per capita trash production? Estimate the yearly per capita trash production for the United States on a basis of 5 pounds per person per day and then estimate what you might expect it to be for India on the basis of the relative affluence between the two countries.

Exercise 12. Ten percent of the budget of a typical community is spent on collection and disposal of solid waste. This adds up to $4.5 billion annually in the United States. An efficient collection and disposal system is vital to a community and the efficiency can be ensured by establishing a collection route and then minimizing the collection time. For a particular system (see "Balancing Waste Collection Routes" by Dean S. Shupe and Richard L. Shell, *Journal of Environmental Systems* **1**, No. 4, 367, Dec. 1971), it is found that the collection time, T_w, is

$$T_w (\text{min}) = 2.57\, T + 0.347\, L - 3.43\, H - 0.00516\, F + 0.0124\, S - 6.93\, Tr + 260.96$$

where T = average number of tons collected, L = average number of loads times distance to incinerator, H = number of helpers, F = number of families, S = number of stops, and Tr = number of trucks.

  a) What are the most important variables in this equation?

b) Why are some of the terms negative and some positive?

c) What is the significance of the constant term 260.96?

d) Since the constant term is so large, to what extent does it affect the reliability of the equation?

Exercise 13.   Aluminum has grown in popularity as a food and beverage container because of its strength, lightness, and durability. It is refined from a compound called bauxite ($Al_2O_3(H_2O)_x$) by a process which requires large amounts of electric energy. It takes the coal equivalent of 56,000 kilowatt-hours to refine 1 ton of aluminum (see "Physics Looks at Waste Management," *Physics Today*, February 1972).

a) Show that this amount of energy amounts to 12.6×10^{29} eV.

b) The atomic mass of aluminum is 27. A mass in grams equal to the atomic mass always contains Avogadro's number (6.023×10^{23}) of atoms. Show that the number of atoms in one ton of aluminum is 2.0×10^{28}. (1 ton = 2000 lb; 1000 g = 2.2 lb)

c) Show that the energy per atom for refining aluminum is about 60 eV/atom.

d) To remelt an aluminum container amounts to about 1 to 2 eV/atom. Make a case for recycling aluminum just on the basis of energy considerations.

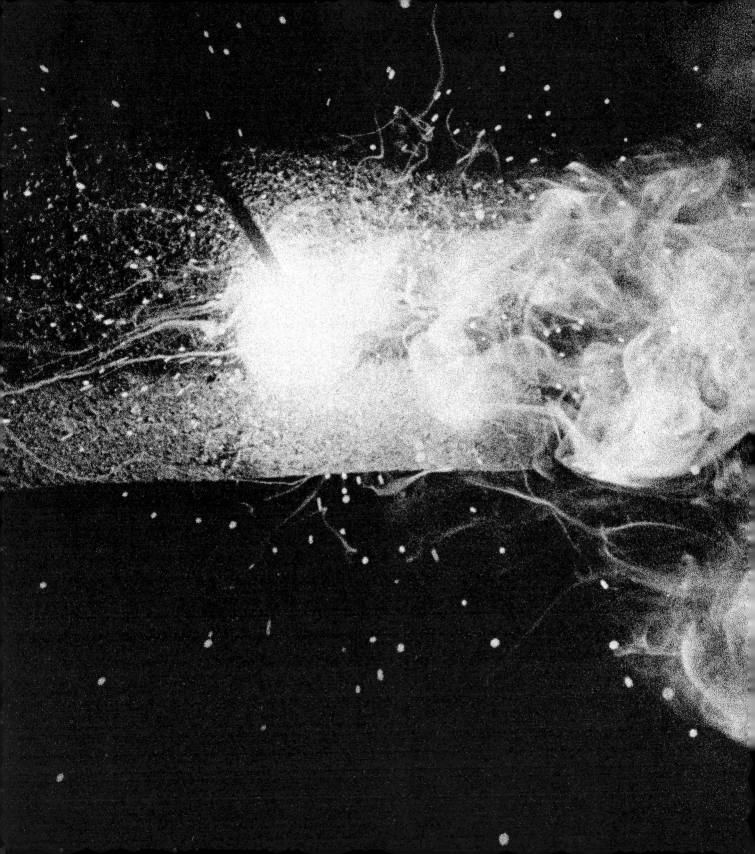

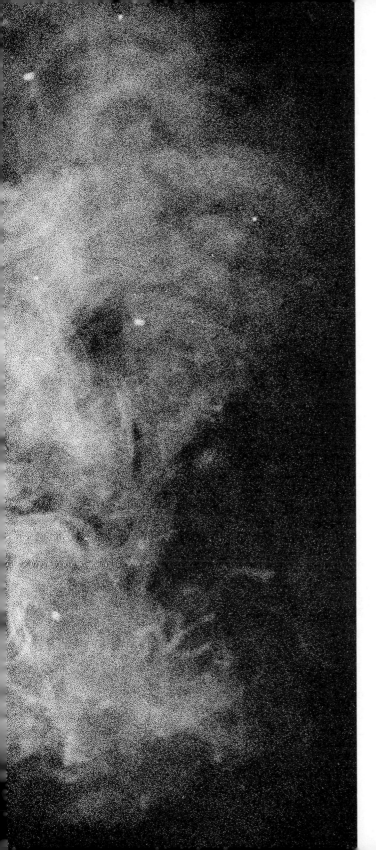

APPENDICES

Photograph courtesy of Bruce Anderson.

Appendix 1

POWERS OF TEN
AND LOGARITHMS

POWERS OF TEN

The range of numbers encountered in assessing the environment is staggering. For example, the radius of the nucleus of the aluminum atom is 0.00000000000036 cm, and the U.S. electrical energy production in 1968 was 1,330,000,000,000 kW-hr. Multiplying, dividing, and just keeping track of numbers like these is tedious without good bookkeeping.

A number like 1,330,000,000,000 represents a measurement or an estimate and there is, therefore, some uncertainty in the number. If the number were known to an accuracy of 1%, which normally would be quite good, it would mean that the true value would, perhaps, be between 1,317,000,000,000 and 1,347,000,000,000. At best, only the first four digits, that is, 1, 3, 3, and 0, are significant. The remaining zeros denote the relative size or magnitude. For example, the number might be written as 1,330 million kW-hr. Million denotes the magnitude and is used as book-keeping procedure, albeit not a particularly scientific one in the sense that it does not lend itself to easy multiplication and division. To illustrate a better scientific method, consider the numbers 0.0015 and 1500. These numbers could also be recorded as

$$1.5 \div 1000 \quad \text{and} \quad 1.5 \times 1000.$$

The 1000 figure denotes the magnitude. Now 1000 can be generated by multiplying $10 \times 10 \times 10$. A shorthand way of symbolizing this operation is 10^3. Using this notation, we can write

$$0.0015 = 1.5 \div 10^3 = \frac{1.5}{10^3}$$

$$1500 = 1.5 \times 10^3$$

It is conventional and meaningful to write $1/10^3$ as 10^{-3} so that $1.5/10^3$ becomes 1.5×10^{-3}. The 3 in 10^3 is called the power of 10. The beauty of this method is that the appropriate power of ten can be obtained by counting the positions the decimal point is moved.

$$0.0015 = 1.5 \times 10^{-3} \qquad \text{Decimal point is moved 3 places to left}$$

$$1500. = 1.5 \times 10^3 \qquad \text{Decimal point is moved 3 places to right}$$

To convince yourself of the utility of the method, express the following numbers as a number times ten to a power.

0.00000001 cm (approximate diameter of an atom)
30,000,000,000 cm/sec (speed of light)

Express the following numbers in decimal form.

9.11×10^{-28} g (mass of the electron)
3.31×10^4 cm/sec (speed of sound in air at one atmosphere pressure and 0°C).

You should be convinced of the usefulness of the method if for no other reason than compactness. The real utility is in multiplication and division because it reduces much of the effort to that of addition and subtraction. To illustrate, consider

$(1000) \times (1000) = 1,000,000$ (long method)

$10^3 \times 10^3 = 10^{3+3} = 10^6$ (short method)

The general rule is $10^a \cdot 10^b = 10^{a+b}$, where a and b are the powers of 10. There are no restrictions on a and b; they can be positive, negative, and even fractional. The reason for writing $1/10^3$ as 10^{-3} can now be seen.

Example:

$$\frac{1,000,000}{1000} = 1000 \text{ (long method)}$$

$$\frac{10^6}{10^3} = 10^6 \cdot 10^{-3} = 10^{6-3} = 10^3 \text{ (short method)}$$

Example:

$$\begin{aligned}
(5.1 \times 10^4) \cdot (3.2 \times 10^{-1}) &= 5.1\,(3.2) \times 10^{4-1} \\
&= 16.32 \times 10^3 \\
&= 1.632 \times 10^4
\end{aligned}$$

Example:

$$\begin{aligned}
(5.1 \times 10^4) \div (3.2 \times 10^{-1}) &= \frac{5.1 \times 10^4}{3.2 \times 10^{-1}} \\
&= \left(\frac{5.1}{3.2}\right) \times 10^{4+1} \\
&= 1.59 \times 10^5
\end{aligned}$$

Perform the following operations using the power of ten notation. Check your results using the conventional method for multiplication and division.

$(0.000303) \cdot (43200) =$
$(250,000) \div 0.0025 =$
$(0.0101) \cdot (0.0005) \div (0.0002) =$

Devise some exercises of your own and practice the method.

LOGARITHMS

Many concepts carry such mysterious and difficult sounding titles that one is frightened and often scared away at the outset. Such is the case with the notion of logarithms. If, however, the power of ten notation is understood, the ideas are simple. To illustrate, let us write:

$1000 = 10^3$ $100 = 10^2$ $10 = 10^1$ $1 = 10^0$ $0.1 = 10^{-1}$ $0.01 = 10^{-2}$
$0.001 = 10^{-3}$

The power required to raise 10 to equal the number in question is called the logarithm of the number to the base 10. The notation for this operation is $\log_{10}$. Thus,

$\log_{10} 1000 = 3$ $\log_{10} 100 = 2$ $\log_{10} 10 = 1$ $\log_{10} 1 = 0$
$\log_{10} 0.1 = -1$ $\log_{10} 0.01 = -2$ $\log_{10} 0.001 = -3$

The logs* of these numbers and all powers of 10 are easy to deduce. We can't help but wonder, though, about the log of an arbitrary number, say 400. This is no simple task. An estimate, if you are willing, is not difficult. For example, 400 is between 100 and 1000; therefore, $\log_{10} 400$ is between 2. and 3.

$\log_{10} 1000 = 3$
$\log_{10} 400 = ?$
$\log_{10} 100 = 2$

With no other information to guide us, we might guess that $\log_{10} 400 = 2.33$. Actually $\log_{10} 400 = 2.602$, so we have missed by about 10%, which is not bad. The calculation of logarithms with a modern computer is quite easy. One can also resort to tables if a computer is not available or appropriate. It would seem that a table with an infinite number of entries would be required because the number system is infinite in extent. However, there is a saving feature.

The decimal part of the logarithm is called the *mantissa* (another ominous word); the integral part, the *characteristic*.

$\log_{10} 400 = 2.602$

characteristic mantissa

Note that the mantissas for the logs of 1000, 100, 1, 0.1, etc., are the same, namely zero. Only the characteristics differ.

---

*Logs is jargon for logarithms.

Now consider some other numbers which differ only by a factor of 10, say 400, 40, 4, 0.4, and 0.004. The logarithms of these numbers are;

$\log_{10} 400 = 2.602$
$\log_{10} 40 = 1.602$ Mantissas are the same.
$\log_{10} 4 = 0.602$
$\log_{10} 0.4 = -0.398$
$\log_{10} 0.04 = -1.398$ Mantissas are the same.
$\log_{10} 0.004 = -2.398$

Note that the mantissas are the same for the numbers greater than one and less than one. It is easy to get the characteristic of a number greater than, or equal to one. It is simply one less than the number of digits left of the decimal point. Likewise, it is easy to get the characteristics of numbers less than one, but another property of logarithms is needed.

As a reminder, the logarithm to the base 10 is the power required to raise 10 to equal the number in question. So if

$$X = 10^a \qquad Y = 10^b$$

then, by definition of the logarithm,

$$\log_{10} X = a \qquad \log_{10} Y = b$$

Now from the rules for multiplying and dividing numbers written as powers of 10 it follows that

$$X \cdot Y = 10^a \cdot 10^b = 10^{a+b}$$

$$\frac{X}{Y} = \frac{10^a}{10^b} = 10^{a-b}$$

Hence,

$$\log_{10} X \cdot Y = a + b$$

$$\log_{10} \frac{X}{Y} = a - b$$

The logarithm of a product of two numbers is the sum of the logarithms of the numbers. The logarithm of the quotient of two numbers is the difference between the logarithms of the numbers.

Now, 0.4 can be written as 4/10. Hence,

$$\log_{10} 0.4 = \log_{10} \frac{4}{10} = \underbrace{\log_{10} 4}_{0.602} - \underbrace{\log_{10} 10}_{1}$$
$$= -0.398$$

Likewise $\log_{10} 0.04 = \underbrace{\log_{10} 4}_{0.602} - \underbrace{\log_{10} 100}_{2}$

$= -1.398$

Tables of logarithms are for numbers greater than one. There is no fundamental problem in getting other logarithms, but it does involve some arithmetic. If a table is desired for a number with, say, 10 significant figures (1,234,567,890, for example), then a huge volume would be required. However, a set for numbers with three significant figures occupies only two pages (Table 1). This table will suffice for our purposes.

Test your facility with logarithms by determining the logs of the following numbers.

326.
5280.
0.634
0.0000082

There are many uses of logarithms. Our interest is only to facilitate handling the enormous range of numbers encountered, especially when graphing data. For example, consider the following set of hypothetical, but not atypical, data.

Year	Amount
1930	1.5
1940	15
1950	150
1960	1500
1970	15,000

The data is to be displayed as a graph of amount versus time. Doing this in a conventional manner, we have what is shown in Fig. 1.

To get all the data on a reasonably sized graph, most of the detail is lost. Now suppose we calculate the logarithms of the amount.

Year	Amount	$\log_{10} A$
1930	1.5	0.18
1940	15	1.18
1950	150	2.18
1960	1500	3.18
1970	15,000	4.18

Note that the amount changes by a factor of 10,000 from 1930 to 1970, but the logarithm of the amount changes by only a factor of 23. Plotting the log of the amount versus time we arrive at Fig. 2. Now there is equal detail for all points, and making an estimate of the amounts for any arbitrary time is equally accurate for all times. The advantages of this method are evident, however it is tedious to look up logarithms, or calculate them. So graph paper has been made with one or both of the axes scaled proportionally to the logarithm of the number. Figure 3 shows how this is accomplished by plotting the log of the number versus the number.

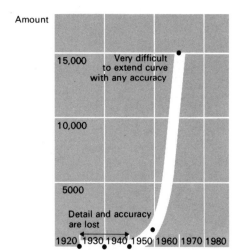

Fig. 1 Conventional plot of amount versus time.

Number	$\log_{10}$ Number
1	0
2	.301
3	.477
4	.602
5	.699
6	.778
7	.845
8	.903
9	.954
10	1.0

Plotting our data on this type of graph paper we arrive at Fig. 4. The same detail is achieved and looking up the logarithms is avoided. For your own benefit, make conventional and logarithmic plots of the following data:

Year	Amount
1900	0.18
1915	1.8
1930	18.
1945	180.
1960	1800.

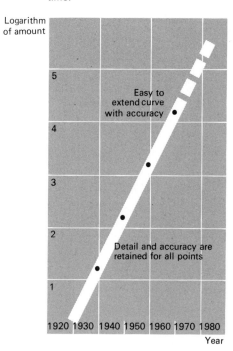

Fig. 2 Plot of $\log_{10}$ (amount) versus time.

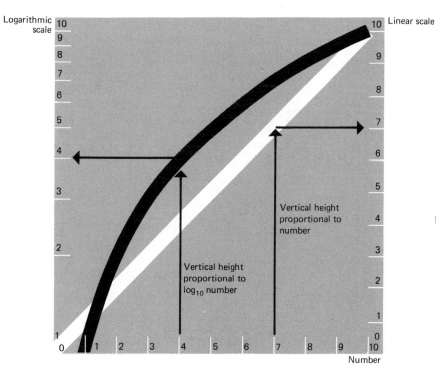

Fig. 3 Illustration of linear and logarithmic scales.

Table 1. Four-place table of logarithms to base 10.

N	0	1	2	3	4	5	6	7	8	9
10	0000	0043	0086	0128	0170	0212	0253	0294	0334	0374
11	0414	0453	0492	0531	0569	0607	0645	0682	0719	0755
12	0792	0828	0864	0899	0934	0969	1004	1038	1072	1106
13	1139	1173	1206	1239	1271	1303	1335	1367	1399	1430
14	1461	1492	1523	1553	1584	1614	1644	1673	1703	1732
15	1761	1790	1818	1847	1875	1903	1931	1959	1987	2014
16	2041	2068	2095	2122	2148	2175	2201	2227	2253	2279
17	2304	2330	2355	2380	2405	2430	2455	2480	2504	2529
18	2553	2577	2601	2625	2648	2672	2695	2718	2742	2765
19	2788	2810	2833	2856	2878	2900	2923	2945	2967	2989
20	3010	3032	3054	3075	3096	3118	3139	3160	3181	3201
21	3222	3243	3263	3284	3304	3324	3345	3365	3385	3404
22	3424	3444	3464	3483	3502	3522	3541	3560	3579	3598
23	3617	3636	3655	3674	3692	3711	3729	3747	3766	3784
24	3802	3820	3838	3856	3874	3892	3909	3927	3945	3962
25	3979	3997	4014	4031	4048	4065	4082	4099	4116	4133
26	4150	4166	4183	4200	4216	4232	4249	4265	4281	4298
27	4314	4330	4346	4362	4378	4393	4409	4425	4440	4456
28	4472	4487	4502	4518	4533	4548	4564	4579	4594	4609
29	4624	4639	4654	4669	4683	4698	4713	4728	4742	4757
30	4771	4786	4800	4814	4829	4843	4857	4871	4886	4900
31	4914	4928	4942	4955	4969	4983	4997	5011	5024	5038
32	5051	5065	5079	5092	5105	5119	5132	5145	5159	5172
33	5185	5198	5211	5224	5237	5250	5263	5276	5289	5302
34	5315	5328	5340	5353	5366	5378	5391	5403	5416	5428
35	5441	5453	5465	5478	5490	5502	5514	5527	5539	5551
36	5563	5575	5587	5599	5611	5623	5635	5647	5658	5670
37	5682	5694	5705	5717	5729	5740	5752	5763	5775	5786
38	5798	5809	5821	5832	5843	5855	5866	5877	5888	5899
39	5911	5922	5933	5944	5955	5966	5977	5988	5999	6010
40	6021	6031	6042	6053	6064	6075	6085	6096	6107	6117
41	6128	6138	6149	6160	6170	6180	6191	6201	6212	6222
42	6232	6243	6253	6263	6274	6284	6294	6304	6314	6325
43	6335	6345	6355	6365	6375	6385	6395	6405	6415	6425
44	6435	6444	6454	6464	6474	6484	6493	6503	6513	6522
45	6532	6542	6551	6561	6571	6580	6590	6599	6609	6618
46	6628	6637	6646	6656	6665	6675	6684	6693	6702	6712
47	6721	6730	6739	6749	6758	6767	6776	6785	6794	6803
48	6812	6821	6830	6839	6848	6857	6866	6875	6884	6893
49	6902	6911	6920	6928	6937	6946	6955	6964	6972	6981
50	6990	6998	7007	7016	7024	7033	7042	7050	7059	7067
51	7076	7084	7093	7101	7110	7118	7126	7135	7143	7152
52	7160	7168	7177	7185	7193	7202	7210	7218	7226	7235
53	7243	7251	7259	7267	7275	7284	7292	7300	7308	7316
54	7324	7332	7340	7348	7356	7364	7372	7380	7388	7396

N	0	1	2	3	4	5	6	7	8	9
55	7404	7412	7419	7427	7435	7443	7451	7459	7466	7474
56	7482	7490	7497	7505	7513	7520	7528	7536	7543	7551
57	7559	7566	7574	7582	7589	7597	7604	7612	7619	7627
58	7634	7642	7649	7657	7664	7672	7679	7686	7694	7701
59	7709	7716	7723	7731	7738	7745	7752	7760	7767	7774
60	7782	7789	7796	7803	7810	7818	7825	7832	7839	7846
61	7853	7860	7868	7875	7882	7889	7896	7903	7910	7917
62	7924	7931	7938	7945	7952	7959	7966	7973	7980	7987
63	7993	8000	8007	8014	8021	8028	8035	8041	8048	8055
64	8062	8069	8075	8082	8089	8096	8102	8109	8116	8122
65	8129	8136	8142	8149	8156	8162	8169	8176	8182	8189
66	8195	8202	8209	8215	8222	8228	8235	8241	8248	8254
67	8261	8267	8274	8280	8287	8293	8299	8306	8312	8319
68	8325	8331	8338	8344	8351	8357	8363	8370	8376	8382
69	8388	8395	8401	8407	8414	8420	8426	8432	8439	8445
70	8451	8457	8463	8470	8476	8482	8488	8494	8500	8506
71	8513	8519	8525	8531	8537	8543	8549	8555	8561	8567
72	8573	8579	8585	8591	8597	8603	8609	8615	8621	8627
73	8633	8639	8645	8651	8657	8663	8669	8675	8681	8686
74	8692	8698	8704	8710	8716	8722	8727	8733	8739	8745
75	8751	8756	8762	8768	8774	8779	8785	8791	8797	8802
76	8808	8814	8820	8825	8831	8837	8842	8848	8854	8859
77	8865	8871	8876	8882	8887	8893	8899	8904	8910	8915
78	8921	8927	8932	8938	8943	8949	8954	8960	8965	8971
79	8976	8982	8987	8993	8998	9004	9009	9015	9020	9025
80	9031	9036	9042	9047	9053	9058	9063	9069	9074	9079
81	9085	9090	9096	9101	9106	9112	9117	9122	9128	9133
82	9138	9143	9149	9154	9159	9165	9170	9175	9180	9186
83	9191	9196	9201	9206	9212	9217	9222	9227	9232	9238
84	9243	9248	9253	9258	9263	9269	9274	9279	9284	9289
85	9294	9299	9304	9309	9315	9320	9325	9330	9335	9340
86	9345	9350	9355	9360	9365	9370	9375	9380	9385	9390
87	9395	9400	9405	9410	9415	9420	9425	9430	9435	9440
88	9445	9450	9455	9460	9465	9469	9474	9479	9484	9489
89	9494	9499	9504	9509	9513	9518	9523	9528	9533	9538
90	9542	9547	9552	9557	9652	9566	9571	9576	9581	9586
91	9590	9595	9600	9605	9609	9614	9619	9624	9628	9633
92	9638	9643	9647	9652	9657	9661	9666	9671	9675	9680
93	9685	9689	9694	9699	9703	9708	9713	9717	9722	9727
94	9731	9736	9741	9745	9750	9754	9759	9763	9768	9773
95	9777	9782	9786	9791	9795	9800	9805	9809	9814	9818
96	9823	9827	9832	9836	9841	9845	9850	9854	9859	9863
97	9868	9872	9877	9881	9886	9890	9894	9899	9903	9908
98	9912	9917	9921	9926	9930	9934	9939	9943	9948	9952
99	9956	9961	9965	9969	9974	9978	9983	9987	9991	9996

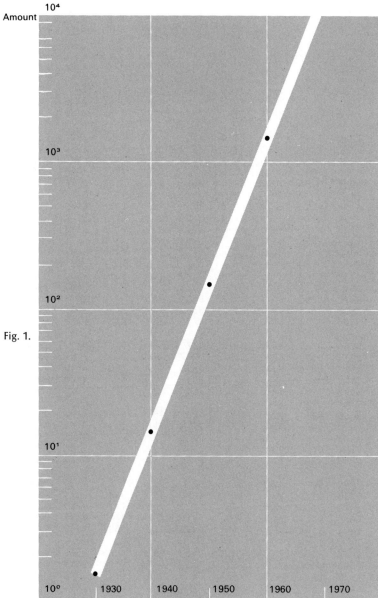

Fig. 4 Semilogarithmic plot of the data shown in Fig. 1.

Predict the amount for the year 1975 and deduce for yourself which plot will yield the most accurate result.

Appendix 2

GENERAL PROPERTIES OF WAVES AND THERMAL RADIATION

GENERAL PROPERTIES OF WAVES

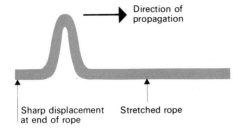

Fig. 1 A sharp vertical snap of the rope at one end produces a disturbance which propagates down the rope.

A wave is defined as "a disturbance or oscillation propagated from point to point in a medium or in space and described, in general, by mathematical specification of its *amplitude, frequency, velocity* (or speed), and *phase*."* A sharp clap of the hands producing an air pressure disturbance which propagates through the air constitutes a wave. Or if the end of a rope is sharply displaced, the disturbance produced propagates along the rope (Fig. 1). Such waves propagate with a speed characteristic of the properties of the medium through which they move. For example, the speed of a wave in a rope depends on how tightly the rope is stretched (tension) and the mass of the string per unit length. A general mathematical description of a wave is very difficult. However, a continuous wave such as that produced by a pure sound from a musical instrument or by continually moving the end of a rope up and down has features which are mathematically tenable. And, fortunately, many waves in nature are of this type.

Figure 2 shows "snapshots" of a rope at various stages in the development of a traveling wave. Viewing the rope from the side, we would see a series of crests or peaks moving past similar to the crests of water waves moving toward the shore. The time it takes for one crest to pass and another to appear is called the period T. (The period would also correspond to the time it takes to execute one oscillation at the end of the rope). The number of oscillations passing each second is called the frequency f. (The frequency also corresponds to the number of oscillations per second at the end of the rope.) So if there is 0.01 second between crests, the number going by in one second is expressed as

$$\text{number of crests} = \frac{1 \text{ sec}}{0.01 \text{ sec/crest}} = 100$$

That is to say, there are 100 crests per second passing the eye. Thus, the period and frequency are related by

$$f = \frac{1}{T}$$

*American Heritage Dictionary of the English Language, American Heritage Publishing Co., New York (1969).

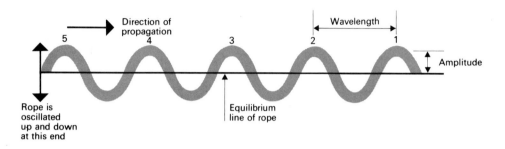

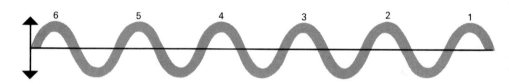

Fig. 2 "Snapshots" of a rope at various stages in the development of a traveling wave. A complete oscillation corresponds to taking the end of the rope up to its maximum height above the equilibrium position, back down to an equal distance below the equilibrium position, and then back to the equilibrium position.

The distance between two adjacent crests of the wave is called the wavelength (λ). So in one period the wave has traveled a distance equal to one wavelength. The speed is then

$$\text{speed} = \frac{\text{distance}}{\text{time}}$$

$$v = \frac{\lambda}{T} \qquad \text{or} \qquad v = f\lambda$$

The distance from a crest to the equilibrium position is called the amplitude A.

It is obvious that if some object gets in the path of the wave, a force will be exerted on the object and energy will be *transmitted* to it. For example, if the ear intercepts the pressure disturbance associated with a sound wave, a force is exerted on the ear drum which triggers a series of responses in the ear and brain. There is also no reason why waves cannot interfere with each other. In this context, the notion of *phase* is fundamental. There are a variety of ways of discussing phase. One commonly used method is—it is the relative separation of the waves at any instance in time. (This is analogous to the meaning of the phase of the moon.) It can be expressed as a fraction of a wavelength.

Suppose that we had two stretched ropes and instigated traveling waves in each, but did not start them at the same time. At any given instance, wave 2 is displaced from wave 1 by one-half wavelength. We would say that they are one-half wavelength out-of-phase. Now suppose that both of these waves are instigated in

Fig. 3 Waves 1 and 2, respectively, would produce an ordinary traveling wave. But if they are present together in the rope, one wave will exactly cancel the effects of the other.

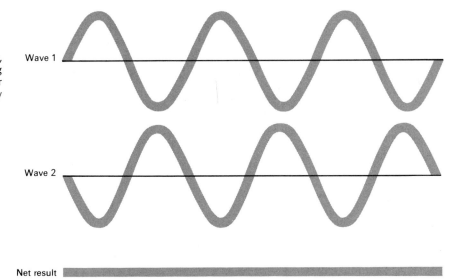

Wave 1

Wave 2

Net result

the same rope (see Fig. 3). (This might be hard to do in a rope, but would be easy for sound waves.) The net result is now the superposition of the effects produced separately. In the situation as outlined, the waves will produce exactly opposite effects. When one tends to push the rope up, the other tends to push it down. The net result is null, and hence this is referred to as complete *destructive* interference. There could also be no relative separation of the waves. We would say that they are exactly in-phase.

When the two waves interfere, they produce identical effects (see Fig. 4). The net result for this case will be just the sum of the two individual effects, and the interference is referred to as complete *constructive* interference. The general interference of waves of different amplitudes, frequencies, and wavelengths is obviously complicated. The net result, though, always lies between the extremes of complete destructive and complete constructive interference.

In addition to superimposing and interfering, waves can also reflect, refract, diffract, and be polarized. The change in direction of a ball after it has bounced from a wall is an example of reflection. Similar effects occur with waves when they suddenly encounter a different medium in their path. A water wave, for example, will reflect off a pier. When we see our reflection in a mirror, it is because light from our body struck the mirror and returned to our eyes by reflection from the surface of the mirror. It is also possible for a wave to penetrate into a different medium that it may encounter and suffer a change in direction and velocity in the process. This is the phenomenon of refraction. Refraction is most commonly observed with light. A narrow beam of light will change its direction when incident on a surface of water as shown in Fig. 5. The direction taken by the reflected and refracted light is of fundamental importance in the study of optics and optical instruments.

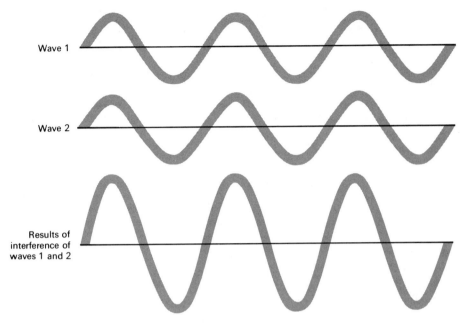

Wave 1

Wave 2

Results of
interference of
waves 1 and 2

Fig. 4 Both waves 1 and 2, separately, produce the same effect in the rope. Therefore the net effect is just twice the effect of either wave.

Fig. 5 Reflection and refraction of light at an air-water interface.

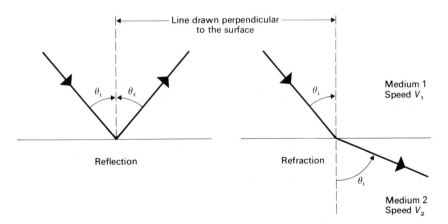

The idea of a "ray" of light is as old as the study of optics. A very narrow beam of light constitutes a ray in the experimental sense. Conceptually, a ray simply denotes the direction in which light is propagating. So if a source is radiating in all directions we would say that rays are emitted in all directions. Just as for waves, the speed of a ray would depend on the physical properties of the medium. Reflection and refraction of a ray of light are shown in Fig. 6.

Fig. 6 Reflection and refraction of a ray (or beam) of light. The refraction example is for the case in which V_2 is greater than V_1. If V_2 is less than V_1, then the angle of transmission, θ_t, is less than the angle of incidence, θ_i.

Fig. 7 Path of least time.

Polarization refers to the direction of the motion associated with the wave disturbance. For example, for the waves described in Fig. 2, the displacement of the rope always lies in the plane of the paper if the oscillations of the mechanism producing the waves lie in this plane. This wave would be said to be plane polarized. If the orientation of the oscillating mechanism were changed, the wave would still be plane polarized, but the plane would not be that of the paper. If several oscillating mechanisms with different orientations were fixed to the end of the rope, the displacements of the rope would not lie in a plane and the wave produced would be unpolarized.

The laws of reflection and refraction which are basic premises in optical technology state that

angle of incidence = angle of reflection

$$\theta_i = \theta_r \text{ (Law of reflection)}$$

(speed in medium 2) × sine of angle of incidence =

(speed in medium 1) × sine of angle of transmission

$$v_2 \sin \theta_i = v_1 \sin \theta_t \text{ (Law of refraction)}$$

The law of reflection is rather obvious; the law of refraction is not. Interestingly, the path taken by a ray in going from a source to a receiver, the eye for example, is just the one that requires the least time. And this is no different from the path chosen by a person bypassing a sidewalk while going from one building to another (Fig. 7) or from the route taken by a cab driver to get to a customer in the least amount of time.

Reflection and refraction of waves are easy to visualize by analogy with the behavior of particles impinging on the interface of two media. Diffraction, the bending of waves around corners, is strictly a wave phenomenon and is not so readily interpretable. It is, in fact, one of the methods used to identify a wave phenomenon. There are, however, many instances where this effect is observed. The most visual display is provided by water waves impinging on a slit in a barrier in their path (Fig. 8). A person standing outside a room can hear sound which is diffracted around the door opening. And light will diffract around the edges of a narrow slit in a manner very much like that shown in the water wave example. The diffraction pattern produced depends on the size of the opening (slit) and the wavelength of the waves. Measurements of these patterns provide an indirect method of measuring the wavelength.

Fig. 8 Water waves in a tank are generated by oscillating a horizontal bar up and down. These waves then impinge on a horizontal barrier with an opening. The waves are seen to bend around the edges of the opening. (Reproduced from *Ripple Tank Studies of Wave Motion* with permission of Clarendon Press, Oxford, England.)

WAVES AND PHOTONS

One has only to get in the path of a water wave rushing to the shore or radiation from the sun to know that both carry energy. But to say that the radiation is due to electromagnetic (EM) waves having characteristic wave properties is not so easily sensed. Abundant evidence supports this conclusion even though it is not the whole story, because other phenomena are induced by this same radiation which cannot be explained by a wave theory. The difficulty is easy to appreciate for we never see the EM waves in the sense that we see water waves. We see only the effects produced by the radiation when it interacts with other radiation or matter. We then try to deduce the nature of the radiation from these observations. Radiation will reflect, refract, and diffract precisely like the water waves which we can actually see. The speed of radiation can be measured and a wavelength deduced from diffraction experiments. These measurements are consistent with the relation $V = f\lambda$ for continuous waves. EM radiation can also be polarized and superimposed like other waves. The natural conclusion, then, is that EM radiation is a wave phenomenon. However, this same radiation will release electrons when it impinges on the surface of many metals (photoelectric effect), and interpretation of this phenomenon by a wave theory fails. Until the advent of quantum physics, the dual behavior of EM radiation left an uncomfortable feeling with physicists.

Quantum physics assumes that the energy associated with an EM wave is comprised of fundamental units or quanta called photons. A photon has energy proportional to the frequency of the wave. It is a fundamental constant of nature and is called Planck's constant.

$$\underbrace{\text{photon energy}}_{E} \quad \underbrace{\text{is proportional to}}_{\propto} \quad \underbrace{\text{frequency}}_{f}$$

or

$E = hf$

$h = 6.62 \times 10^{-34}$ joule-sec

The wave aspects emerge when many photons act together. So when we see reflection, refraction, and diffraction of EM radiation, we are observing the cumulative effect of a large number of photons, and the resulting phenomenon is one in which the wave aspects overwhelm the particle aspects.

There still remain the questions of "What is the disturbance associated with the waves?" and "How are the photons generated?" A water or sound wave possesses ordinary mechanical energy, that is, energy associated with particles in motion. Through actual collision, the waves can transmit energy. An electromagnetic wave, as the same suggests, possesses electric and magnetic energy by virtue of associated electric and magnetic fields (Fig. 9). The electromagnetic wave transfers energy through the interaction of its associated electric and magnetic fields with the electric and magnetic fields that it encounters. For example, if a charge gets in the path of an EM wave, the electric field of the charge interacts with that of the wave and the charge experiences a force which sets it into motion.

Photon emission is associated with the loss of energy that accompanies the deceleration of an electric charge. For example, the EM radiation from the antenna

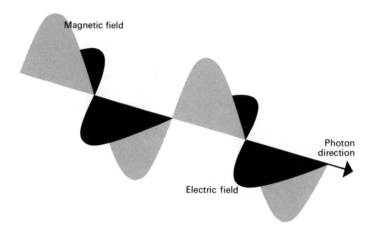

Fig. 9 Wave representation of electromagnetic radiation. Note that the electric and magnetic fields are perpendicular to each other and that both are perpendicular to the direction of propagation.

of a TV station is induced by accelerating electrons in the metallic conductors of the antenna. A free electron which is accelerated by entering the EM field of an atom will radiate photons. The radiation from an incandescent light bulb is due to random accelerations of electrons in the heated filament. As a result of this activity, a broad, continuous spectrum of wavelengths is emitted. Photons are also emitted when atoms, molecules, and nuclei make transitions from higher energy states to lower energy states. This is analogous to an energy transformation that evolves when a ball makes a transition from the top of an inclined plane (higher energy state) to the bottom (lower energy state). (See Fig. 10a.) But unlike the inclined plane example, the changes in energy occur discontinuously, more like those which would occur with a ball on a staircase (Fig. 10b). The radiation from a neon sign is not a continuous distribution of wavelengths (or colors) like that from an incandescent light bulb because the radiation is due to transitions between atomic states in neon atoms. Such a discontinuous change in energy is accompanied by the emission of a photon with energy equal to the transition energy

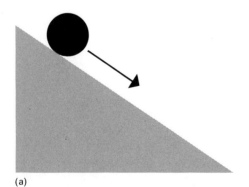

(a)

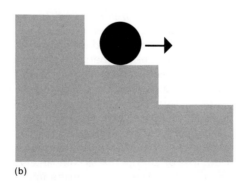

(b)

Fig. 10 (a) A ball rolling down an inclined plane continuously loses potential energy and gains kinetic energy. (b) A ball rolling down a staircase loses potential energy in "jumps" each time it rolls over the edge of a step.

(the recoil energy of the atom which is usually negligible is not taken into account.) See Fig. 11. The energy given up by the system is transformed into electromagnetic energy carried by the photon. The frequency of the radiation is given by

$$f = \frac{E_2 - E_1}{h}$$

Thus, the greater the energy given up in the transition, the higher the frequency of the radiation. The range of energies covered by these radiations, as well as the range of their associated frequencies, is absolutely staggering (Fig. 12). Interestingly, the speed of all these waves in vacuum is 2.9979×10^8 meters per second *regardless* of the *frequency*.

Several experiments verify the photon concept of EM radiation. One is the photoelectric effect mentioned earlier. Another is the study of thermal radiation from a heated object. In fact, the thermal radiation from the sun, the most important heated object, and from other objects as well is widely exploited in the technology of remote sensing.

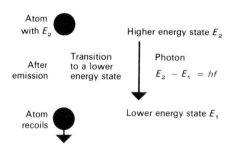

Fig. 11 Energy transformations in photon emission by an atom.

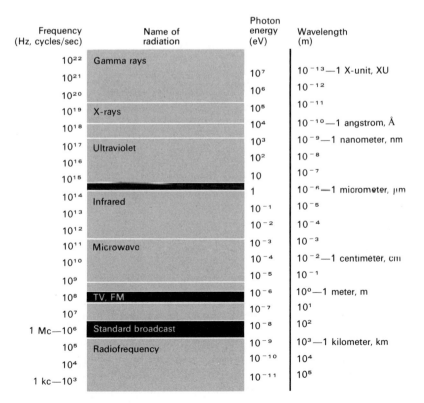

Fig. 12 The ranges of frequency, energy, and wavelength in the spectrum of electromagnetic waves.

Frequency (Hz, cycles/sec)	Name of radiation	Photon energy (eV)	Wavelength (m)
10^{22}	Gamma rays		
10^{21}		10^7	10^{-13}—1 X-unit, XU
10^{20}		10^6	10^{-12}
10^{19}	X-rays	10^5	10^{-11}
10^{18}		10^4	10^{-10}—1 angstrom, Å
10^{17}	Ultraviolet	10^3	10^{-9}—1 nanometer, nm
10^{16}		10^2	10^{-8}
10^{15}		10	10^{-7}
10^{14}	Infrared	1	10^{-6}—1 micrometer, μm
10^{13}		10^{-1}	10^{-5}
10^{12}		10^{-2}	10^{-4}
10^{11}	Microwave	10^{-3}	10^{-3}
10^{10}		10^{-4}	10^{-2}—1 centimeter, cm
10^9		10^{-5}	10^{-1}
10^8	TV, FM	10^{-6}	10^0—1 meter, m
10^7		10^{-7}	10^1
1 Mc—10^6	Standard broadcast	10^{-8}	10^2
10^5	Radiofrequency	10^{-9}	10^3—1 kilometer, km
10^4		10^{-10}	10^4
1 kc—10^3		10^{-11}	10^5

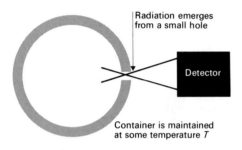

Fig. 13 Schematic diagram of a cavity radiator.

Radiation emerges
from a small hole

Detector

Container is maintained
at some temperature T

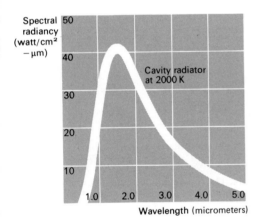

Spectral radiancy (watt/cm² – μm)

Cavity radiator
at 2000 K

Wavelength (micrometers)

Fig. 14 Experimental measurement of spectral radiancy for a cavity resonator.

THERMAL RADIATION

Experiments with thermal radiation involve the measurement of spectral radiancy which is the power per unit area associated with each frequency (or wavelength) of the radiation. Typical units for power, area, and wavelength are watts, square centimeters, and micrometers (10^{-6} m), respectively, so the measured quantity would be expressed in watts per square centimeters per micrometer. Now if measurements are taken of different materials at the *same* temperature, different results will be obtained. Hence the measurements include not only information about the fundamental properties of the radiation, but also information about the properties of the radiating material. The difference in the emission characteristics of various materials can be used in a practical sense to distinguish one object from another. If we are primarily interested in the fundamental properties of the radiation, the dependence of the emission on the material of the radiator complicates matters. Thus we need to find a radiating source which does not depend on the nature of its construction material. A so-called cavity radiator provides such a source. A cavity radiator consists of a hollow container of arbitrary shape with a tiny hole in the container wall (Fig. 13). The interior and walls are maintained at the same temperature and only the radiation from the hole is measured. Figure 14 shows results of such a measurement. If the measurement is repeated at different temperatures similar results are obtained, but the peak in the curve shifts (Fig. 15) and the total amount of radiation changes. The wavelength for the peak in the spectral radiancy curve obeys a relation of the form

$$\lambda_m T = \text{constant}$$
$$= 2900 \text{ micrometer} \cdot \text{Kelvins}$$

Consequently, as the temperature increases, the peak in the spectral radiancy curve shifts to smaller wavelengths. For example, if the temperature of two cavity radiators were 300K (about the temperature of the earth) and 6000K (about the temperature of the sun), the peaks would occur at about 10 and 0.5 micrometers, respectively. A wavelength of 10μm is in the infrared region of the electromagnetic spectrum and one of 0.5μm is in the visible region. This explains the difference in emission characteristics of the earth and sun alluded to in Chapter 3 in the discussion of the "greenhouse" effect.

The radiancy obeys a relation of the form

$$e_c = \sigma T^4 \ \text{W/cm}^2$$

Here the subscript c is for cavity where $\sigma = 5.67 \times 10^{-12} \text{ W/(cm}^2 - K^4)$

Hence the higher the temperature, the greater the power output of the radiator.

The theoretical explanation of the spectral radiancy curve was a major triumph in physics and is due to Max Planck. It provided one of the major links between classical and quantum physics. Planck assumed that the atoms in the walls of the radiator behaved as oscillators emitting electromagnetic radiation. However, an oscillator cannot have just any energy, but only those that satisfy the relation $E = nhf$, where n is an integer (0,1,2,3,...), h is Planck's constant mentioned earlier,

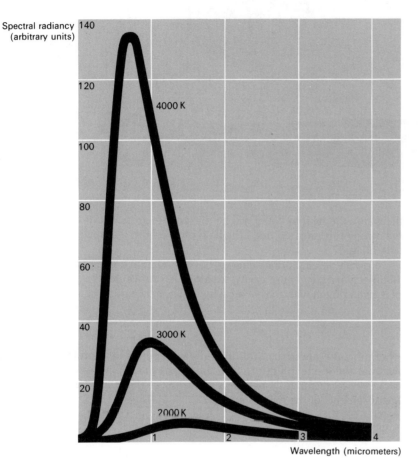

Spectral radiancy (arbitrary units)

Wavelength (micrometers)

Fig. 15 Illustration of how spectral radiancy changes with temperature.

Fig. 16 Planck's theoretical curve for spectral radiancy superimposed on an experimental result.

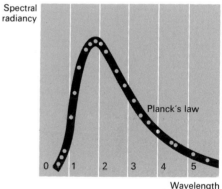

Spectral radiancy

Planck's law

Wavelength (micrometers)

and f is the frequency of the oscillator. Secondly, the oscillators do not radiate continuously but only in "jumps" or quanta. He then proceeded to derive the theoretical curve shown in Fig. 16.

The major importance of the relation $R_c = \sigma T^4$ lies in its revelation of the fact that any object at a temperature greater than 0 K will radiate EM energy. The amount of energy radiated depends very strongly on the temperature. This strong temperature dependence allows us to determine differences in temperature through measurements of the radiant energy. For example, the temperature of a river will be higher in the vicinity of a power plant dumping hot water from its turbine condenser. Although there is no physical difference in the appearance of

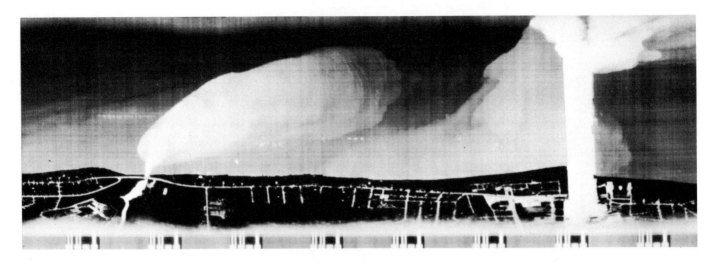

Fig. 17 Aerial thermal image of Genessee River, Lake Ontario, New York. (Courtesy of the U.S. Geological Survey.)

the water, the heat radiated will be different in areas of different temperature. By measuring this radiated heat, the heat plume from the power plant can be delineated (Fig. 17).

Now all objects radiate energy, but our relation does not always apply because all objects are clearly not cavity radiators. It is convenient to compare the radiancy of a given object with that of a cavity radiator by writing

$$\frac{R}{R_c} = e$$

where e is the emissivity and, in general, depends both on the type of radiating material and the temperature. The difference in emissivities of different materials allows us to distinguish between materials at the *same temperature* by taking measurements of the radiant energy.

GLOSSARY

Acceleration: Time rate of change of velocity.

Adiabatic process: A thermodynamic process that involves neither a gain nor loss of thermal energy.

Aerosol: Literally, a gaseous suspension of solid or liquid particles. Environmentally, a maximum size such as 10^{-4} m is sometimes included.

Albedo: The fraction (or percentage) of incident electromagnetic radiation reflected by a surface.

Alpha particle: Originally designated as a radioactive particle emitted naturally by several heavy elements such as radium and uranium. It is indistinguishable from the nucleus of the helium atom having two protons and two neutrons.

Alternating current (a.c.): An electric current that alternates direction at regular intervals.

Angstrom (Å): A unit of length equal to one-hundredth-millionth (10^{-8}) of a centimeter. Named for Anders Jonas Angstrom, Swedish physicist 1814–1874.

Atom: A unit of matter bound together by electric forces consisting of a dense, central, positively-charged nucleus surrounded by a system of electrons equal in number to the number of nuclear protons.

Atomic mass unit (amu): A unit of mass equal to $\frac{1}{12}$ the mass of the carbon isotope with six protons and six neutrons. 1 amu = 1.6604×10^{-24} g.

Bay Area Rapid Transit (BART): A modern mass transit system serving the San Francisco bay area.

Beta particles: Charged particles emanating from the nucleus of the atom which are created at the time of emission. The negative beta particle has all the characteristics of an electron which orbits the nucleus. The positive beta particle (positron) is identical to the negative beta particle except that it is positively charged.

Binding energy: In nuclear physics, binding energy is the energy required to separate a nucleus into its constituent neutrons and protons.

Boiling Water Reactor (BWR): A reactor in which water in thermal contact with the reactor core is brought to a boil. The vapor from the boiling water is then used to drive a steam turbine.

Bound system: A system such as a nucleus, atom, or molecule bound (or held) together by some force(s).

Breeder reactor: A reactor which generates fissionable fuel by bombarding appropriate nuclei with neutrons formed in the routine operation of the reactor.

British Thermal Unit (Btu):   The engineering unit of heat. It is the amount of heat required to raise the temperature of one pound of water one degree Fahrenheit.

Calorie (Cal):   The amount of heat required to raise the temperature of one gram of water one degree Celsius. The food calorie is equivalent to one thousand calories defined in this manner.

Carbon dioxide (CO_2):   A molecule containing one carbon and two oxygen atoms. It is the product of combustion of carbon in fossil fuels and is of concern because of the "greenhouse effect."

Carbon monoxide (CO):   A molecule containing one carbon and one oxygen atom. It is a product of incomplete combustion of carbon in fossil fuels and is of concern because of its effect on human body functions.

Carnot cycle: A thermodynamic cyclic process involving two isothermal and two adiabatic changes.

Catalyst: A substance which modifies the rate of a chemical reaction without being consumed in the process.

Catalytic converter: A device used to convert molecules regarded as hazardous to harmless species using catalytic chemical reactions.

Chain reaction: A self-sustaining, multi-stage nuclear reaction instigated by neutrons and sustained by neutrons generated in the reactions.

Condenser: A device at the exit of a steam turbine used to extract heat from the water vapor so that it will condense to a liquid.

Conduction: The transmission of something through a passage or medium without motion of the medium. Heat is conducted between adjacent parts of a medium if their temperatures are different. Moving electrons in a metal constitute electric conduction.

Control rod: A rod often made of boron or cadmium which can be inserted into the reactor core to control the fission reactions by selective absorption of neutrons.

Convection: Transfer of heat from one place to another by actual motion of the heated material.

Cooling tower: A tower used to collect the heated water effluent from the condenser of a steam turbine and transfer the thermal energy to the atmosphere.

Criticality: A condition in a nuclear reactor in which the neutron generation and loss rates are equal.

Critical mass: The minimum mass of fissionable material that will sustain a nuclear chain reaction.

Crankcase blowby: The leakage of gas or liquid between a piston and its cylinder during operation.

Curie (Ci):   A unit of radioactivity amounting to 3.7×10^{10} disintegrations per second.

Cyclonic collector: A particulate collecting device in which a whirlwind of air throws particles out of an air stream to the walls of a container on whose surface they collect or adhere.

Decibel (dB): A unit used to express relative differences in power (or intensity). It is defined as ten times the logarithm to base ten of the ratio of some power (or intensity) to a reference power (or intensity). The reference intensity is 10^{-16} W/cm^2.

Density: The amount of something divided by the volume it occupies. Mass density is the mass of a substance divided by the volume it occupies.

Direct current (d.c.): An electric current flowing in one direction.

Disintegration: A transformation involving the nucleus of an atom which results in a less massive configuration by emission of radiation which may be either particle or electromagnetic.

Dry-bulb temperature: The temperature achieved by a dry thermometer.

Earth Resources Technology Satellite (ERTS): A highly sophisticated earth-orbiting satellite for remotely evaluating earth resources and environmental quality. Launched in 1972.

Efficiency: The useful output of any system divided by the total input.

Electric current: Electric charges in motion.

Electric force: A force between two objects each having the physical property of charge.

Electricity: In the popular sense, it is electric current used as a source of power.

Electric potential difference: The work done on a charge to move it between two points divided by the strength of the charge.

Electron capture: A process in which the nucleus of an atom captures an electron orbiting the nucleus.

Electron volt (eV): A unit of energy. It is the energy acquired by an electron accelerated through a potential difference of one volt. 1 eV $= 1.602 \times 10^{-19}$ joules.

Electrostatic precipitator: An apparatus for the removal of suspended particles from a gas by charging the particles and precipitating them through application of a strong electric field.

Endoergic: A reaction such as a nuclear reaction which requires energy to proceed is said to be endoergic.

Endothermic: A chemical reaction which requires thermal energy to proceed is said to be endothermic.

Energy: The capacity for doing work.

Entropy: A measure of the disorder (or chaos) in a system.

Environmental Protection Agency (E.P.A.): A federal agency officially established on December 2, 1970 under a Presidential reorganizational plan. Placed in E.P.A. were the Interior Department's Federal Water Quality Administration; the HEW Department's National Air Pollution Control Administration, Bureau of Solid Waste, and Bureau of Water Hygiene; the pesticide registration, research, and regulation functions of the Agriculture, Interior, and HEW Departments, and certain radiation functions of the Atomic Energy Commission; the Federal Radiation Council; and HEW's Bureau of Radiological Health. The 1970 Clean Air amendments also authorized the formation of an Office of Noise Abatement with E.P.A.

Epilimnion: The bottom stratum (layer) of a lake or sea.

Equilibrium: A condition involving no change in the translational and rotational motion of a system.

Eutrophic: Designating a body of water in which the increase of mineral and organic nutrients has reduced the dissolved oxygen, producing an environment that favors plant over animal life.

Exoergic: A reaction such as a nuclear reaction which releases energy is said to be exoergic.

Exothermic: A chemical reaction which releases thermal energy is said to be exothermic.

Exponential growth: The growth in some quantity characterized by doubling of its value at regular intervals of time. For example, if some quantity doubles its value every ten years, it is growing at an exponential rate.

Extrapolate: To infer information by extending known information. Information obtained from a graph extended beyond known values is an example of extrapolation.

Fast neutrons: A loosely-defined classification of neutrons according to speed (or energy). A neutron with energy greater than 10 keV is called a fast neutron.

Feedback: The return of a portion of the output of any process or system to the input.

Fertile nucleus: A nucleus which serves as the target for generating fissionable fuel in a breeder reactor.

Fission: The splitting of an atomic nucleus into fragments, two of which are usually of comparable mass. Energy is released in the process.

Fly ash: Fine particles that are generated from noncombustible products in the burning of coal.

Fossil fuel: Fuel such as coal and petroleum derived from remnants of organisms of a past geological age.

Fuel cell: An electrochemical cell in which the energy of a reaction between a fuel such as liquid hydrogen and an oxidant such as liquid oxygen is converted directly into electric energy.

Fusion: A nuclear reaction in which nuclei are combined to form other nuclei. Energy is released in the process.

Gamma ray: A quantum (or photon) of electromagnetic radiation. It is distinguished from other electromagnetic radiation by its energy. The energy limits are loosely defined, but typically photons with energies greater than 100 keV are called gamma rays.

Gas Cooled Fast Breeder Reactor (GCFBR): A fast breeder reactor which uses a gas to transfer the heat from the reactor core.

Gravitational collector: A device for collecing noncombustible products released in the burning of coal. It utilizes the gravitational attraction between the earth and the particles to effect the separation.

Gravitational force: A force between two objects each having the physical property of mass.

Gravity Vacuum Transport (GVT): A subway-type mass transit system which utilizes the gravitational force between the earth and the vehicle for acceleration.

Greenhouse effect: An effect which produces a warming of a system by trapping electromagnetic radiation within the system. A common greenhouse warms primarily by trapping infrared radiation within the structure.

Gross National Product (GNP): The total market value of all the goods and services produced by a nation during a specified period.

Half-life: The time required for one-half of a given number of radioactive nuclei to disintegrate.

Heat: A form of energy associated with the motion of atoms or molecules.

Heat plume: The hot water discharge of a steam turbine into a body of water is confined to a "feather-shaped" region called a heat plume.

Heat sink: A system to which thermal energy can be transferred.

Hertz (Hz): A unit of frequency equal to one cycle per second. Named for Heinrich Hertz, German physicist 1857–1894.

Horsepower (hp): A unit of power equivalent to 746 watts.

Humidity: A qualitative term pertaining to the moisture in air.

Hydrocarbon: A compound containing only hydrogen and carbon.

Hypolimnion: The upper stratum (layer) of a lake or sea.

Impedance: A measure of the opposition to some flow of energy. Mathematically, it is the magnitude of the stimulus producing the flow divided by the flow. Electrical impedance, for example, is voltage divided by current.

Impedance matcher: A device used to equalize the impedance of a source and receiver.

Internal combustion (IC) engine: An engine, such as an automotive gasoline piston or rotary engine, in which fuel is burned within the engine proper rather than in an external furnace as in a steam engine.

Internal conversion: A process in which the internal energy of a nucleus is given up to an orbiting electron.

Internal energy: Thermodynamically, it is the kinetic and potential energy of the atomic or molecular constituents of a system.

Isothermal process: A process for which the temperature does not change.

Isotopes: Atoms with nuclei having the same number of protons, but a different number of neutrons.

Joule (J): Unit of work or energy. The work done by a force of one newton moving an object one meter in the direction of the force. Named for James Prescott Joule, British physicist 1818–1889.

Kelvin (K): Unit of temperature based on a freezing point of water of 273 and a boiling point of water of 373. Named for Lord Kelvin, William Thomson, British physicist 1824–1907.

Kilowatt-hour (kW-hr): A unit of energy equal to a power of one kilowatt (1000 W) acting for one hour.

Kinetic energy: One-half the product of the mass of a body times the square of its velocity. It is energy due to motion.

Lapse rate: The variation of temperature with respect to height above the earth.

Laser: A word coined from the first letters of the main words of "*l*ight amplification by *s*timulated *e*mission of *r*adiation." It is a device which produces an intense beam of electromagnetic radiation of very uniform wavelength.

Liquid Metal Fast Breeder Reactor (LMFBR): A breeder reactor using fast neutrons to produce the fissionable fuel and a liquid metal, typically sodium, as the heat transferrant.

Logarithm: If a number *X* raised to power *a*, equals a number *Y*, then *a* is said to be the logarithm of *Y* to base *X*. For example, the logarithm of 100 to base 10 is 2 because

$$\underset{X}{10}\overset{a}{\underset{}{2}} = \underset{Y}{100}$$

London smog: A heavily polluted atmosphere consisting mainly of particulates and sulfur dioxide emitted from the combustion of coal and petroleum products.

Macroscopic: Large enough to be characterized without instrumentation other than the eye.

Magnetic field: A region of space where a magnet would experience a force.

Magnetohydrodynamics: The science concerned with the motion of electrically conducting fluids through electric and magnetic fields.

Mass: The measure of a body's resistance to acceleration.

Micrograms per cubic meter ($\mu g/m^3$): One microgram (10^{-6} g) of a given substance in a one cubic meter volume of air.

Micrometer (μm): One millionth (10^{-6}) of a meter.

Microscopic: Small enough to require instrumentation other than the eye for characterization.

Model: A quantitative, numerical description of a physical system.

Moderator: A mechanism used to reduce the speed of neutrons in a nuclear reactor.

Molecule: A bound system of two or more atoms.

Neutrino: A massless, electrically neutral particle emitted in beta decay.

Neutron: An electrically neutral subatomic particle, which, with the proton, constitutes the atomic nucleus.

Neutron activation analysis: A technique for determining contaminants as low as 10^{-12} g by making a substance radioactive by neutron bombardment.

Newton (N): Unit of force. The force required to accelerate one kilogram, at one meter per second, each second. Named for Sir Isaac Newton, English scientist 1642–1727.

Nitrogen oxides (NO_x): For example, molecules produced by the oxidation of atmospheric nitrogen in the combustion of fossil fuels.

Nuclear reactor: A device for converting nuclear energy to thermal energy through nuclear fission reactions.

Nucleon: A proton or a neutron.

Nucleus of atom: The positively-charged central region of an atom, composed of neutrons and protons, containing nearly all the mass of the atom.

Octane rating: A numerical rating of the ability of a gasoline to reduce an audible knock (noise) in an internal combustion engine.

Ozone (O_3): A highly reactive, unstable molecule containing three atoms of oxygen.

Particulate: Existing in the form of minute separate particles.

Parts-per-million (ppm): A measure of the concentration of a substance. On a molecular basis, a 1 ppm concentration of a given molecule means that one of every one million molecules is a molecule of the specified type.

Photochemical (Los Angeles) smog: A heavily-polluted atmosphere consisting mainly of eye-irritating chemicals produced from the interaction of automobile exhaust gases with sunlight.

Photosynthesis: The process by which plants convert electromagnetic (solar) energy, carbon dioxide, and water to carbohydrates and oxygen.

Polarization: Denoting alignment or the assumption of two, usually conflicting, positions. Separation of oppositely-charged particles is called polarization. A wave is said to be linearly polarized if its displacement always lies along a line.

Pollutant: A constituent of some substance (such as air, water, and land, for example) that alters its value.

Potential energy: Energy that is potentially converted to another form, usually kinetic, of energy.

Power: Rate of doing work.

Pressure: The force on a surface divided by the area over which the force acts. $P = F/A$.

Pressurized Water Reactor (PWR):   A nuclear reactor in which the water in contact with the reactor core is allowed to become pressurized to prevent boiling and thereby allows a higher operating temperature.

Proton: A subatomic particle containing one unit of positive charge. It and the neutron are the constituents of the atomic nucleus.

Radiation: The emission and propagation of waves (such as light and sound) or particles (such as beta and alpha).

Radiation absorbed dose (Rad):   A unit of energy absorbed from ionizing radiation. One Rad is equal to 100 ergs per gram of irradiated material.

Raman effect: The alteration in frequency of light scattered by molecules in a medium. Named for Sir Chandrasekhara Venkata Raman, Indian physicist, 1888–present.

Rate: The change in some quantity divided by the time required to produce the change.

Relative Biological Effectiveness (RBE):   A measure of the capacity of a specific ionizing radiation to produce a specific biological effect.

Relative humidity: The amount of water vapor in the air at a specific temperature divided by the maximum capacity of water vapor in the air at that temperature.

Remote sensing: The art of acquiring information about objects or systems using remote measurements or observations. A camera is a remote sensor.

Roentgen equivalent man (Rem):   A biological unit of absorbed ionizing radiation which accounts for biological damage. It is the radiation dose measured in Rads multiplied by the relative biological effectiveness (RBE).

Semi-logarithmic graph paper: Graph paper with one axis scaled logarithmically and the other axis scaled arithmetically.

Settling velocity: The velocity (or speed) with which a particle in the atmosphere settles to the ground.

Shock wave: A large-amplitude wave produced by an object traveling faster than the characteristic speed of a wave in the medium in which the object is traveling.

Smog: A word derived from smoke and fog describing a polluted atmosphere that has resulted from the combustion of fossil fuels.

Sonic boom: A loud explosive-like sound caused by an object traveling faster than the local speed of sound.

Sound Pressure Level (SPL):   A relative measure of sound pressure. It is twenty times the logarithm (to base ten) of the ratio of the sound pressure to a reference pressure taken as 0.0002 dynes/cm^2.

Specific energy: The total available energy of a motive vehicle divided by the mass (or weight) of the vehicle.

Specific heat: The amount of heat required to raise the temperature of a unit mass (or weight) of a substance by one degree. Typical units are calories per gram per degree Celsius and Btu per pound per degree Fahrenheit.

Specific power: The power developed by a motive vehicle divided by the mass (or weight) of the vehicle.

Subsidence: The sinking or subsiding of an air mass.

Sulfur dioxide (SO_2): A stable molecule consisting of one atom of sulfur and two atoms of oxygen. It is emitted as an unwanted by-product in the combustion of most coal.

Supersonic: Traveling faster than the local speed of sound.

Synergism: A situation in which the combined action of two or more agents acting together is greater than the sum of the actions of the agents acting separately.

Temperature: A measure of the sensation of hotness or coldness referred to a standard scale such as the height of a column of mercury in a tube.

Temperature inversion: An atmospheric condition in which the temperature increases with increasing altitude.

Thermal pollution: The addition of thermal energy to the environment, usually the water environment, to the extent that the value of the environment is altered.

Thermal reactor: A device used in conjunction with an internal combustion engine to complete the combustion of hydrocarbons and carbon monoxide.

Thermal stratification: The layering of a body of water into sections of nearly constant temperature.

Thermodynamics: The science of the relationship between heat and other forms of energy.

Threshold of pain: A noise level, typically 120–140 dB, which produces a painful response in the ear.

Tokamak: A device developed by the Russians for converting nuclear energy using nuclear fusion reactions.

Tracked Air Cushion Vehicle (TACV): A vehicle supported by a cushion of air and constrained by some appropriate track.

Troposphere: The lower region of the earth's atmosphere extending up to an average altitude of about 40,000 ft (8 miles).

Turbidity: In meteorology, any condition in the atmosphere which reduces its transparency to radiation, especially to visible radiation.

Turbine: A device in which the kinetic energy of a fluid is converted to rotational kinetic energy of a shaft by impulses exerted on vanes attached to the shaft.

Visual range: The distance, under daylight conditions, at which the apparent contrast between an object chosen as the target of observation and its background becomes equal to the threshold of contrast of the observer.

Wankel engine: An internal combustion engine in which the kinetic energy of the exploding gas is converted directly to rotational kinetic energy.

Wet-bulb temperature: The temperature achieved by a thermometer covered with a cloth moistened with water.

Work: The product of the distance an object is moved times the force operating in the direction in which the object moves.

X-ray: A quantum (or photon) of electromagnetic radiation. It is distinguished from other electromagnetic radiation by its energy. The energy limits are loosely defined but typically photons with energies between about 1 and 100 keV are called X-rays.

X-ray fluorescence: The emission of X-rays resulting from the absorption of radiation. It is a sensitive technique for determining trace amounts of contaminants.

ANSWERS TO SELECTED EXERCISES

Chapter 1

1. (a) -5 mi/hr per sec   (b) 21.8 mi/hr per sec

2. Units are mi/hr per second. The physical quantity is an acceleration.

3. Acceleration would be ft/sec^2. From Newton's second law $m = F/a$. Thus the units are (lb/ft) (sec)^2. This unit is called a slug.

4. $$\frac{\text{mi}}{\text{hr}} \div \frac{\text{mi}}{\text{gal}} = \frac{\text{gal}}{\text{hr}}$$

   Thus

   $$45 \frac{\text{mi}}{\text{hr}} \div 15 \frac{\text{mi}}{\text{gal}} = 3 \frac{\text{gal}}{\text{hr}}$$

8. Eight flushes. Assuming that a shower takes 5 min, we have,

   amount = rate · time
   $$= 4 \frac{\text{gal}}{\text{min}} \cdot 5 \text{ min} = 20 \text{ gal}$$

10. (a) About 10 years   (b) 15 million tons

11. 0.6 passengers per trip

12. (a) Assume that it takes $\frac{1}{2}$ hour to dry a load of clothes. Then
    energy = power × time = 6000 W · $\frac{1}{2}$ hr
    $$= 3000 \text{ W-hr} = 3 \text{ kW-hr}$$

    cost $= $ energy × rate $= 3 \text{ kW-hr} \times \dfrac{8\cent}{\text{kW-hr}}$

    $\quad\quad = 24\cent$

    (b) energy $= 5 \text{ W} \times 365 \text{ days} \times 24 \dfrac{\text{hr}}{\text{day}}$

    $\quad\quad\quad = 43.8 \text{ kW-hr}$

    cost $= 43.8 \text{ kW-hr} \times \dfrac{8\cent}{\text{kW-hr}} = \$3.50$

13. 610 mi

15. Dimensionally k is length divided by time squared. The constant is the acceleration due to gravity, usually symbolized by g.

16. (a) 1650 billion tons   (b) between 1930 and 1931

17. The ultimate speed is believed to be the speed of light in vacuum— 3×10^8 m/sec. The distance traveled in 3 hr by something moving at 3×10^8 m/sec is 2×10^9 mi.

22. (a) 10 N   (b) The spring tension is relieved.

23. About 90%

25. (a) 2.2 billion tons/year   (b) 14 years

26. (b) 2.6 min

27. (a) 14,750,000   (c) 15,500,000. The population has grown.

Chapter 2

1. 100 lb

2. (a) $12{,}720 \, \frac{\text{Btu}}{\text{lb}} \cdot 5000 \, \text{lb} = 63{,}600{,}000 \, \text{Btu}$

   (b) heat into turbine = conversion efficiency multiplied by energy input
   $$= 0.85 \, (63{,}600{,}000)$$
   $$= 53{,}550{,}000 \, \text{Btu}$$

   (c) Energy escapes as thermal energy to the environment.

4. $10{,}000 \, \dfrac{\text{tons}}{\text{day}} \div 100 \, \dfrac{\text{tons}}{\text{car}} = 100 \, \dfrac{\text{cars}}{\text{day}}$

5. 41%

6. $I = \dfrac{P}{V} = \dfrac{6000 \, \text{W}}{115 \, \text{V}} = 52 \, \text{amp}$

7. 17% and 38%

9. radius = 0.00985 cm; length of side = 0.159 cm

10. About 4 in.

11. About 20 tons

14. 4.2×10^8 tons of coal; 16.4×10^6 tons of SO_2

15.
Diameter (micrometers)	Settling time (seconds)
0.1	1.25×10^9
1.	2.5×10^7
10	3.3×10^5
100	4×10^3
1000	260

18. (a) Charge

   (b) 100 amp · hr = current · time
   $$\text{time} = \frac{100 \, \text{amp} \cdot \text{hr}}{0.25 \, \text{amp}} = 400 \, \text{hr}$$

19. (c) $R = \dfrac{V^2}{P} = \dfrac{(115)^2}{1000} = 13.2 \, \text{ohms}$; an ohm is the unit of resistance.

Chapter 3

1. $C = \frac{5}{9}(F - 32) = 27°C$; $K = C + 273 = 300K$

2. $-40° C = -40° F$

3. $°M = °C + 100$

5. $135 \text{ lb} \div 15 \frac{\text{lb}}{\text{in}^2} = 9 \text{ in}^2$. The square would be 3 in. on a side.

6. (b) $P = \frac{F}{A} = \frac{20,000}{20} \frac{\text{lb}}{\text{ft}^2} = 1000 \text{ lb/ft}^2$

   (c) 110,000,000 lb/ft^2

7. (a) 0.000314 m^3 (b) 400 J

8. (a) 30,000 Btu   (b) 60%   (c) First law of thermodynamics

10. 15 lb/in^2

11. $\epsilon = (1 - \frac{273}{373}) \times 100\% = 27\%$. Temperatures must be in Kelvins.

15. It does not say how long it will take.

17. (a) 540 cal to vaporize 1 g of water. 1 cal to raise the temperature of 1 g by 1°C.
    (b) Water boiling vigorously vaporizes faster.

18.

T_{cold}	T_{hot}	$\varepsilon = 1 - \frac{T_{cold}}{T_{hot}}$
300	900	0.667
275	900	0.695
300	925	0.676

19. (a) 5000 cal   (b) $Q = mc \triangle T$
    (c) From part (b), $\triangle T = Q/mc$

    For copper:
    $c = 0.093 \text{ cal/g} = °C$; $\triangle T = \frac{20,000}{1,000(.093)} = 215°C$

    For water:
    $c = 1 \text{ cal/g} = °C$; $\triangle T = \frac{20,000}{1,000(1)} = 20°C$

20. (a) 7.5%   (b) 6.2%   (c) 17.7%   (d) 8.0%

26.

	Nitrogen gas	Water
Volume	$22.4 \times 10^3 \text{ cm}^3$	18.015 cm^3
Volume shared by each molecule	$3.73 \times 10^{-20} \frac{\text{cm}^3}{\text{molecule}}$	$3.0 \times 10^{-23} \frac{\text{cm}^3}{\text{molecule}}$
Length of side	3.3×10^{-7} cm	3.1×10^{-8} cm

Chapter 4

1. $\text{number molecules} = \dfrac{1,000,000 \text{ eV}}{10 \text{ eV/molecule}} = 100,000 \text{ molecules}$

   60 yr

3. $Z = 38, N = 67, A = 105$

4. Two alpha particles

8. 7.7 MeV/nucleon

10. About 25 blocks

11. (a) $\dfrac{1}{4}$ (b) $\dfrac{3}{4}$ (c) 18 hr

12. ^{91}Sr has a much shorter half-life.

13. 0.25 Rad

17. $^{14}_{6}C_8 \rightarrow\ ^{14}_{7}N_7 + \beta^- + \bar{\nu}$

19. Complete fission of ^{235}U produces 4.3×10^{10} Btu/lb. Burning of coal produces about 13,100 Btu/lb.

20. (a) $Z = 54, N = 77$ (b) 77 millicuries

23. (b) 100 units   (c) 1.4 cm for 50%. In principle an infinite thickness is required to absorb all the gamma rays.

    (d) Lead is the better absorber. One reason is that it is much denser than aluminum.

30. (a) $^{12}_{6}C_6$

    (b) Endoergic by 2.7 MeV

31. 2×10^7 kW-hr

32. 7.3 cycles

35. (a) 4.5×10^{-23} g   (b) 2.0×10^{-37} cm^3 (c) 2.25×10^{14} g/m^3
    (d) 5.1×10^7 tons/cm^3

Chapter 5

1. EBR-II: 26.4%; Fermi: 33.5%; Demo No. 1 and Demo No. 2: 40%

2. The square would be 10 m on a side.

3. 10^{11} W

4. $\text{density} \cdot \text{time} = 3 \times 10^{11} \dfrac{\sec}{\text{cm}^3}$;

   $\text{Lawson criterion} = 10^{14} \dfrac{\sec}{\text{cm}^3}$

5. Force exerted on the person by the merry-go-round.

9. Assume that the United States is a rectangle 3000 mi × 1000 mi.
   (a) 5.5×10^7 MW

(b) At 1% conversion, power $= 5.5 \times 10^5$ MW

$$= 550{,}000 \text{ MW}$$

Generating capacity of the United States is about 300,000 MW.

11. (a) 1   (b) 2

12. (a) 1.8×10^4 cm^2 corresponding to a circle about 4.5 feet in diameter.

(b) 50 min

19. About 10,000

Chapter 6

1. 50

2. Endoergic. Energy is required for the breakup.

3. $C_7H_{16} + 11O_2 \rightarrow 7CO_2 + 8H_2O$

4. No NO and NO$_2$ are consumed. They function as catalysts.

9. $\varepsilon = 0.4(0.9)(0.9)(0.8)(0.9) = 0.23$

$= 23\%$

15. (a) $\varepsilon = 0$ if $r = 1$

(b)

r	$\varepsilon(\%)$
1	0
4	50
9	67
16	75

(c) From $r = 4$ to $r = 9$, change $= 17\%$. From $r = 9$ to $r = 16$, change $= 8\%$.

16. $6NO + 4NH_3 \rightarrow 5N_2 + 6H_2O$

$6NO_2 + 8NH_3 \rightarrow 7H_2 + 12H_2O$

17. $H_2 + 2OH^- \rightarrow 2H_2O + 2e^-$

$2K^+ + O_2 + H_2 + 2e^- \rightarrow 2KOH$

Chapter 7

1. $\lambda = v/f$; $f = 20$ Hz, $\lambda = 55$ ft; $f = 20{,}000$ Hz, $\lambda = 0.055$ ft $= 0.66$ in.

2. 10^{-6} W/cm^2

3. 60 dB

5. 110 dB

6. No.

7. $t = \dfrac{\text{distance}}{\text{speed}} = \dfrac{100(3) \text{ ft}}{1100 \text{ ft/sec}} = 0.27 \text{ sec}$

    The time would be significant.

8. $\dfrac{I_{\text{outside}}}{I_{\text{inside}}} = 10^{-4} = 0.0001$

Chapter 8

1. 750 mi/hr

2. $M = \dfrac{1750}{750} = 2.34$

3. 60

4. 3 hr for the SST; 6 hr for the jet. Including time to and from the airports:

    500 mi/hr for the SST; 375 mi/hr for the jet.

5. (a) 0.0008 gal   (b) 1300

7. Assume that the United States is 1000 miles wide and that there are 200,000,000 people in the United States.
    Then number affected is approximately

    $\dfrac{20 \text{ mi}}{1000 \text{ mi}} \times 200{,}000{,}000 = 4{,}000{,}000 \text{ people}$

8. $a = \dfrac{v}{t} = \dfrac{80 \text{ mi/hr}}{1/2 \text{ min}} = 160 \dfrac{\text{mi}}{\text{hr-min}}$

9. About 1975.

10. (a) Number of trains per hour $= \dfrac{3600 \text{ sec}}{90 \text{ sec/train}} = 40 \text{ trains/hr}$

    (b) Capacity $= 40 \dfrac{\text{trains}}{\text{hr}} \times 6 \dfrac{\text{cars}}{\text{train}} \times 72 \dfrac{\text{passengers}}{\text{car}}$
    $= 17{,}280 \text{ passengers/hr}$

11. (a) SST: $t = 1.5$ hr; B-707: $t = 5$ hr

    (b) SST: exposure $= 1.5$ mrem; B-707: exposure $= 2.5$ mrem

12.

Average speed (mi/hr)	(a) Time (sec)	(b) Average speed (mi/hr)
20	180	18
30	120	26
40	90	33
50	72	39
60	60	45

13.

Speed (mi/hr)	Time (hr)	Time saved (hr)
20	12	4 (20 mi/hr to 3 mi/hr)
30	8	2 (30 mi/hr to 40 mi/hr)
40	6	1.2 (40 mi/hr to 50 mi/hr)
50	4.8	0.8 (50 mi/hr to 60 mi/hr)
60	4	

14.

(a) $t = \dfrac{distance}{speed} = \dfrac{1 \text{ mi}}{30 \text{ mi/hr}} = 120 \text{ sec}$

(b) $v = \dfrac{distance}{time} = \dfrac{1 \text{ mi}}{180 \text{ sec}} = 20 \text{ mi/hr}$

(c)

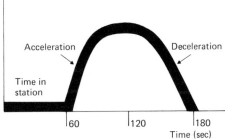

15.

(a)

Time (sec)	Speed (mi/hr)	Distance (mi)
5	20	.014
10	40	.056
15	60	.13
20	80	.22
25	100	.35
30	120	.5

(b)

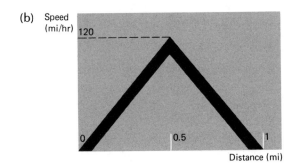

(c) $\text{speed} = \dfrac{\text{distance}}{\text{time}} = \dfrac{1\ \text{mi}}{60\ \text{sec}} = 60\ \text{mi/hr}$

(d) Speed (mi/hr)

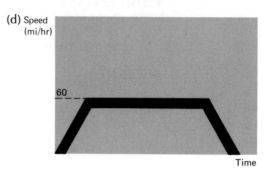

Chapter 9

1. $^{76}_{33}\text{As}_{43}$

3. Assuming that $\lambda = 0.03\mu\text{m}$, we determine that the temperature would be about 100,000K.

4. 10.5 yr

6. About 77,000 tons/day

7. (a) 143,000   (b) 81 mi

9. The power radiated from the source at 400K is about 32 times greater than the power radiated from the source at 300K.

11. United States: 1800 pounds per person per year; India: 200 pounds per person per year

INDEX

INDEX